DPL™ 4.0
Professional Decision Analysis Software

Academic Version

Applied Decision Analysis LLC
a wholly owned subsidiary of
PriceWaterhouseCoopers LLP

Duxbury
Thomson Learning™

Pacific Grove • Albany • Belmont • Boston • Cincinnati • Johannesburg • London • Madrid
Melbourne • Mexico City • New York • Scottsdale • Singapore • Tokyo • Toronto

Sponsoring Editor: *Curt Hinrichs*
Marketing: *Karin Sandberg*
Editorial Assistant: *Sarah Kaminskis*
Production Editor: *Tessa McGlasson Avila*

Cover Design: *Vernon Boes*
Print Buyer: *John Cronin*
Printing and Binding: *Webcom, Limited*

DPL Team:		
	Adam Borison	*Bill Miller*
	Justin Claeys	*Julia Langel*
	Jeff Cloutier	*Darlyn Linka-Pettenati*
	Chris Dalton	*Pavel Mizenin*
	Christy Downs	*Dan Smith*

Duxbury is an imprint of Brooks/Cole, a division of Thomson Learning
The Thomson Learning logo is a trademark used herein under license.

For more information, contact:
DUXBURY
511 Forest Lodge Road
Pacific Grove, CA 93950 USA
www.duxbury.com

For permission to use material from this work, contact us by
Web: www.thomsonrights.com
fax: 1-800-730-2215
phone: 1-800-730-2214

Printed in Canada

10   9   8   7   6   5   4   3   2   1

Library of Congress Cataloging-in-Publication Data

DPL 4.0 professional decision analysis software : academic version /
   Applied Decision Analysis LLC.
        p.   cm.
   Includes bibliographical references.
   ISBN 0-534-35368-1
   1. Decision support systems—Data processing.   2. DPL.
 I. Applied Decision Analysis, Inc.
 T58.62.D65   1999
 658.4' 03' 02855369—dc21

                                                            99-36299

Contents

The following chapters are in PDF format on the supplied CD-ROM

Preface

Welcome to DPL, professional decision analysis software by Applied Decision Analysis LLC, a wholly-owned subsidiary of PricewaterhouseCoopers LLP. DPL requires a computer with Windows 95, 98, 2000, or NT 4.0 or higher. Installation requires approximately 4MB of disk space. For best performance, we recommend a 133MHz or faster CPU and at least 32MB of RAM.

DPL is a powerful, flexible modeling environment with many features and options. To get started, we recommend that you read through the first two chapters, Overview of Decision Analysis and System Overview, then work through the introductory tutorial in Chapter 3, which will walk you through all the steps of building and evaluating a simple model. If you are familiar with decision analysis, these three chapters will probably be all you need to get started with DPL.

Chapters 1 to 9 document DPL's modeling features in detail, including influence diagrams, decision trees, and spreadsheet links. Chapters 10 to 13 discuss DPL's evaluation methods, most frequently used outputs, and most commonly used type of sensitivity analysis. Each chapter includes a description of the feature, guidelines and tips for using it, and tutorials demonstrating its most common uses.

Chapters 14 to 19 are included on the CD-ROM in PDF format. These chapters cover the features of DPL which are used less frequently, including DPL Programs, advanced sensitivity analyses, and additional outputs.

In this manual, we have attempted to give you the equivalent of having an expert modeler and analyst guiding you through a decision analysis project. If you need more information than this book provides, you have two additional resources. The On-line Help system has a complete technical reference for all features, functions, named distributions, and syntax. Our website, at www.adainc.com, contains technical notes, user applications, and additional tutorials.

Good luck, and happy modeling!

The DPL Team

Decision analysis is much more than using a decision tree to calculate an expected value. As it is currently practiced, decision analysis is a powerful and flexible set of analytic tools that support and drive a structured evaluation process. Decision analysis is used widely by businesses, government agencies, universities, and research institutes. It is especially useful for addressing difficult decisions that have one or more of the following characteristics:

- multiple and complex alternatives
- uncertain outcomes
- high stakes
- multiple stakeholders
- conflicting objectives
- multiple sources of relevant information

Chapter 1

OVERVIEW OF DECISION ANALYSIS

Chapter 1: Overview of Decision Analysis

What is Decision Analysis?

Decision analysis is a structured, quantitative approach for evaluating decisions with complex alternatives, uncertain outcomes, and competing objectives. First developed at Harvard and Stanford universities in the 1960s, decision analysis has become especially popular in the last five or ten years because of the presence of three factors:

- a solid quantitative foundation
- a flexible, practical and simple decision making process
- fast, reliable computer tools, like DPL, which can perform the numerous calculations quickly and present the results graphically.

The decision analysis process has four main steps:

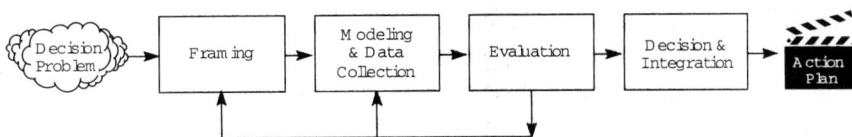

One advantage of the decision analysis process is that each step can produce a great deal of insight, so you can start benefiting from it very early in a project. In fact, some people find the first step, framing, so useful that they don't feel a need to go any further.

Another benefit of the process is its scalability. You can conduct a useful decision analysis in as little as two or three days, if that's all the time you have. Or, you can spend three to six months on an analysis. The process is the same; the difference is in the level of detail you can reach.

In the process diagram, you will see that the arrows suggest looping back to earlier phases. We strongly recommend that you allow time in your project schedule for this iteration, for this is where decision analysis can really add value to your decision making. When you take what you learn during an early pass through the project, use that as a starting point for creative problem solving, then make a second pass through the phases to test and refine your new ideas, you will usually finish with a strategy much better than any of the alternatives you started with.

The rest of this chapter will discuss the phases in more detail and describe how DPL's tools can help.

Phase I: Framing

In the framing phase, your goal is to get the right people evaluating the right problem. Although it's tempting to move on to the quantitative phases, rushing through the framing phase will do more harm than mistakes in any later phase.

The Right People

In any decision situation, there are three roles people play:

The Decision-Maker is the person (or group) who has the final word on a decision. The decision-maker is the person with the authority and responsibility to commit resources and start (or stop) the organization's activities. Usually, but not always, the decision-maker is not the one conducting a decision analysis, but will see the results and recommendations.

The Evaluation Team includes the people who are conducting the evaluation. This core team usually includes a project or department manager, some individuals experienced in the affected areas, and some people who are analyzing data, building spreadsheets, and running DPL.

Information Experts have the best available data, experience, and judgment relevant to specific topics and issues relevant to the evaluation.

It is crucial that the evaluation team communicate regularly with the decision-makers and information experts. Ideally, this should happen at least once in each phase of the decision analysis.

Occasionally, you may find yourself playing two or even all three of these roles. Take care to spend some time wearing each hat during every phase — looking at a problem from a different point of view can provide useful insights.

The Right Problem

Once the evaluation team is assembled, it's time to do some brainstorming. Your goals are to produce a problem statement and thorough lists of objectives, decisions, and uncertainties. From these, you can create an influence diagram and decision tree that will serve as the basis of your quantitative model, and a project management plan that will help you plan your schedule, reserve resources, and allocate responsibilities.

Objectives

Before you can select a strategy, you need to know the criteria by which the strategy will be judged. Be honest and complete. Will success be judged purely on financial considerations, or are there other objectives such as environmental impact or employee morale? If there is a financial objective, how is it measured — total project over five years? Capital and maintenance costs over ten years? If there are multiple objectives, how will you resolve trade-offs?

Decisions

Now that you know what you want to accomplish, think about how to get there. What decisions need to be made immediately? Two years from now? After the new environmental regulations are final? Think about each of your objectives. If employee morale was your only concern, what might you do? Now, what if environmental impacts were paramount? What might you do if you were very daring and aggressive? Very cautious? If a new competitor enters the market, how might you respond?

Uncertainties

Your last brainstorming task is to think of all the uncertainties that make choosing a strategy difficult. Think locally — could your costs be higher than expected? Could construction take longer than planned? Think about your market — will a new product cannibalize sales from the rest of the product line? Will your competitors lower their prices? Think about low probability, high consequence events — what if someone develops a new technology that makes your product obsolete? What if this year's teen heartthrob uses your product on TV and everyone between twelve and sixteen has to have it? What if there's an earthquake?

Starting the Model

Once you've finished brainstorming, start your decision model. Most DPL models have two parts — an influence diagram and a decision tree.

Influence Diagram

An influence diagram is a graphic representation of the elements in a decision problem and the relationships among them.

- Decisions are represented with yellow rectangles
- Uncertainties are represented with green ovals
- Values (objectives, and the numbers and formulas needed to calculate them) are represented by blue rounded rectangles
- Relationships are indicated with arrows

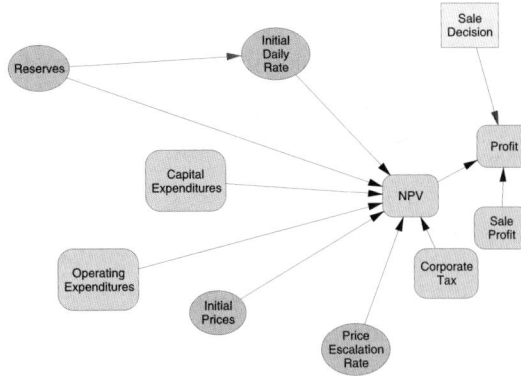

Decision Tree

A decision tree displays the time of the events in your model. It defines the range of future scenarios.

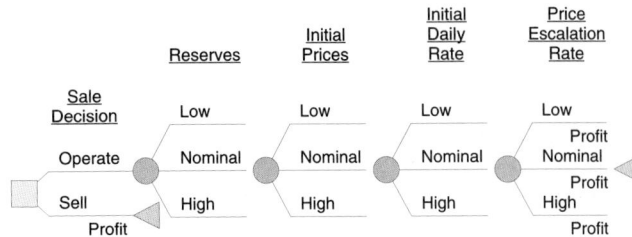

These two tools are useful both as the basis of your mathematical model and a communication aide when you present the results of your framing efforts with the decision makers, information experts and others. You will continue to refine and modify these diagrams as your move through the process.

DPL provides many tools for drawing influence diagrams and decision trees. For complete information and tutorials, refer to:

- Chapter 4: Decisions and Strategy Tables
- Chapter 5: Chance Events
- Chapter 6: Values, Get/Pay Expressions and Objective Functions
- Chapter 7: Multiple Objective Models
- Chapter 8: Timing and Structure

Phase II: Building a Model and Gathering Data

During the model building phase, you will put numbers behind your influence diagram and decision tree. There are four steps:

- Build a deterministic model
- Conduct deterministic sensitivity analysis
- Assess probability distributions
- Build the full probabilistic decision model

Deterministic Model

A deterministic model assigns an outcome value to any scenario in your decision tree. It can be as simple as a table of values or as complex as a multi-sheet spreadsheet. A good place to start is with a base, or nominal case, then add any logic necessary to handle other scenarios.

For information and tutorials, refer to Chapter 9: Spreadsheets.

Deterministic Sensitivity Analysis

When you first frame a decision problem, you may have as few as three or four uncertainties or as many as thirty or forty. It is important to understand which uncertainties have a significant impact on your choice of strategy and which are less important. This helps you gain insight into the decision and prioritize your analysis efforts.

To perform a deterministic sensitivity analysis, set all variables to their nominal, or base case, values and evaluate each alternative to determine the best one (called the nominal policy). Then vary each uncertain variable over its range of uncertainty to see the impact on outcome value and choice of policy. The usual format for presenting the results is a tornado diagram.

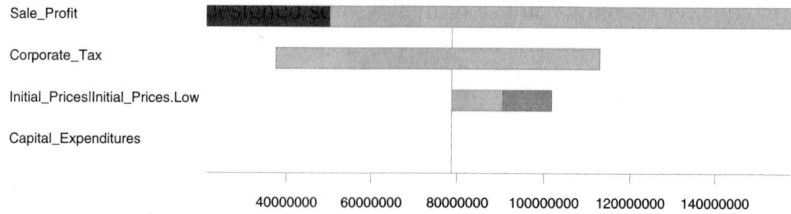

Sale_Profit					
Corporate_Tax					
Initial_Prices|Initial_Prices.Low					
Capital_Expenditures					

40000000 60000000 80000000 100000000 120000000 140000000

The most sensitive variables have the widest bars. Variables whose uncertainty have an impact on strategy choice have color changes.

For information and a tutorial, refer to Chapter 13: Expected Value Tornado Diagrams.

Probability Distributions

For each of the most significant uncertainties, you need to assess a probability distribution. The probability for an uncertainty represents your best state of knowledge about the uncertainty. You will probably need to interview an information expert to assess the distribution; the expert should draw on all available data, experience, history, forecasts, and judgment necessary.

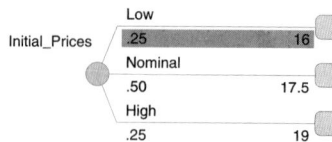

For information and tutorials, refer to Chapter 5: Chance Events.

The Probabilistic Decision Model

The final step is to build a probabilistic decision model. Modify the influence diagram and decision tree to reflect the significant uncertainties, and enter the data for their probability distributions. Link the influence diagram to the deterministic model, so it can calculate outcome values for each scenario in the decision tree. Now your model is ready for analysis.

Phase III: Analysis and Evaluation

A complete decision model can be evaluated in a number of ways. DPL provides several types of decision and sensitivity analysis, each of which gives you a slightly different perspective on the problem. Of course, you may not need each analysis feature on every model.

You will probably find that you don't proceed directly through this phase. Instead, you will run a decision analysis, notice something interesting, run some sensitivity analysis to explore it, then decide to refine your model or add a new alternative or uncertainty. Then you'll come back to the analysis features, learn something else interesting, and pursue that. DPL's speed and flexibility make it easy to continue to explore, test and modify until you have a thorough understanding of your strategy.

Decision Analysis

A decision analysis is a single evaluation of the complete decision tree. Chapter 10: Evaluation Methods discusses this in detail. DPL has four ways to examine the results:

- The Policy Tree graphically displays every path in the decision tree, complete with probabilities, outcome values, and expected values. Most importantly, it also displays the optimal strategy.

Sale_Decision [8.14025e+007]

Reserves: Opera [8.14025e+007]

- Initial_Prices
  - Low [4.34136e+007] .250
  - Nomir [7.39538e+007] .500
    - Initial_Daily_Rate
      - Low [6.21992e+007] .250
        - Price_Escalation_Rate
          - Low [5.48471e+007] .167
            - Low .250 5.01858e+007 [5.01858e+007]
            - Nomir .500 5.37438e+007 [5.37438e+007]
            - High .250 6.17151e+007 [6.17151e+007]
          - Nomir [6.3018e+007] .667
            - Low .250 5.86827e+007 [5.86827e+007]
            - Nomir .500 6.20039e+007 [6.20039e+007]
            - High .250 6.93813e+007 [6.93813e+007]
          - High [6.62764e+007] .167
            - Low .250 6.29803e+007 [6.29803e+007]
            - Nomir .500 6.55353e+007 [6.55353e+007]
            - High .250 7.10547e+007 [7.10547e+007]
      - Nomir [7.39538e+007] .500
        - Price_Escalation_Rate
          - Low [6.55449e+007] .167
            - Low .250 6.04466e+007 [6.04466e+007]
            - Nomir .500 6.43382e+007 [6.43382e+007]
            - High .250 7.30567e+007 [7.30567e+007]
          - Nomir [7.48737e+007] .667
            - Low .250 7.0132e+007 [7.0132e+007]
            - Nomir .500 7.37646e+007 [7.37646e+007]
            - High .250 8.18336e+007 [8.18336e+007]
          - High [7.86834e+007] .167
            - Low .250 7.50783e+007 [7.50783e+007]
            - Nomir .500 7.78728e+007 [7.78728e+007]
            - High .250 8.39096e+007 [8.39096e+007]
      - High [8.57085e+007] .250
        - Price_Escalation_Rate
          - Low [7.62427e+007] .167
            - Low .250 7.07074e+007 [7.07074e+007]
            - Nomir .500 7.49326e+007 [7.49326e+007]
            - High .250 8.43984e+007 [8.43984e+007]
          - Nomir [8.67294e+007] .667
            - Low .250 8.15813e+007 [8.15813e+007]
            - Nomir .500 8.55253e+007 [8.55253e+007]
            - High .250 9.42859e+007 [9.42859e+007]
          - High [9.10903e+007] .167
            - Low .250 8.71763e+007 [8.71763e+007]
            - Nomir .500 9.02103e+007 [9.02103e+007]
            - High .250 9.67645e+007 [9.67645e+007]
  - High [1.34289e+008] .250

Sell [7.875e+007] / 7.875e+007

Refer to Chapter 11: Policy Trees for a thorough discussion and tutorials.

- Risk Profiles graph the range of outcomes under alternative strategies. They help you understand which strategies offer the most risk and which offer the most potential.

Refer to Chapter 12: Risk Profiles for a thorough discussion and tutorials.

• The Policy Summary provides an overview of the optimal policy and helps you determine which scenarios lead to very good or very bad outcomes.

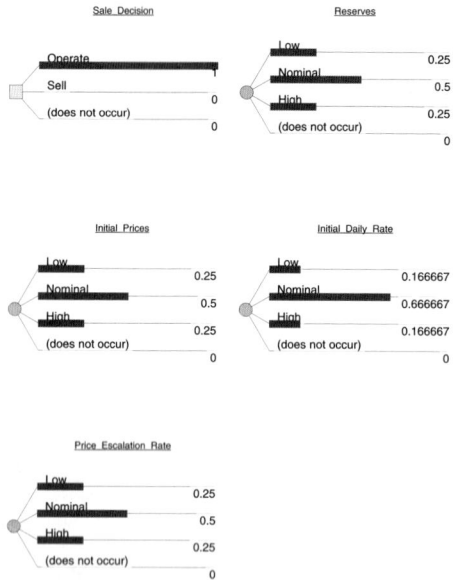

Refer to Chapter 15: Policy Summaries and Comparisons for a thorough discussion and tutorials.

- The Value of Information and Control graph shows what it would be worth to gain additional information about an uncertainty before making a decision or to try to obtain control over the uncertainty or its impact.

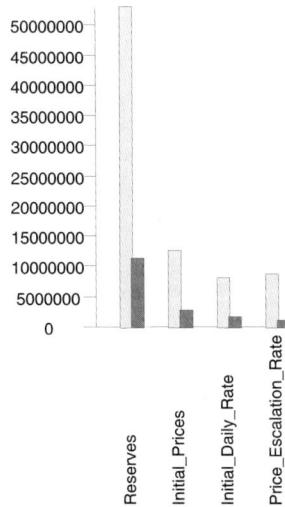

Refer to Chapter 16: Value of Perfect Information and Control for a thorough discussion and tutorials.

Sensitivity Analysis

Sensitivity analysis contributes to your understanding of the decision problem by estimating how important the data and assumptions in your model are. They can increase your confidence in your recommendations, and suggest directions for creative alternatives, additional modeling or data gathering. DPL has four kinds of sensitivity analysis.

- Rainbow Diagrams show you have the expected value and optimal decision policy of the model change as you vary one value through a range.

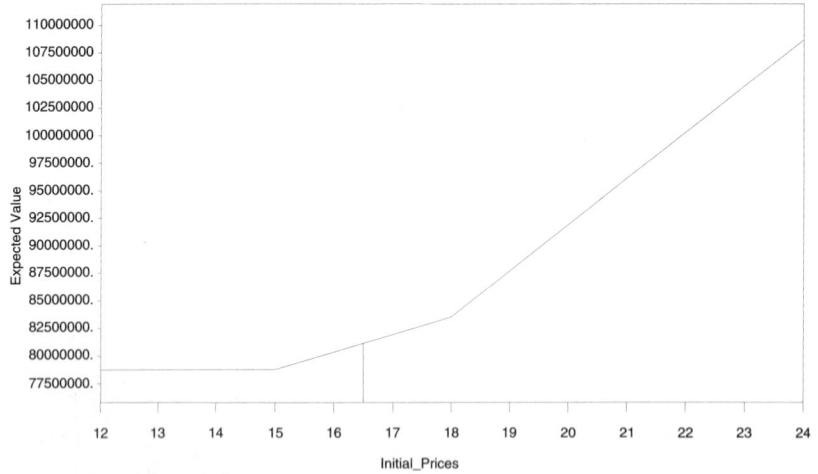

Refer to Chapter 17: Rainbow Diagrams for a thorough discussion and tutorials.

- Expected Value Tornado Diagrams also show the effects of varying values in the model, but they present the results for several variables at once, allowing comparisons.

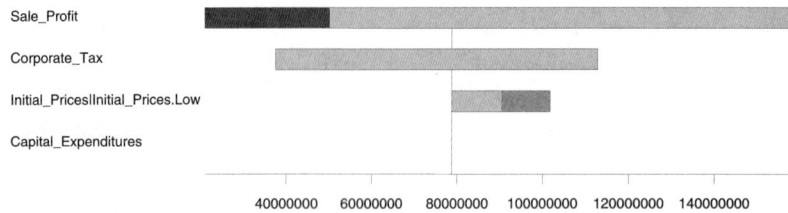

Refer to Chapter 13: Expected Value Tornado Diagrams for a thorough discussion and tutorials.

• Base Case Tornado Diagrams show the amount of uncertainty each chance event in the model contributes, compared to a base case where all uncertainties are set to their base, or nominal, settings.

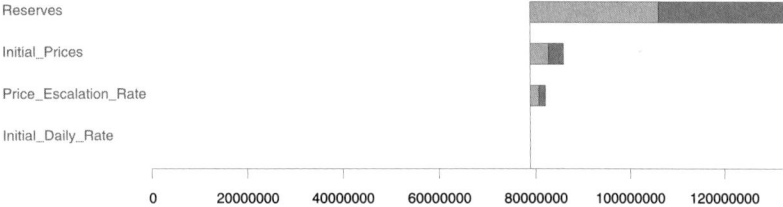

Refer to Chapter 18: Base Case Tornado Diagrams for a thorough discussion and tutorials.

• Event Tornado Diagrams also show the relative contribution each uncertainty makes to the uncertainty in the optimal policy. In this analysis though, the effects are compared to an expected value, rather than nominal, base case.

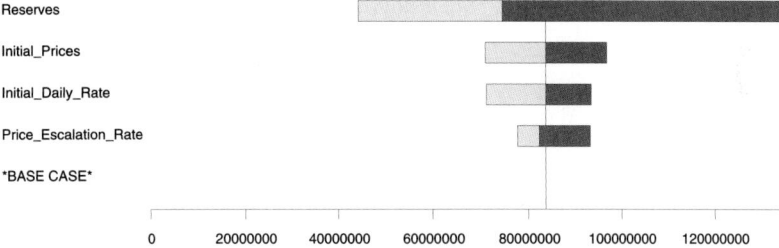

Refer to Chapter 19: Event Tornado Diagrams for a thorough discussion and tutorials.

Phase IV: Making the Decision

With a thorough evaluation completed, it is time to make a decision (or recommendation). In addition to summarizing the insight gained from analyzing the model, this should include an action plan for moving forward into implementation.

Many of DPL's graphical features, such as influence diagrams, decision trees, and tornado diagrams, can be helpful for communicating your results to others. They can all be pasted into other applications, such as work processors and presentation packages.

Tip: Don't make your audience work too hard — before using an analytic output directly from DPL, ask yourself whether it would be better to summarize the insights from the output in a bullet list or other format.

Tip: Try to avoid using the phase "The model says we should..." People hate taking orders from a computer! And well they should; the model isn't making the decision — you are making the decision based, in part, on insights gained from the model.

DPL is designed to help you construct, manage, and analyze your project efficiently and effectively. A DPL project can contain a number of different elements, such as models, programs, outputs, and results. All of the elements of a DPL analysis are contained in separate windows within a single DPL project, and managed with a Project Manager.

Chapter 2

Chapter 2: System Overview

DPL Windows

Each element of a DPL decision analysis, including models, programs, and outputs has its own window in a DPL project. These windows can be maximized, minimized, tiled, or cascaded to allow you to customize your view. Whichever window you're working in is the active window, and has a distinctive border and title bar. The active window is always in the forefront (i.e. on top), and making a window active brings it to the forefront. You can select any window from the Windows menu to make a specific window the active window.

Project Manager

The Project Manager is the control center of a DPL project. The Project Manager is essentially a list of all of the windows and data contained in the project. Double-clicking on the name of a window will jump you to that window. Double-clicking on data will create a new window to display that data. Whenever you get lost in a project, the Project Manager is the place to go to find your way again!

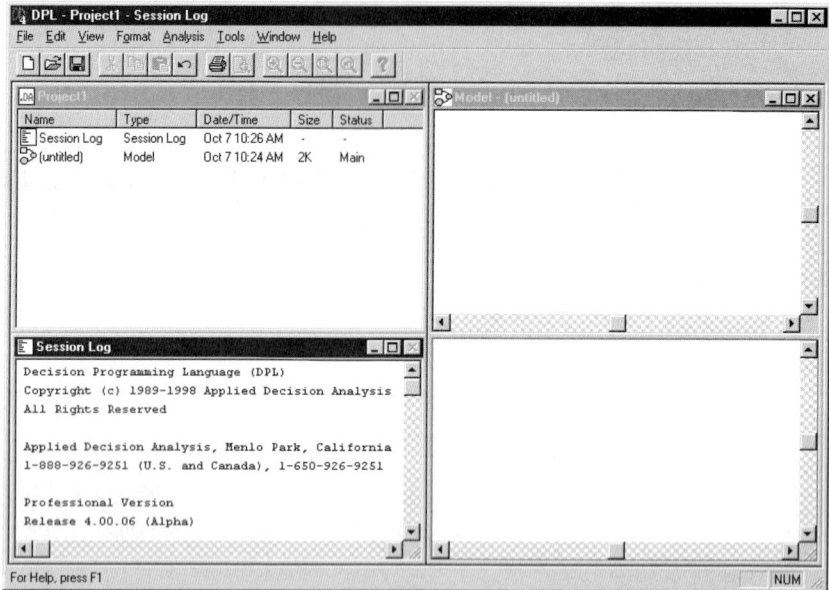

As the name implies, the Project Manager is also the place to manage your project, beginning with renaming and deleting windows. The Date/Time column tells you when each element was last changed, so you will always know which is the most current information. The Project Manager is also where you designate which model is the "main" model (more on this later).

The Model Window

The Model Window is the heart of a DPL project, and is where you construct your model using Influence Diagrams and Decision Trees. The Model Window is divided into two panes by a splitter bar. The top (or left) pane is the Influence Diagram, and the bottom (or right) pane is the Decision Tree. Only one pane is active at a time, and you can tell which pane is active because it has a thin magenta outline. You can use the splitter bar between the two panes to resize the panes. If you prefer to maximize the view of a pane, double-click on the splitter bar. With a pane maximized in the window, pressing the Tab key will switch to a maximized view of the other pane.

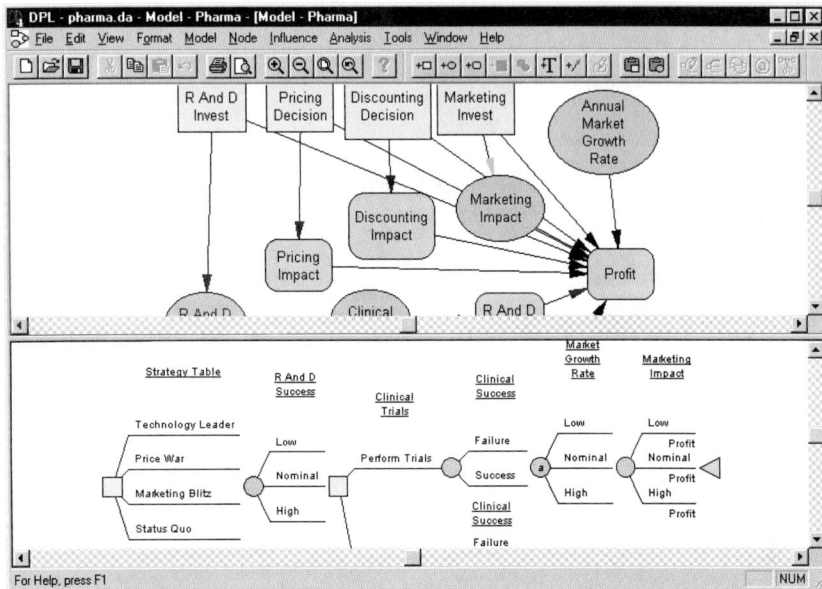

A new DPL project contains only one Model Window, but you can add a Model Window to the project by selecting File Add Model. While you can have as many models as you like, DPL can only run one model at a time. Multiple models should be used to store old versions or different variations of your model, not component parts of a single model which you want to run in the same analysis.

Before running an analysis, you need to tell DPL which model is the one it should analyze. This is done in the Project Manager. Simply right-click on the model you want to analyze and select "Make Main". DPL will identify the Main model by the word "Main" in the "Status" column of the Project Manager. If you only have one Model Window, DPL will designate it as the Main model by default.

Running An Analysis

You can run a Decision Analysis from any window in the DPL project. Regardless of which window is active, DPL will run the analysis on the Main model.

Endpoint Database

When you run a Decision Analysis, DPL creates an Endpoint Database containing the numerical results of the analysis. From this database, DPL can create a Policy Tree, Risk Profile Chart, Policy Summary, and Value of Information/Control Chart without having to re-run the model.

When you select Analysis Decision Analysis, DPL displays the Decision Analysis Options dialog box. The dialog box asks you which of the following four outputs you would like to generate: Policy Tree, Risk Profile Chart, Policy Summary, and Value of Information/Control Chart. DPL will automatically generate the selected outputs at the completion of the run. In addition, DPL generates an Endpoint Database, containing the numerical results of the analysis. From this database, DPL can generate any of the four outputs without re-running the analysis, which can be a big time-saver!

To generate outputs from the Endpoint Database, simply select Analysis Decision Analysis as you would if you were going to re-run the model. At the bottom of the Decision Analysis Options dialog box is a check box labeled "Reuse endpoint database", which is checked by default. If this box is checked, DPL will not re-run the model, and use the results from the Endpoint Database to generate the outputs. To force DPL to re-run the model, simply uncheck the box. (**Note:** The current model must match the model which was used to generate the Endpoint Database. If you have changed the model since running the analysis, DPL will not allow you to re-use the Endpoint Database.)

Because the Endpoint Database can be extremely large, you can only have one Endpoint Database in a project. Each time you re-run the model, DPL overwrites the existing Endpoint Database with data from the new run. Because the Policy Tree and Policy Summary rely on the data from the Endpoint Database, you can have only one Policy Tree and one Policy Summary in the model.

Saving

When you select File Save Project in any window, DPL saves all the elements of the project. DPL projects have a file extension of ".da". When you save the project, you save everything except the Session Log. (If you really want the Session Log saved, you'll have to export it as a file to save it.)

Some elements are not saved when you re-run the model:

- The Endpoint Database is overwritten each time you run the model (this is because of its size). Because the Policy Tree and Policy Summary rely on data from the Endpoint Database, they are also erased each time you run the model. **Note:** When you open a file, if the current Main model is not the model which was used to generate the Endpoint Database (e.g., you changed the model between the time you last ran it and the time you saved the project), DPL will not be able to regenerate the Policy Tree and Policy Summary.
- A trickier issue is what happens to Risk Profile Data. If you generate a Risk Profile for the Objective Function only, DPL saves it in the Project Manager with the name "Expected Value". When you run another Risk Profile on the Objective Function, DPL again saves it with the name "Expected Value", overwriting the previous one. To keep the Risk Profile Data from being overwritten, simply rename it to something other than "Expected Value" in the Project Manager before running another Risk Profile. (**Note:** If you generate Risk Profile Data for "Initial decision alternatives", they must be renamed as well in order to avoid being overwritten.)
- When you run Rainbow Diagrams, Tornado Diagrams, and Value of Information/Control Charts, none of the existing elements are erased. You can keep running them and they will keep being added to the Project Manager. The same is true for Risk Profile Charts, which contain the chart formatting information for Risk Profile Data, such as titles, legends, etc.

Importing and Exporting Elements of a DPL Project

DPL allows you to import and export elements of DPL projects. This can be useful if you would like to have an influence diagram from one project included in another project. It is also the means for exchanging models and outputs with previous versions of DPL.

To import an element into your project, go to (or create) a window of the same type. For example, to import a DPL influence diagram and decision tree model, go to the Model Window and select File Import. Enter the path and filename of the element you wish to import, or click Browse to search for it, then click Open.

Warning! The contents of the active window will be overwritten by the file being imported. To prevent this from happening, create a new window into which you can import the element.

To export an element from your project, go to the window which contains the element. For example, to export an influence diagram and decision tree model, go to the Model Window and select File Export. DPL will prompt you to specify a path and assign the element a filename. A copy of the elements will be written to a file. (**Note:** The element will remain in your model.)

Elements of DPL projects have various file extensions when they are saved separately from a DPL project. These extensions are as follows:

Element	File Extension
DPL Project	.DA
Influence Diagram/Decision Tree	.INF
DPL Program	.DPL
Tornado Diagram	.CYC
Distribution points	.CSV
Value of Information/Control	.CYC
Rainbow Diagram	.CSV
Session Log	.LOG
Policy Tree (Text)	.POL

Printing

You can print the contents of any window you're in simply by selecting File Print.

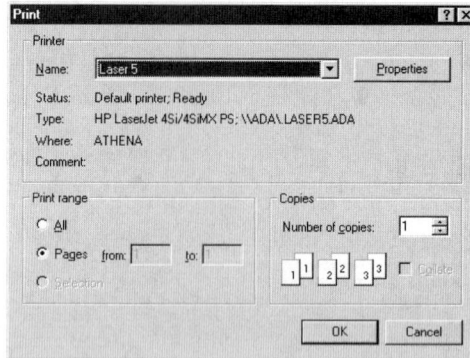

You can control the appearance your printed documents by selecting File Page Setup.

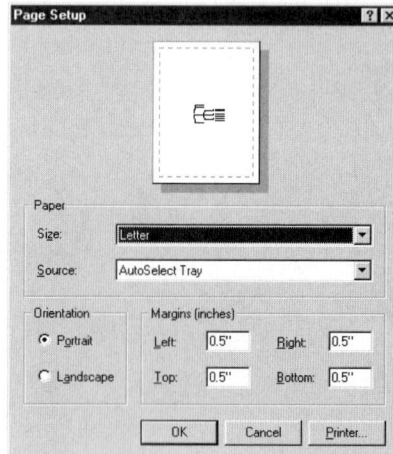

Finally, you can preview what will be printed by selecting File Print Preview.

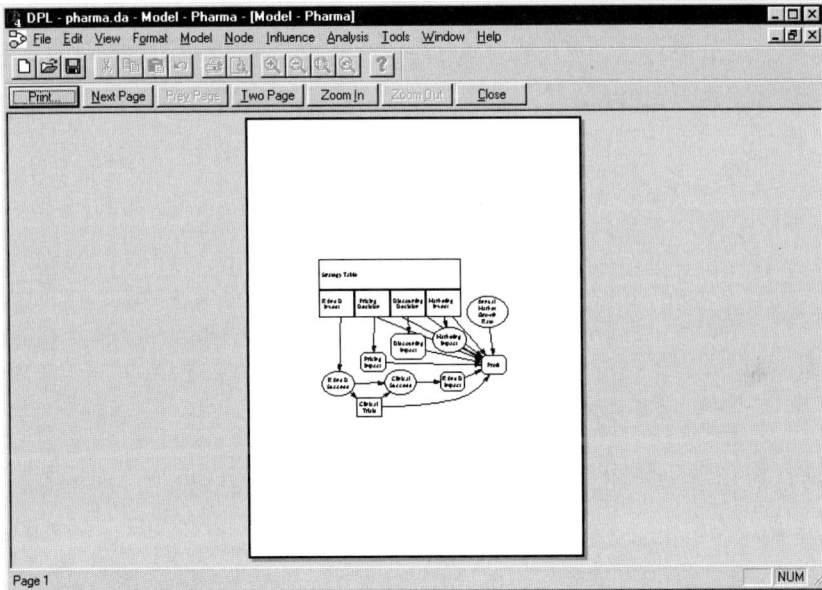

In the Model Window, DPL will print the visible contents of the active pane (either the Influence Diagram or the Decision Tree, but not both at once). To print the entire contents of the pane, simply select Zoom Full before printing to make sure the entire model is visible.

Because printing an entire model on one sheet of paper may make it difficult to read, DPL also allows you to zoom in on a section of the model and print the entire model (on multiple pages) to the same scale. To do this, zoom in on a section of the model. After you select File Print, go to the "Print range" section of the Print dialog box and click "All" instead of "Pages".

Session Log

The Session Log maintains a record of your DPL session, beginning with basic
information on the version number, copyright statement, and current date and time.
When you run an analysis, the starting and ending times as well as the results are
written to the Session Log. These results vary based on the type of analysis you run.
For example, the results of a decision analysis include the number of paths in the
model, the expected value (and certain equivalent if applicable), and the 10%, 50%,
and 90% points of the outcome distribution. As another example, the results of an
Expected Value Tornado include the name of the variables analyzed, the values at
which each variable was analyzed and the corresponding expected value of the model
for each value.

```
DPL - pharma.da - Session Log - [Session Log]
File  Edit  View  Format  Analysis  Tools  Window  Help

Decision Programming Language (DPL)
Copyright (c) 1989-1998 Applied Decision Analysis
All Rights Reserved

Applied Decision Analysis, Menlo Park, California
1-888-926-9251 (U.S. and Canada), 1-650-926-9251

Professional Version
Release 4.00.06 (Alpha)


17:43:23  Compiling: (untitled)
Number of paths = 324
17:43:24  Complete

17:43:26  Analyzing... Fastest exact
17:43:28  Complete

Expected value = 1442.2
Percentiles 10/50/90:
906.015657224278,1483.79772527333,2302.36159709784

For Help, press F1                                                   NUM
```

The Session Log is not saved with the project. To save the contents of the Session
Log, you must export them to a file, or copy them to a Program Window.

Options

You can customize many of the default settings in DPL in the Options dialog box, which is accessed by selecting Tools Options. This dialog box contains six tabs, each of which allows you to set global parameters for DPL. The Options that pertain to the current project are stored in the .DA file, whereas preferences and computer dependent settings are stored in the system registry.

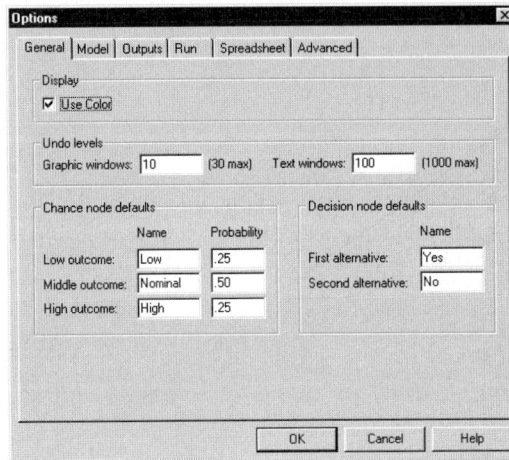

The first five tabs contain general options, from default node state names to optimization options. The sixth tab contains switches which turn on seldom-used advanced features in DPL. The switches are off by default, which prevents the features from being used. The purpose of this is to prevent new DPL users from getting into trouble with some of the more advanced or potentially confusing features of DPL. The settings for these switches are saved with each project.

On-Line Help

To get help at any time in DPL, use the Help Menu or press F1. DPL offers context-sensitive help, which means that clicking on the Help button or pressing F1 will open Help to the topic corresponding to what you're doing in DPL. The On-Line Help is the most complete source of "How-to" documentation.

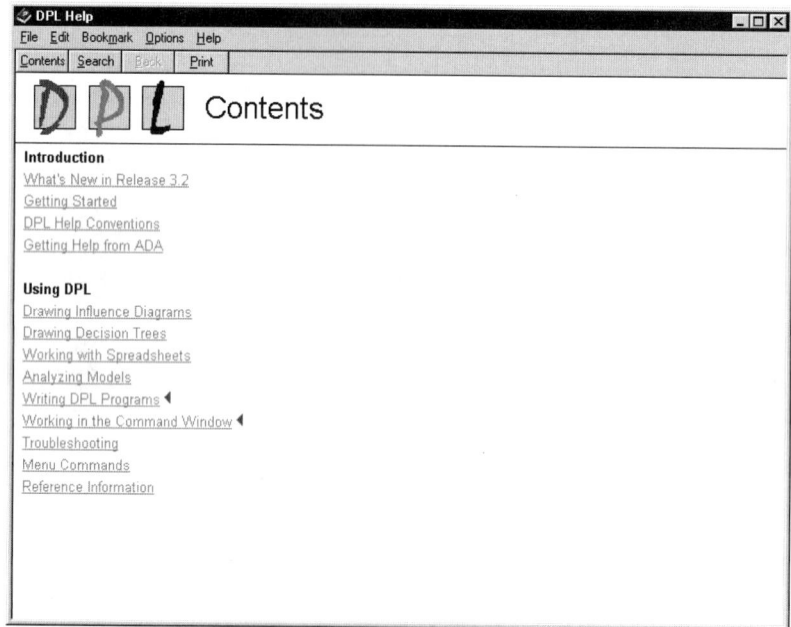

If you are new to decision analysis, DPL's many features can make it a bit difficult to know where to start. This tutorial walks you through building and analyzing a simple model. The tutorial can also be used as a demonstration if someone asks you what this "decision analysis" is all about.

You can download other complete tutorials from our website.

Chapter *3*

Chapter 3: Decision Analysis Tutorial

Introduction

Imagine that you are in charge of developing a business strategy for a product of HealthDrug, a large pharmaceutical company. The business strategy involves making decisions about four key aspects of your business: Marketing, pricing, research and development, and discounts for large customers. You have identified four strategies, each encompassing these four decisions, which you think might be successful for HealthDrug, and have named them Technology Leader, Price War, Marketing Blitz, and Status Quo. Technology Leader positions you as a premium pharmaceutical company, with emphasis on leading-edge R&D and high-priced products. Price War attempts to increase market share by selling your products for less than your competitors', placing little emphasis on R&D. Marketing Blitz relies on strong advertising, holding steady on pricing, and sacrificing R&D. Status Quo has you continuing to do what you have been doing.

You want to pick the strategy that maximizes the net present value of future revenues, which have been modeled in a spreadsheet. However, the model has a number of uncertainties, and you would like to incorporate these uncertainties into your analysis.

Which strategy should you choose?

The Spreadsheet

A simple cashflow spreadsheet has been built to model the revenues generated by your drug. To review this spreadsheet:

▸ Start Microsoft Excel (version 5.0 or higher).

Excel.exe

▸ Open "c:\DPL\tutorials\pharma.xls".

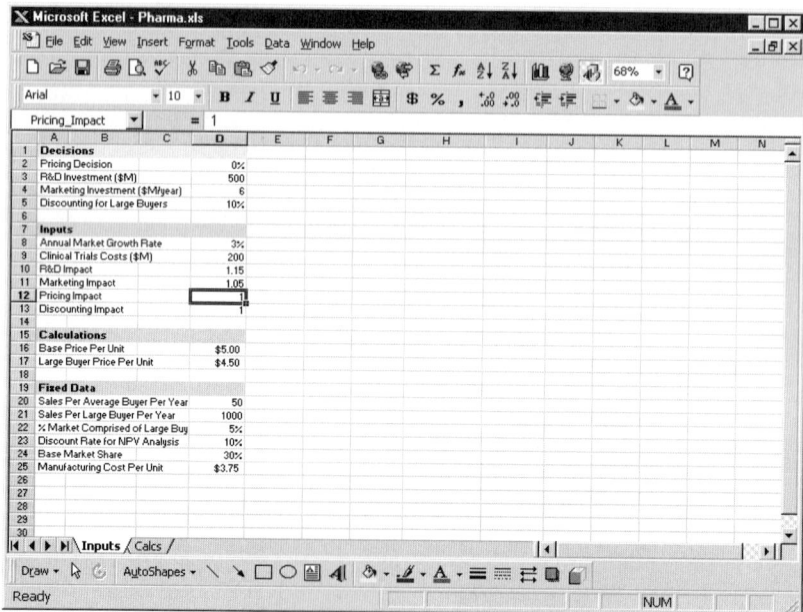

This is a greatly simplified six-year cashflow model, but it will be sufficient for your needs. The first worksheet, titled "Inputs", contains the decisions, inputs, simple calculations, and fixed data. The second worksheet, titled "Calcs", contains the cashflow model.

The four decisions consider the following: the amount by which you are going to raise or lower the price of the drug, how much you are going to invest in R&D on a new version of the drug, how much you are going to invest in marketing the drug, and the level of discount you'll give to large customers.

The inputs section contains all the variables you are going to model in DPL. The first two inputs deal with the annual growth rate of the market and the costs of performing clinical trials. The four "impact" variables are estimated multipliers to market share based on the four decisions.

The third section contains calculations, which you can choose to model in DPL or Excel. To keep the DPL model as simple as possible, leave them in the spreadsheet. The fourth section, fixed data, contains known data and inputs about which there is not a significant amount of uncertainty.

▶ Click on the Calcs tab.

The Revenues section of the cashflow model calculates the revenues generated by your drug based on the market size, market share, and sales.

▸ Scroll down to view the Costs section of the worksheet.

The Costs section analyzes both the capital and operating costs associated with the drug. The profit for each year is also calculated here. Below the Costs section, you can see that the Profit calculation for the base case is $1.058 billion. The spreadsheet model, however, assumes a particular value for all the input variables. You need to examine the potential impact of uncertainty in these inputs.

▸ Select File Close to close the spreadsheet. Do not exit Excel.

Welcome to DPL!

Now that you have a way to calculate Profit for any scenario, build a decision model in DPL.

▸ Start DPL.

DPL opens to a view of the Project Manager, Session Log, and Model Window.

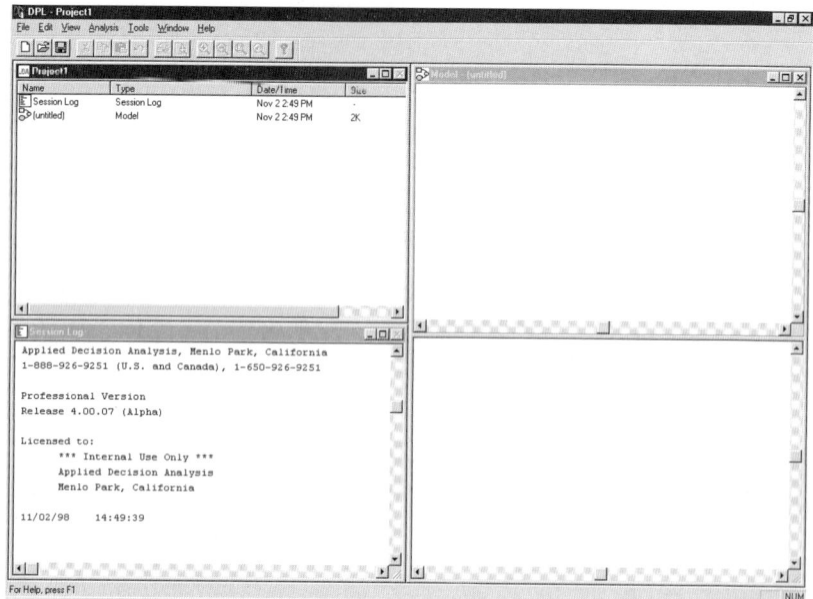

DPL begins with the Project Manager as the active window. (You can tell it is the active window because the window bar and title are the same color as the DPL bar and title at the top of the screen, while the bars and titles of the Session Log and Model Window are grayed out.) The Project Manager is a listing of all of the windows and files in your DPL project. Here you can make windows active, rename them, or delete them. You can also move to any window in the list by double-clicking on the window name or icon in the Project Manager, by clicking on the window itself, or by selecting the window from the Window menu.

▸ Double-click on the Model (untitled) icon in the Project Manager window.

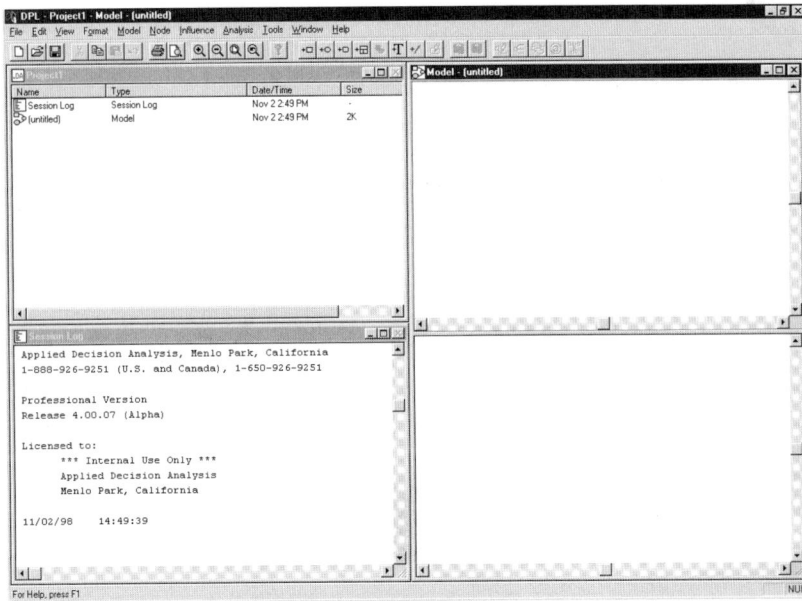

The Model Window is divided into two panes by a splitter bar. The top pane is the Influence Diagram, and bottom pane is the Decision Tree. You can tell that the Influence Diagram is the active pane because it has a thin magenta outline. You can make the Decision Tree active by clicking on it, by selecting View Decision Tree from the menu, or by pressing the Tab key.

▸ Press Tab to make the Decision Tree pane active.
▸ Press Tab again to make the Influence Diagram pane active.

You'll be using the Model Window first, so maximize it.

▸ Click the Maximize icon in the upper-right hand corner of the Model Window.

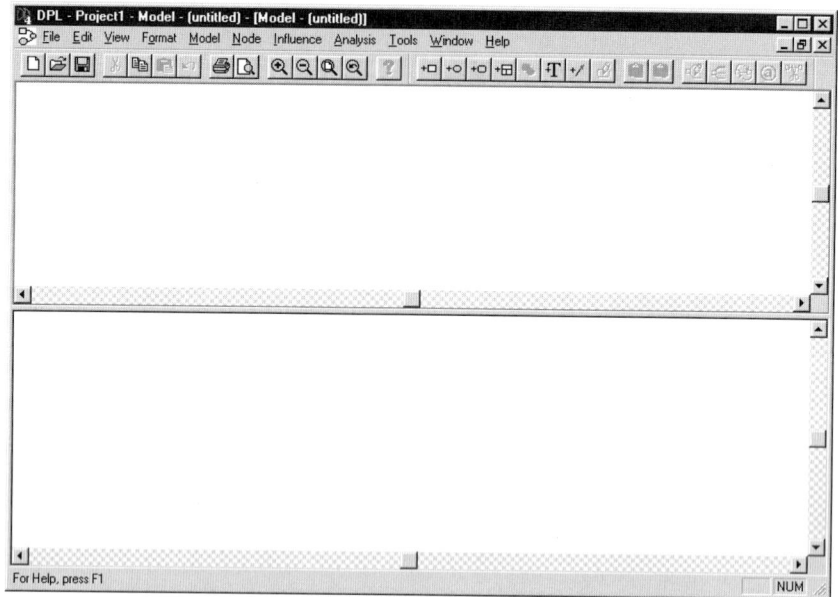

You're going to build the model in the Influence Diagram, so you'll maximize the size of the Influence Diagram pane. The splitter bar between the two panes resizes the panes. You can maximize the view of a pane by moving the splitter bar all the way to the edge of the window. With a pane maximized in the window, pressing the Tab key will switch to a maximized view of the other pane.

▸ Click-and-drag the splitter bar down to the bottom edge of the window.

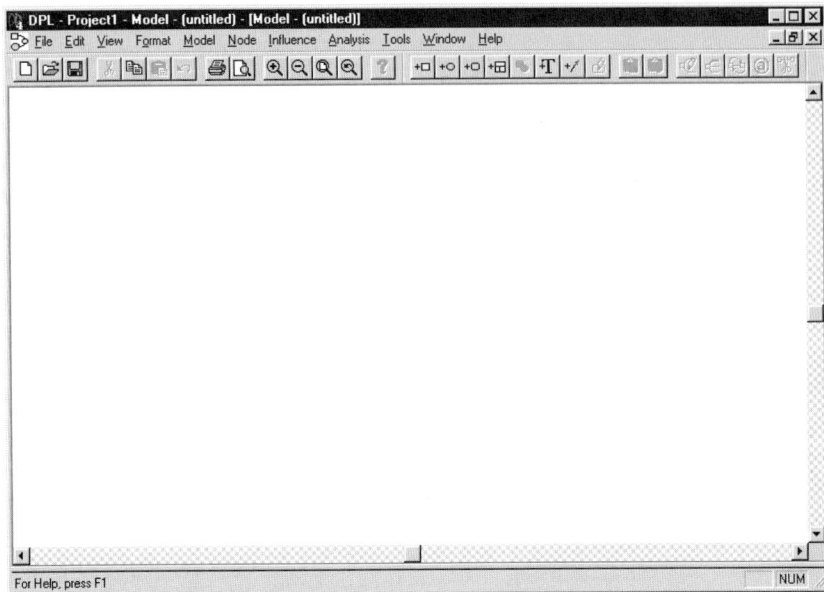

You're now ready to begin building the DPL model.

The DPL Model

You're now ready to turn the spreadsheet model into a DPL Model.

▶ Select Tools Create Model from Excel.

DPL displays the Create Model from Spreadsheet dialog box.

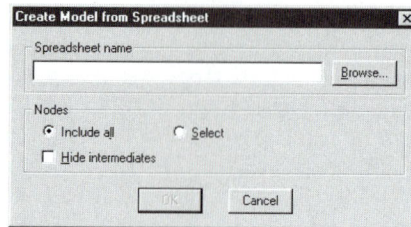

▶ Click on Browse to find the spreadsheet file.

DPL displays the Input File Name dialog box.

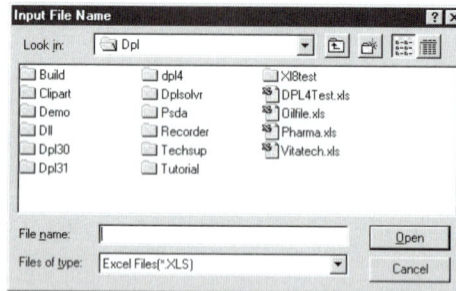

▶ In the Input File Name dialog box, select "c:\DPL\Tutorials\Pharma.xls", then click Open.

where you saved it to in your home directory

DPL displays the Create Model from Spreadsheet dialog box again.

▶ Check the "Hide intermediates" box to keep DPL from importing nodes which represent intermediate calculations.

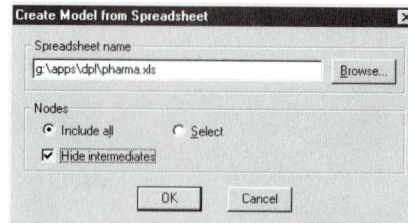

▶ Click OK.

DPL creates value nodes from the named cells in the spreadsheet.

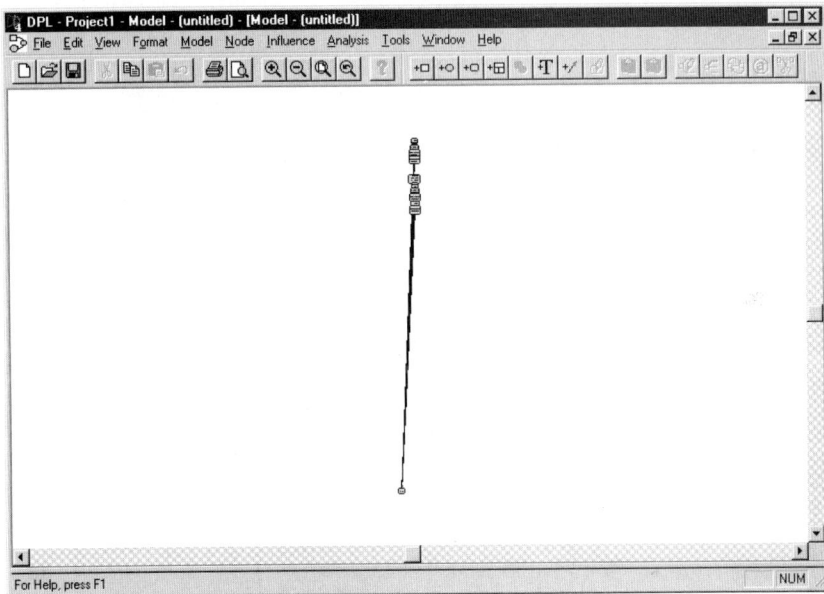

The nodes are placed in the influence diagram in roughly the same order as they appear in the spreadsheet. You can tell DPL to re-arrange the nodes to make the diagram readable.

‣ Select Format Arrange Diagram Left-to-Right.

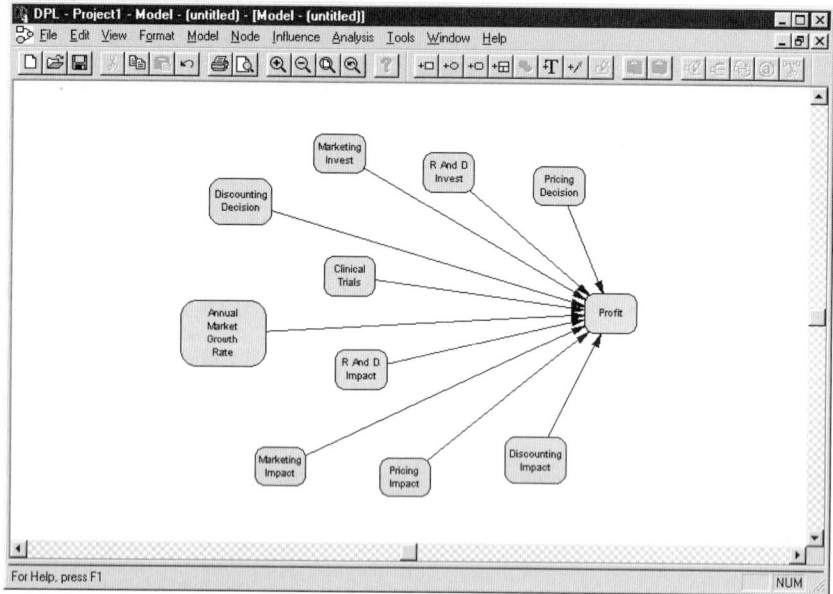

Each node has been assigned a name corresponding to its cell name in the spreadsheet, and is linked to the spreadsheet cell for which it was named. The nodes are all blue rounded rectangles, indicating that they are value nodes. Value nodes can represent either a constant or a formula.

You can view or change a node's name, data, or links in the Node Definition dialog box.

▶ Double-click on the node Marketing Invest, or click once on Marketing Invest and then select Node Edit Definition.

DPL opens the Node Definition dialog box to the Data tab. You can see that Marketing Invest has been assigned the value of 6 it has in the spreadsheet.

▸ Click on the General tab.

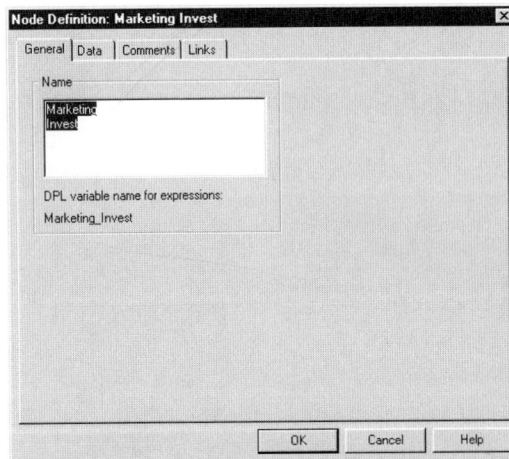

The General tab contains the name and states of the node. (Because Marketing Invest is a value node, it does not have any states.) In the Name section of this tab DPL displays the variable name you can use in formulas to reference Marketing Invest, which is Marketing_Invest.

▶ Click on the Comments tab.

The Comments tab is a place to write special information about the node. Comments are particularly useful if someone else will be using the model.

▶ Click on the Links tab.

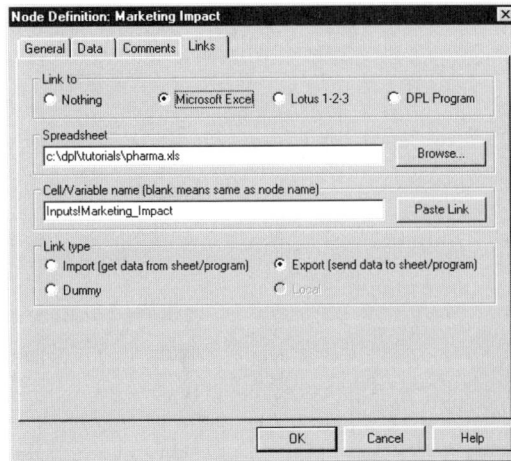

In the Links tab you can see the application, file name, sheet name, and cell name of the spreadsheet to which the variable is linked. You can also see whether it is an import, export, dummy, or local variable. Nodes with data for DPL to send to the spreadsheet are Export variables. In this model, all nodes are Export variables except Profit, which is an Import variable.

▶ Click OK.
▶ Double-click on the Profit node.

The Data tab does not show a value for Profit since the calculated value for Profit will be imported from the spreadsheet during the DPL analysis.

▶ Click OK.

Now that DPL is linked to the spreadsheet model, you can run the spreadsheet calculations from DPL. DPL will open your spreadsheet in Excel for you.

▶ Select Analysis Decision Analysis.

DPL displays the Value Function dialog box, which asks you which spreadsheet variable to calculate.

Value Function

Enter a variable or an expression for the model's value function (the "bottom line"):

OK   Cancel

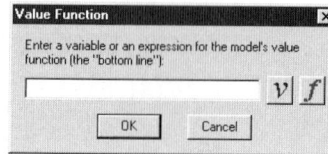

You want to calculate Profit, which can be selected from the list of variables.

▶ Click on the Variable icon.

v

▶ In the dialog box, double-click on Profit.

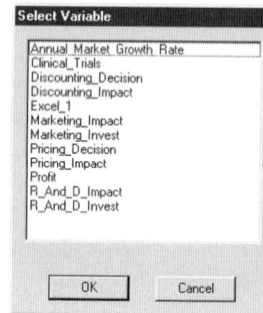

Select Variable

Annual Market Growth Rate
Clinical_Trials
Discounting_Decision
Discounting_Impact
Excel_1
Marketing_Impact
Marketing_Invest
Pricing_Decision
Pricing_Impact
Profit
R_And_D_Impact
R_And_D_Invest

OK   Cancel

▶ Click OK.

DPL displays the Decision Analysis Options dialog box. In this dialog select which outputs you wish DPL to generate, the evaluation method to use, and whether or not you wish to see a status display. For a model without decisions or uncertainties, Risk Profile is the only output available.

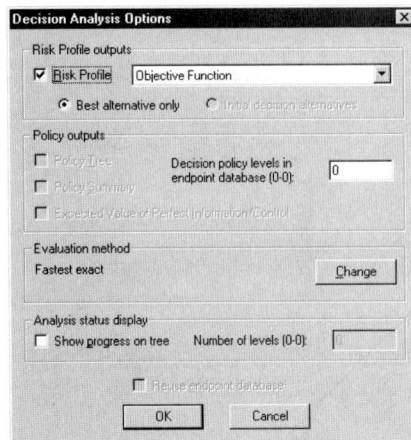

▶ Click OK.

DPL sends the input variable settings to the spreadsheet, calculates the spreadsheet, and gets the value for Profit. The Risk Profile Chart shows the Profit calculation graphically.

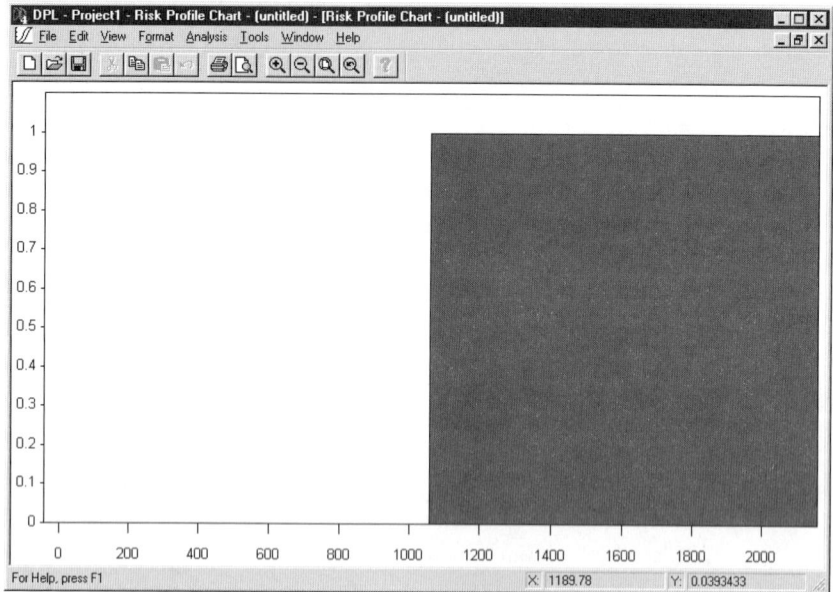

It appears that the value for Profit is just over $1 billion, which is the value seen in the spreadsheet. The links to the spreadsheet are working correctly. (The precise value calculated for Profit is written in the Session Log. You could check it if you wanted to verify the accuracy more carefully.)

This distribution is not very useful, so you can delete the Risk Profile Chart. In the top-right-hand corner of the window, there are two sets of icons. The top set minimizes, tiles, and closes DPL. The lower set minimizes, tiles, and closes the active DPL window. To delete the Risk Profile Chart, click on the "X" icon in the lower set.

▶ Delete the Risk Profile Chart.

Modeling Decisions and Strategy Tables

Next you will change the four value nodes representing your decisions into decision nodes. You will then incorporate these four decisions into a Strategy Table.

▶ Click on Marketing Invest.

Now you want to perform the Change Type command. In the Model Window, frequently-used commands such as Change Type can be accessed in four ways: Menus, shortcut keys, toolbar icons, or context menus (context menus appear when you right-click on something). Therefore, to change the node type, you can select Node Change Type, type Ctrl-T, click on the Change Node Type icon in the toolbar, or right-click on Marketing Invest to get the context menu, then select Change Node Type. To keep the tutorial simple, only one method for each command will be referred to, but if you prefer to use another method you may wish to do that instead.

▶ Click on the Change Node Type icon.

DPL displays the Node Type dialog.

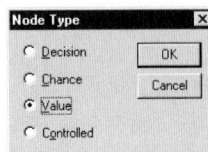

▶ Select Decision and click OK.

DPL displays the General tab of the Node Definition dialog box.

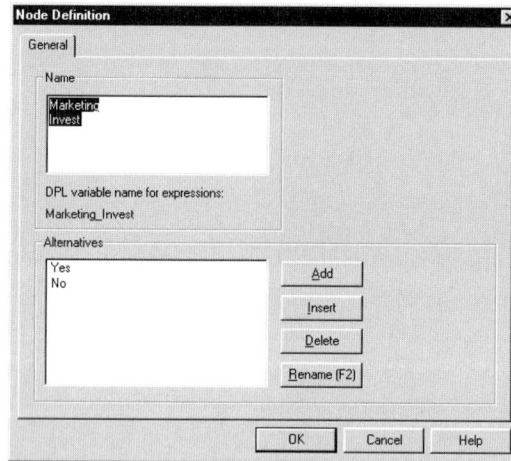

You have three alternatives for your marketing investment decision: invest a lower amount of funding than the current level, the same amount, or a higher amount. The default number of alternatives is two, so insert another alternative and then rename the three alternatives.

▸ Click on the Yes alternative, then click the Insert button.
▸ Type "Lower" and press Enter.

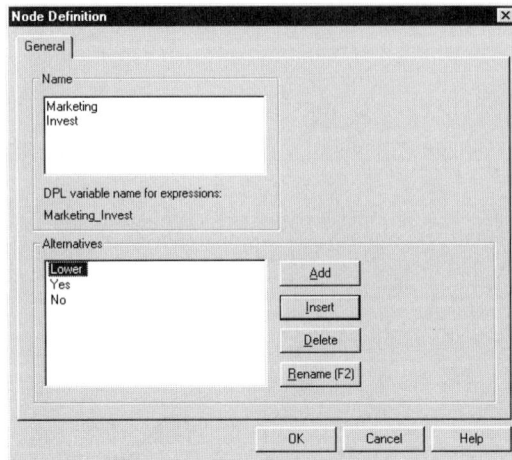

▸ Double-click on the Yes alternative (or click on it once and then click the Rename button).

▸ Type "No change" and press Enter.

▸ Repeat to rename the No alternative as "Higher".

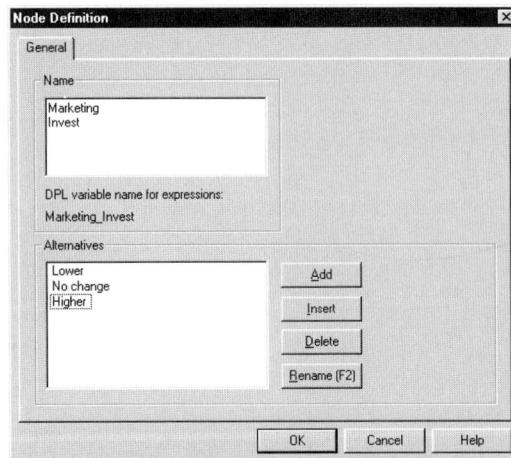

▸ Click OK.

DPL displays decisions in the Influence Diagram as yellow rectangles. (Right now it is magenta because it is selected.)

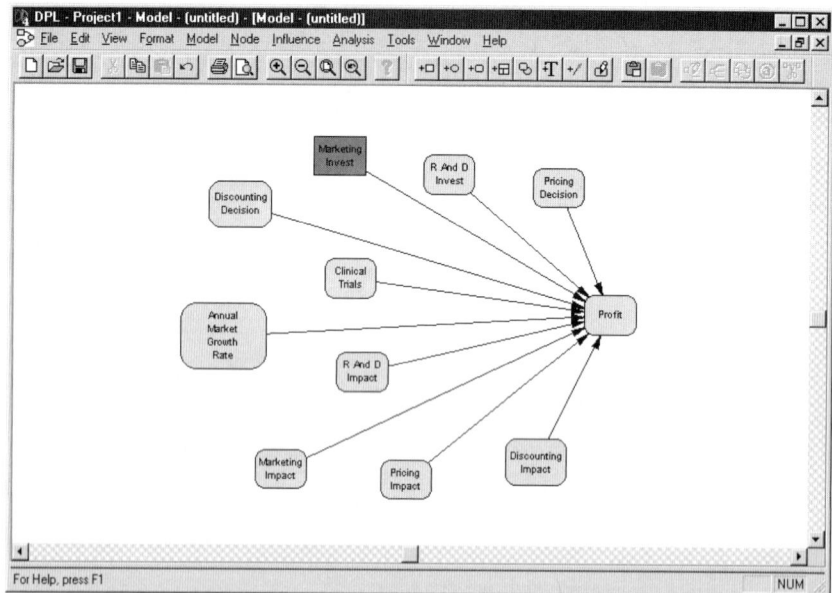

You can now enter values for the different alternatives of Marketing Invest that reflect the different levels of investment.

- ▸ Double-click on Marketing Invest.
- ▸ Click on the "Lower" branch.
- ▸ Type "4", then press Enter.
- ▸ Press Enter again to move to the "Higher" branch.
- ▸ Type "8", then press Enter.

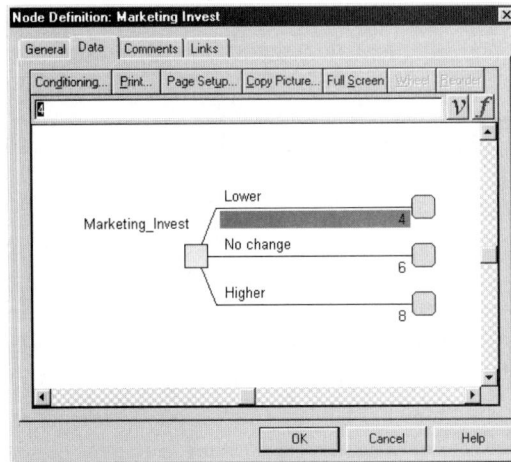

▶ Click OK.

Now, repeat the same process to change the other three decision variables into decision nodes. R And D Invest indicates the amount of funding you'll spend on R&D (in $ Millions), Pricing Decision is the percentage by which you'll change the price of your drug, and Discounting Decision reflects the percent by which you discount the drug for large buyers.

▶ Repeat the previous steps to turn R And D Invest, Pricing Decision, and Discounting Decision into decision nodes. The states and values are as follows:

R And D Invest:	Pricing Decision:	Discounting Decision:
Lower: 200	Lower: -0.05	More: 0.2
No Change: 500	No Change: 0	No Change: 0.1
Higher: 800	Higher: 0.05	Less: 0.05

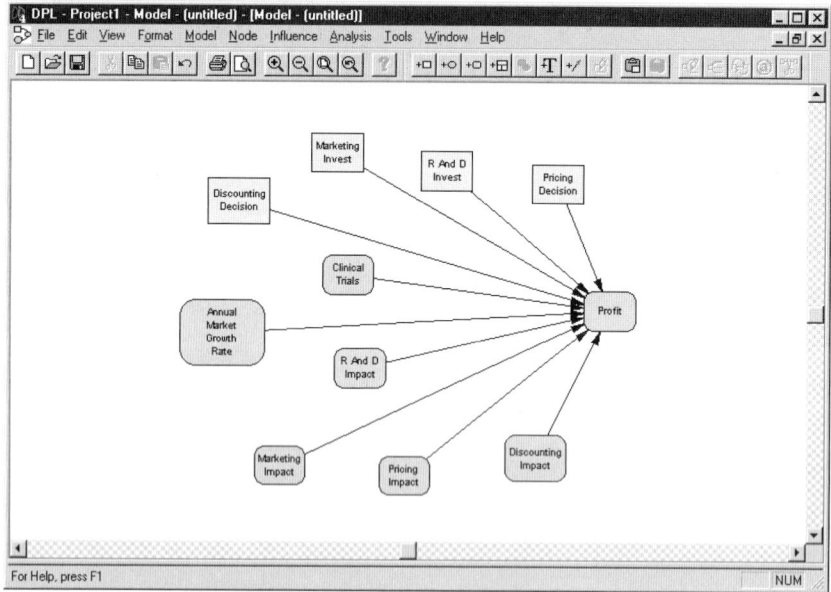

Next, create a strategy table from these four decisions.

▶ Click on the Create Strategy Table icon.

Place the strategy table above Annual Market Growth Rate. (If you want to move the node once it has been placed, simply click-and-drag it to the desired location.)

▶ Click OK to accept the default name for the Strategy Table.

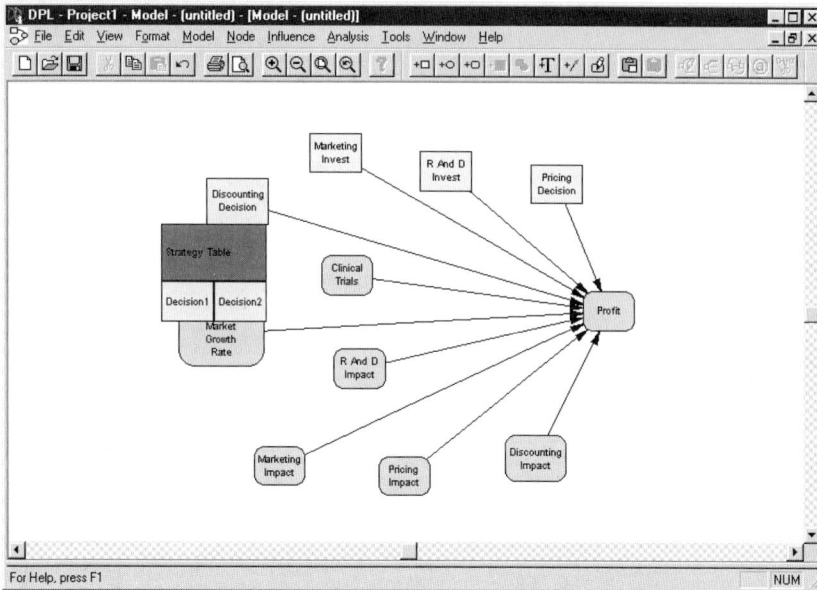

The strategy table node contains one section denoting the strategy table, and additional sections for each of the component decisions. New strategy nodes are created with two new component decisions.

▶ Double-click on the Strategy Table section of the Strategy Node.

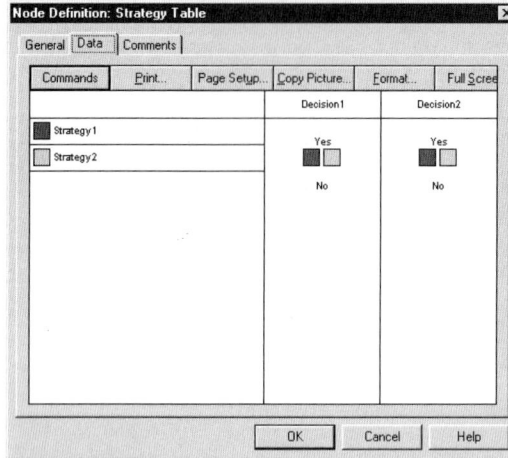

You can make the strategy table easier to see by changing the view to full screen.

▶ Click Full Screen.

Strategy table commands are contained in the context menu. You will use these commands to add your decisions to the strategy table.

▶ Click the right mouse button.
▶ Select Add Decision from the context menu.

DPL displays a list of the decisions.

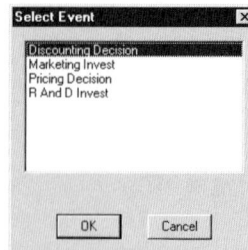

▶ Click OK to select Discounting Decision.

▶ Repeat the previous steps to add the remaining three decisions to the strategy table.

Node Definition: Strategy Table							⊠
Commands	Print...	Page Setup...	Copy Picture...	Format...	Full Screen		

	Decision1	Decision2	Discounting Decision	Marketing Invest	Pricing Decision	R And D Invest
■ Strategy1	Yes ■□	Yes ■□	More ■□	Lower ■□	Lower ■□	Lower ■□
□ Strategy2	No	No	No change	No change	No change	No change
			Less	Higher	Higher	Higher

Next, delete the two decisions which were created with the strategy table.

▶ Click on Decision1.
▶ Click on the right mouse button, then select Delete from the context menu.
▶ Repeat to delete Decision2.

Node Definition: Strategy Table				□ ×
Commands	Print...	Page Setup...	Copy Picture...	Format...	Full Screen

	R And D Invest	Pricing Decision	Discounting Decision	Marketing Invest
■ Strategy1	Lower ■□	Lower ■□	More ■□	Lower ■□
□ Strategy2				
	No change	No change	No change	No change
	Higher	Higher	Less	Higher

Now enter your four strategies into the strategy table.

▸ Click on the right mouse button, then select Create Strategy from the context menu.

▸ Repeat.

▸ Click on Strategy1.

▸ Click the right mouse button, then select Rename from the context menu.

▸ Type "Technology Leader", then press Enter.

▸ Repeat the previous steps to rename the remaining three strategies "Price War", "Marketing Blitz", and "Status Quo".

The icons in the strategy table indicate which decision alternatives are chosen in each strategy. You need to assign each strategy an alternative to choose for each decision. By default, all strategies are assigned the first alternative of each decision. You want to change the assignments to reflect your strategies. For example, you want the strategy Technology Leader to select R And D Investment to be Higher, Pricing Decision to be Higher, Discounting Decision to be Less, and Marketing Invest to be Lower. Now change the strategy table to reflect this.

▸ Click-and-drag the icon next to Technology Leader (a red square).
▸ Place the icon under the Higher alternative of R And D Invest.

This tells DPL that the strategy Technology Leader will choose the "Higher" alternative of R And D Invest.

▶ Repeat the previous steps for all decisions in Technology Leader and then for all strategies until the strategy table has the following structure:

	R and D Invest	Pricing Decision	Discounting Decision	Marketing Invest
Technology Leader	Higher	Higher	Less	Lower
Price War	Lower	Lower	More	No Change
Marketing Blitz	Lower	No Change	No Change	Higher
Status Quo	No Change	No Change	No Change	No Change

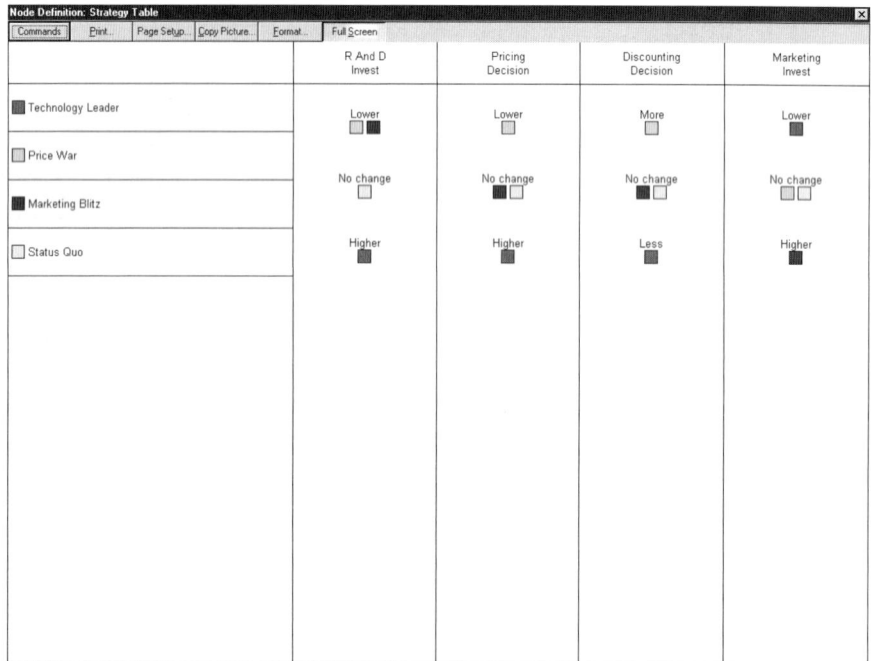

▶ Click Full Screen.
▶ Click OK.

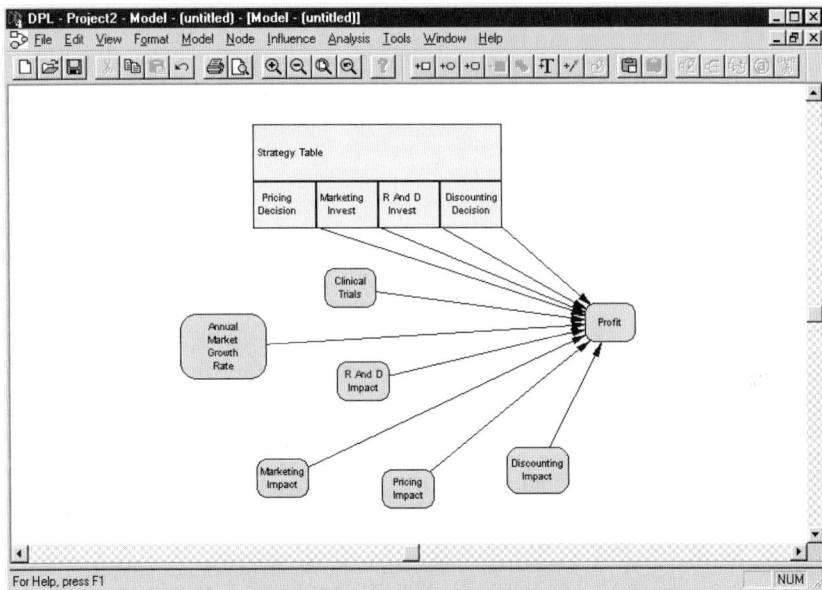

As you built the influence diagram, DPL automatically constructed a decision tree for you.

▸ Select View Decision Tree.

The decision tree has four branches, one for each strategy in the Strategy Table.

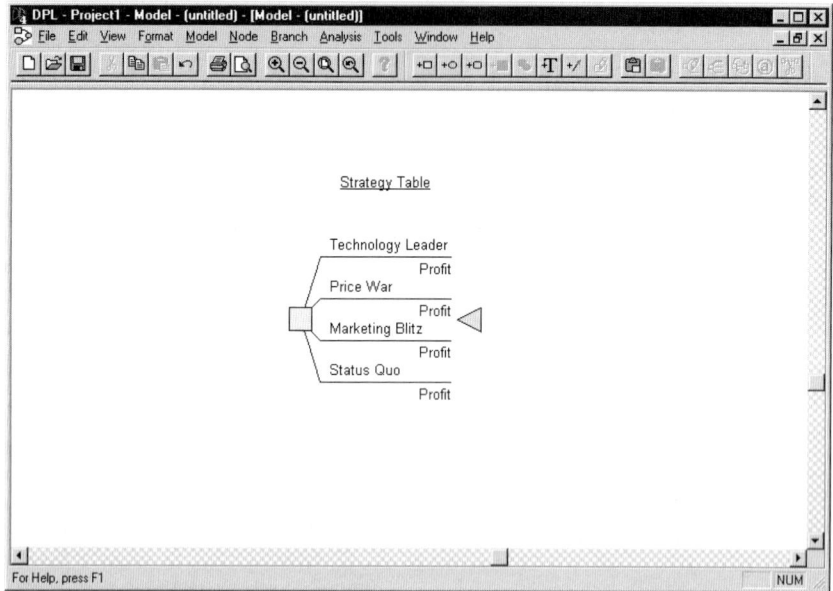

DPL has constructed this decision tree based on your construction of the influence diagram. It has four paths, one for each strategy in the Strategy Table. The blue triangle on the right represents an endpoint, which tells DPL that the Strategy Table is the last point in the tree. The "Profit" underneath each branch is the "get/pay expression", which tells DPL to calculate Profit at each endpoint.

If you wanted to change the decision tree, you could do so, and DPL would leave future updating of the decision tree to you. This default tree will give you the results you want, so continue to leave it in DPL's control.

▸ Select View Influence Diagram.

Now give DPL a label for your outputs. This is done in the Objective dialog box.

▸ Select Model Objective.

DPL displays the Objective dialog box. If the model had multiple attributes, you would identify them here along with an objective function. Because the model only has one attribute, DPL simply assumes the objective function to be Attribute1. All you have to do is provide a units label for the outputs, which is "$ Millions."

▶ In the "Units for output labels" input box, type "$ Millions".

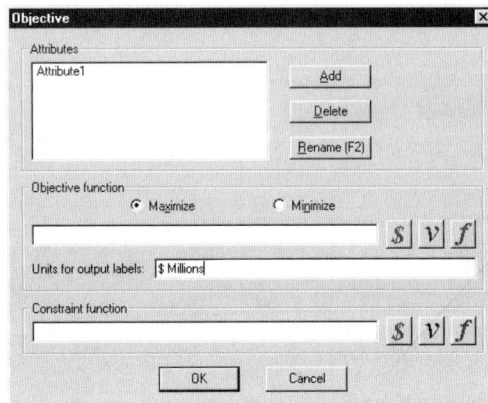

▶ Click OK.

Now you're ready to run a decision analysis. First save the model.

▶ Select File Save Project.
▶ Name the file Pharma.da and click Save.
▶ Select Analysis Decision Analysis.

Accept the default output selections, which are a Risk Profile for your objective function and a Policy Tree.

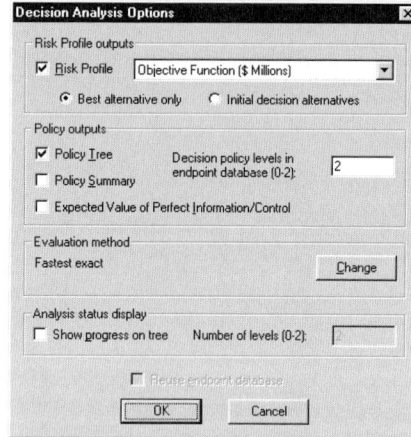

▶ Click OK.

DPL runs the analysis and opens the Policy Tree Window.

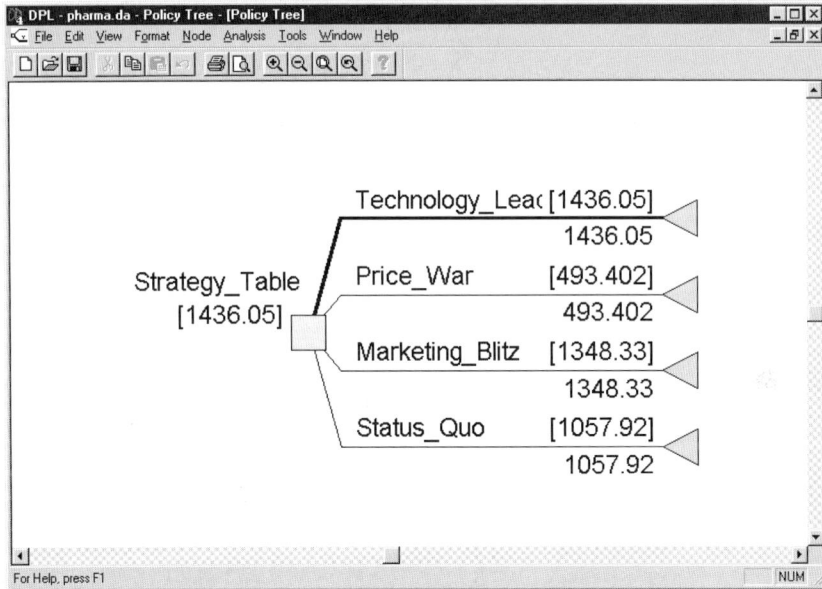

```
DPL - pharma.da - Policy Tree - [Policy Tree]                              _ □ ×
File  Edit  View  Format  Node  Analysis  Tools  Window  Help              _ 8 ×

          Technology_Lea([1436.05]
                              1436.05

Strategy_Table  Price_War      [493.402]
   [1436.05]                      493.402

          Marketing_Blitz  [1348.33]
                              1348.33

          Status_Quo      [1057.92]
                              1057.92

For Help, press F1                                            NUM
```

The Policy Tree shows the four strategies and the value for Profit calculated for each strategy. As in the decision tree, the blue triangles represent endpoints. The number below each branch is the value for the get/pay expression of the branch. The number above each branch is the expected value of the tree from that point onwards (to the end of the tree). Because you only have one get/pay expression in your tree, the value below each branch is the same as the number above each branch.

You can see that the strategy Technology Leader returns the highest expected value — over $1.4 billion. DPL has identified this as the optimal policy by highlighting the Technology Leader branch. The expected value of a decision node is the expected value of the optimal alternative, and is displayed to the left of the yellow square representing the node.

Marketing Blitz has the second-highest value for Profit, followed by Status Quo and Price War.

Next you will look at the Risk Profile.

▸ Select Window Risk Profile Chart.

The Risk Profile Chart displays a cumulative probability distribution for Profit under the optimal policy (the Technology Leader strategy).

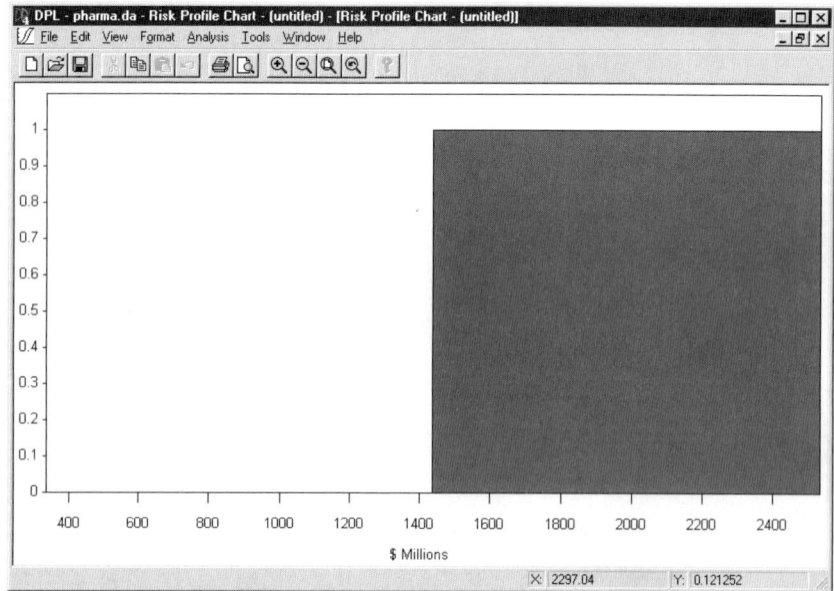

The chart states that you have a probability of 1 of achieving a value of $1.4 billion for Profit. The value matches what was seen in the Decision Policy; it has a probability of 1 because the model has no uncertainty.

By placing the cursor over the left edge of the distribution and looking at the status bar in the bottom right-hand corner of the window, you can see the value for Profit reported in the Decision Policy (represented by the x-coordinate). The y-coordinate displays the cumulative probability at the cursor location.

Again, the distribution is not very useful, so delete this Risk Profile Chart as you did the last.

▶ Delete the Risk Profile Chart by clicking on the Close icon.

Sensitivity Analysis

Now you are ready to incorporate uncertainty into the model. An Expected Value Tornado Diagram is a sensitivity analysis which helps you select which variables to model as uncertainties.

▶ Select Analysis Expected Value Tornado Diagram.

DPL displays the Select Value for Sensitivity Analysis dialog box.

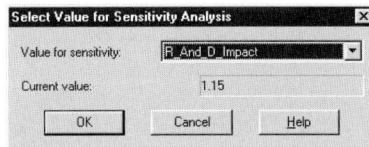

▶ Select Annual Market Growth Rate from the drop-down menu.
▶ Click OK.

DPL displays the Run Sensitivity Comparison dialog box.

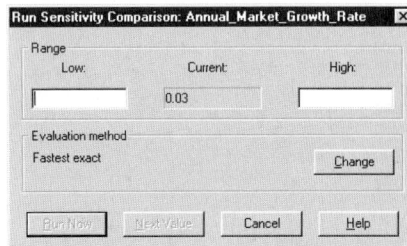

▶ Enter -0.09 for the Low value and 0.15 for the High value.
▶ Click Run Now.

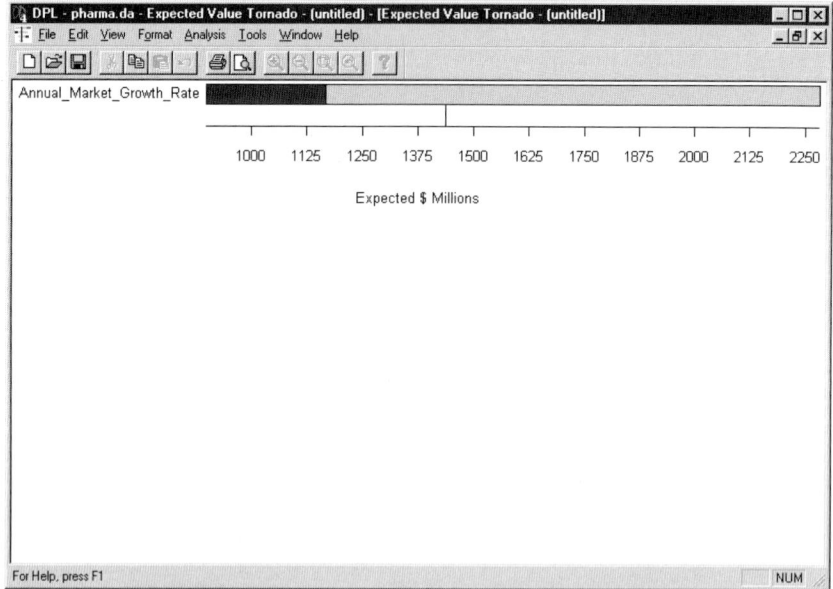

DPL has run the model twice, once each with Annual Market Growth Rate set to the low value and high value, and displays the expected values for the two runs at the ends of the bar. The vertical line in the middle of the bar shows the value of the model you have already run, which was with Annual Market Growth Rate set to its nominal value. The variance in Annual Market Growth Rate appears to have a large effect on Profit, as it ranges from below $1 billion to over $2.2 billion. The color change in the bar indicates that the value of Annual Market Growth Rate affects the strategy you select.

Next add bars for the other variables.

▸ Select Analysis Add Bar.
▸ Select Discounting Impact from the drop-down menu, then click OK.
▸ Enter 0.98 for the Low value and 1.02 for the High value.
▸ Click Run Now.

Instead of running the variables one at a time, you can define ranges for all the variables and then run them all at once. To do this, click "Next Value" instead of "Run Now".

▸ Repeat the previous steps to add bars for Marketing Impact, Pricing Impact, and R And D Impact. The ranges are as follows:

    Marketing Impact: 0.95 to 1.2

    Pricing Impact: 0.98 to 1.03

    R And D Impact: 0.97 to 1.15

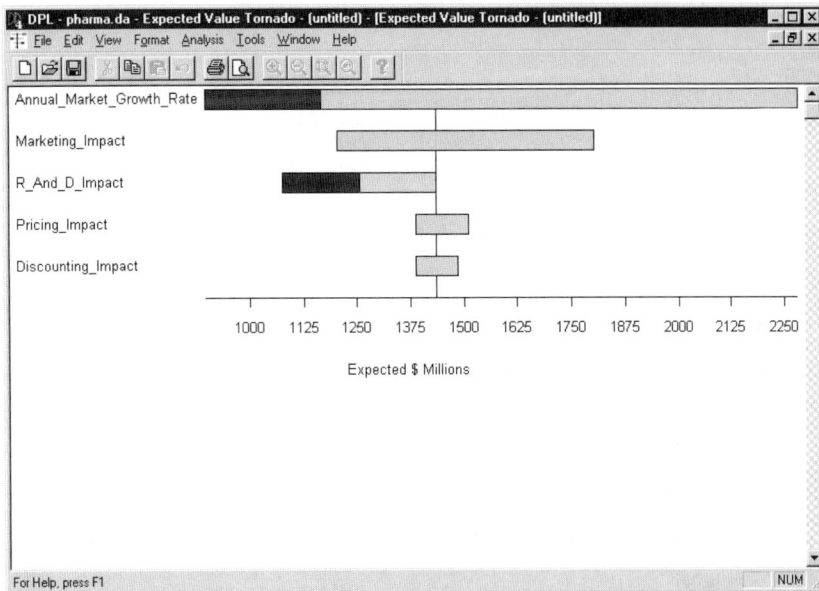

The tornado shows that Annual Market Growth Rate, Marking Impact, and R And D Impact have the greatest affect on Profit. In addition, Annual Market Growth Rate and R And D Impact are decision-sensitive. You will choose to incorporate uncertainty into these three variables.

Now is a good time to save the project. By saving the Project in any window, all windows in the project are saved.

▶ Click the Save icon.

▶ Select Window Model (untitled) to return to the Influence Diagram.

Probabilistic Modeling

The deterministic analysis given by the tornado diagram has identified the variables critical to Profit. Now you can add probability to these variables to look more carefully at the risks you face. From the tornado diagram, you know that Annual Market Growth Rate is the most important variable to consider further. The next step is to change Annual Market Growth Rate into a chance node.

You can change value nodes into chance nodes using the Change Node Type function.

▶ Click on Annual Market Growth Rate, then click on the Change Node Type icon.

▶ Select Chance from the Node Type dialog box, then click OK.

DPL displays the General tab of the Node Definition dialog box.

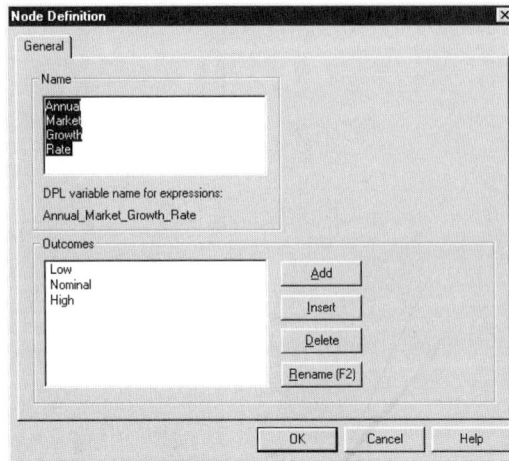

▶ Click OK to accept the default state names of Low, Nominal, and High.

DPL displays chance nodes in the Influence Diagram as green ovals and has changed Annual Market Growth Rate into a green oval to indicate that it is an uncertainty.

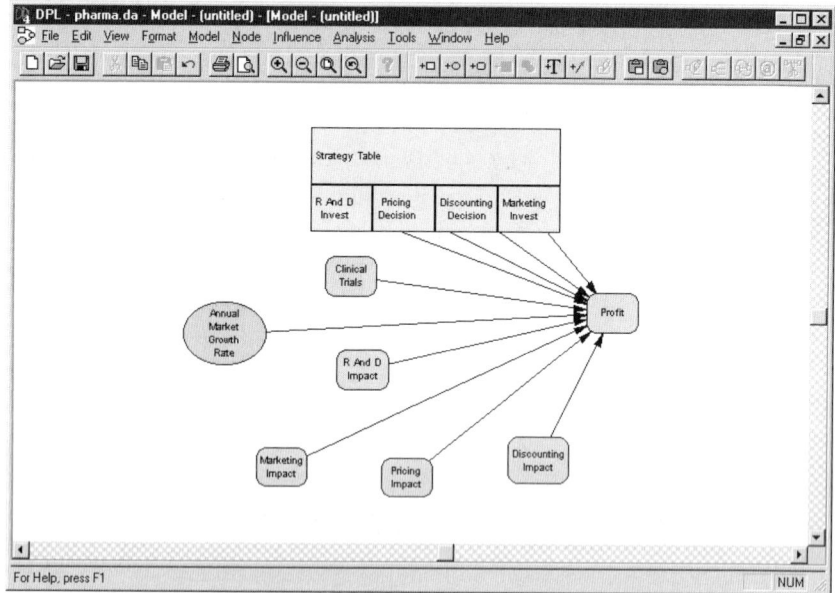

Now you need to edit the data for Annual Market Growth Rate.

▸ Double-click on Annual Market Growth Rate.

Data for a chance node consists of both probabilities and values for each outcome. By default, DPL assigns the Low, Nominal, and High states probabilities of 0.25, 0.5, and 0.25, respectively. DPL has copied the low and high values from the tornado diagram to the Low and High states of the node.

Instead of using the default distribution, you would like to represent the probabilities and values of Annual Market Growth Rate as a Normal Distribution with mean of 0.03 and standard deviation of 0.1.

▸ Click the check box labeled Named Distribution.

DPL displays a list of Named Distributions.

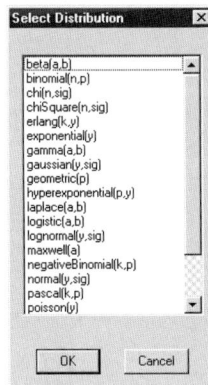

▸ Click on "normal(y,sig)", then click OK.

Notice that each branch of the data tree has "normal(y, sig)" as the probability expression and an asterisk as the value expression. DPL is indicating that the probability and value assigned to each branch is determined by a Normal Distribution with the parameters y (mean) and sig (standard deviation). Next, enter values for the parameters.

▸ Change the probability expression to "normal(0.03,0.1)" and press Enter.

DPL updates all three branches to reflect the new distribution.

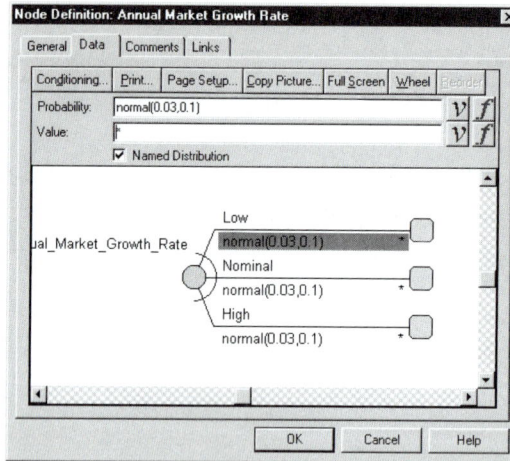

[handwritten note in left margin: Expected value tornado diagram]

Node Definition: Annual Market Growth Rate

General | **Data** | Comments | Links

Conditioning... | Print... | Page Setup... | Copy Picture... | Full Screen | Wheel | Reorder

Probability: normal(0.03,0.1)

Value:

☑ Named Distribution

ual_Market_Growth_Rate

Low
normal(0.03,0.1)

Nominal
normal(0.03,0.1)

High
normal(0.03,0.1)

OK | Cancel | Help

▶ Click OK.

The next variable you want to change into a chance node is Marketing Impact.

▶ Change Marketing Impact into a chance node, accepting the default state names.

▶ Double-click on Marketing Impact to view its data.

As with Annual Market Growth Rate, DPL has copied the low and high values from the tornado diagram to the Low and High states of the node. You want to condition the data for Marketing Impact on Marketing Invest, so leave the data as it is until you input the conditioning.

▶ Click on Conditioning.

DPL displays the Conditioning dialog box.

▶ Check the box labeled Marketing Invest.

▶ Click OK.

▶ Click Full Screen.

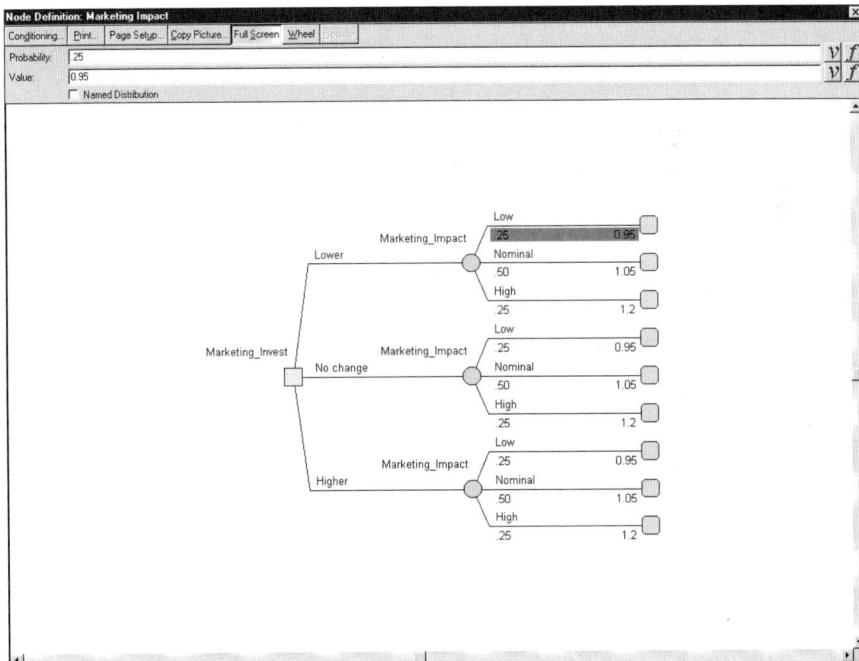

You can now enter three sets of data for Marketing Impact, each corresponding to a different alternative of Marketing Invest. The data currently in Marketing Impact corresponds to the "No Change" alternative of Marketing Invest, so you only have to change the data corresponding to the "Lower" and "Higher" alternatives.

▶ Click on the top branch of Marketing Impact.

DPL highlights the probability associated with this outcome, which is 0.25. You want to change it to 0.6.

▶ Type "0.6", then press Enter.

DPL now prompts you to enter a new value for the outcome. The values of Marketing Impact will not change based on the state of Marketing Invest, so leave this value as 0.95.

▶ Press Enter.

DPL has moved the cursor to the "Nominal" branch of Marketing Impact, and updated the data for the "Low" branch. You can now enter the probabilities for the remaining branches, leaving the values unchanged.

▶ Enter the following probabilities for the Marketing Impact outcomes corresponding to the Lower and Higher states of Marketing Invest:

Lower:	Higher:
Low: 0.6	Low: 0.1
Nominal: 0.3	Nominal: 0.3
High: 0.1	High: 0.6

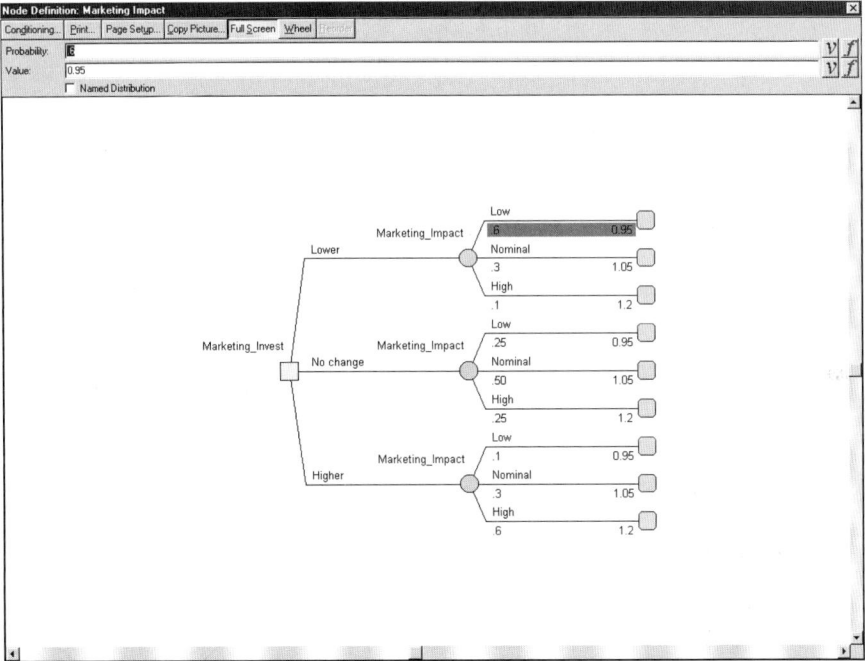

▸ Click Full Screen.

▸ Click OK.

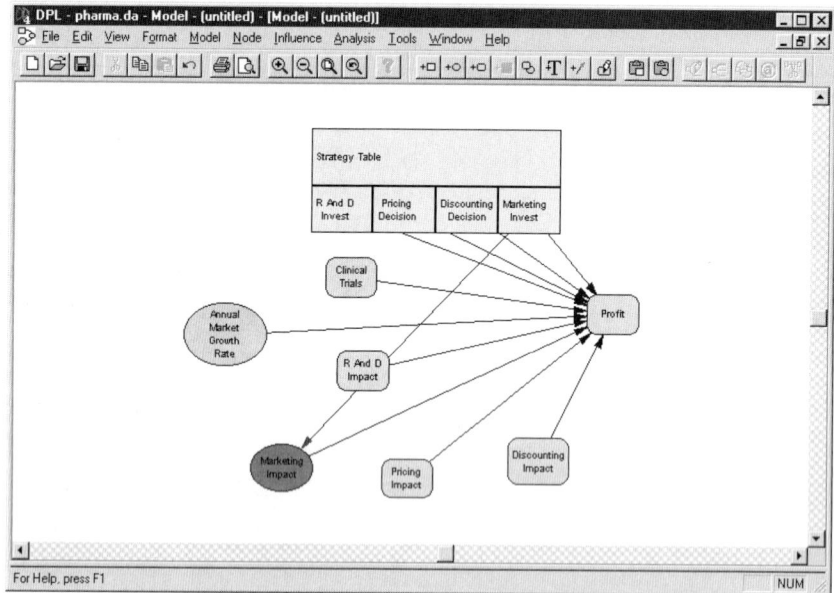

DPL has drawn an arrow from Marketing Invest to Marketing Impact. The burgundy arrowhead indicates that data for Marketing Impact depends on Marketing Invest.

Next you are going to incorporate uncertainty into R And D Impact. The success of your R And D effort is dependent on the amount you spend, and your success in clinical trials is dependent on the success of your R And D effort. Finally, the boost to your market share from R And D is dependent on the success of the drug in clinical trials. Model these dependencies by creating two uncertainties: R And D Success and Clinical Success.

▶ Click on the Create Chance icon.

DPL displays a node and a set of crosshairs to help place the node.

▸ Place the node to the left of Annual Market Growth Rate.

DPL displays the General tab of the Node Definition dialog box.

▸ Name the node R And D Success, pressing Enter after "R And D" to put the name on two lines.
▸ Click OK to accept the default state names for R And D Success.

Next, create a node for Clinical Success.

▸ Click on the Create Chance icon.
▸ Place the node to the left of R And D Impact.
▸ Name the node Clinical Success, pressing Enter after "Clinical" to put the name on two lines.
▸ In the "Outcomes" section, select the "Low" state, then click Delete.
▸ Use the Rename button to rename the remaining two outcomes "Failure" and "Success".

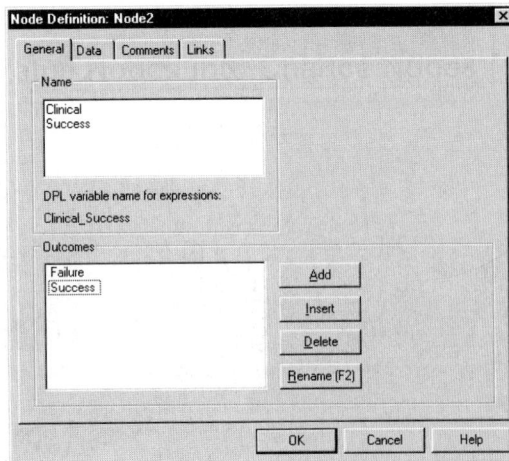

▸ Click OK.

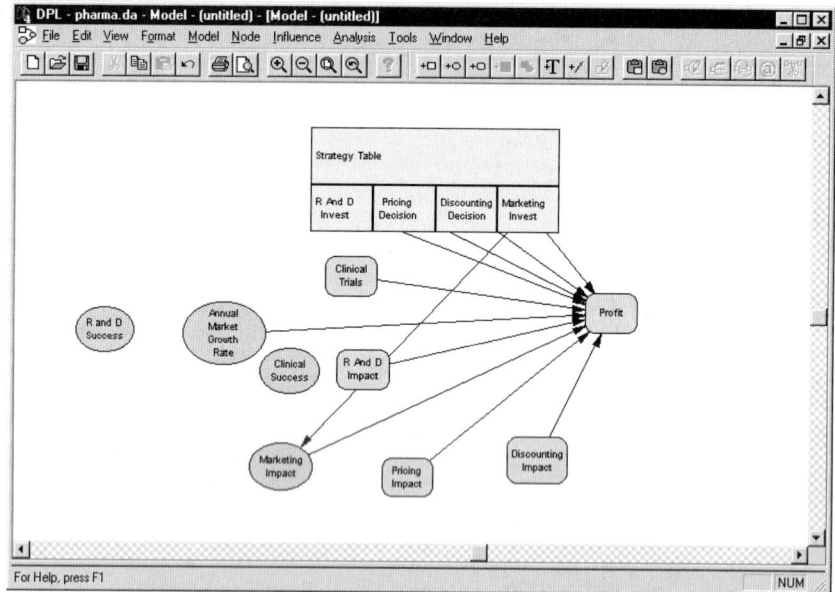

Now condition R And D Success on R And D Invest.

- ▸ Double-click on R And D Success.
- ▸ Click Conditioning, and condition it on R And D Invest.
- ▸ Change the probability distributions for R And D Success corresponding to the
  Lower and Higher states of R And D Invest as follows, leaving the values blank
  (you probably want to click Full Screen to make it easier to read):

Lower:	Higher:
Low: 0.6	Low: 0.1
Nominal: 0.3	Nominal: 0.3
High: 0.1	High: 0.6

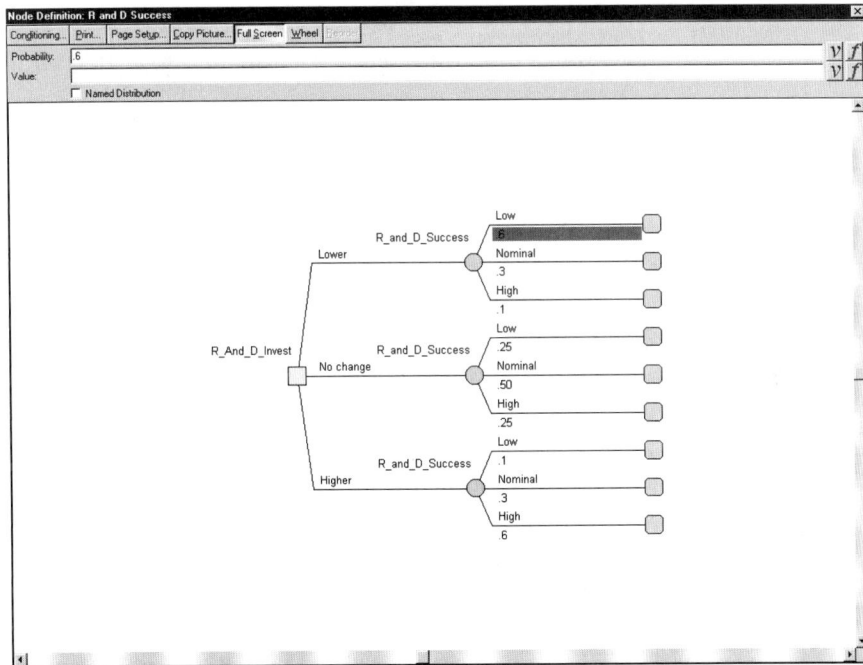

▸ Click OK.

Now condition Clinical Success on R And D Success.

▸ Double-click on Clinical Success.
▸ Condition it on R And D Success.
▸ Change the probability distributions for Clinical Success corresponding to the Low, Nominal, and High states of R And D Success:

Low:	Nominal:	High:
Failure: 0.9	Failure: 0.5	Failure: 0.2
Success: 0.1	Success: 0.5	Success: 0.8

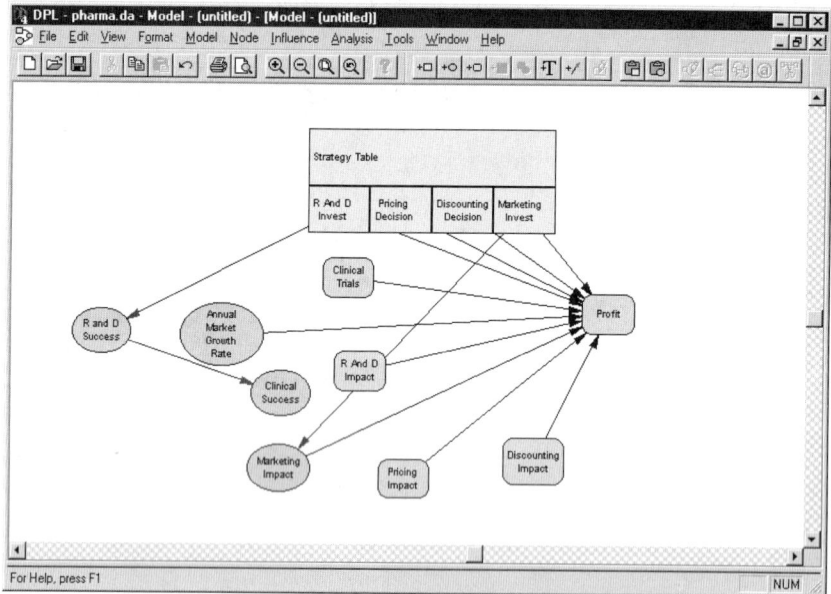

Notice that DPL has drawn an arrow from R And D Invest to R And D Success, and another arrow from R And D Success to Clinical Success.

Now you're ready to condition R And D Impact on Clinical Success.

  ▸ Double-click on R And D Impact.
  ▸ Condition it on Clinical Success.
  ▸ Change the value corresponding to the "Failure" state of Clinical Success from 1.15 to 0.97.

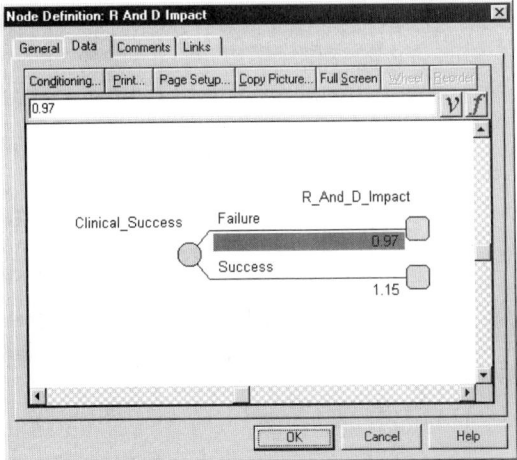

▸ Click OK.

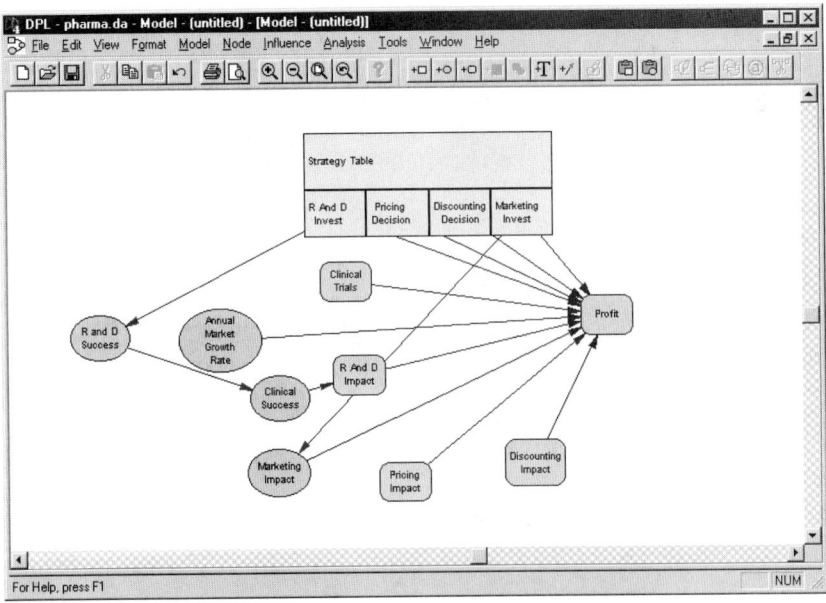

DPL has drawn an arrow from Clinical Success to R And D Impact.

The last thing to do is to condition Pricing Impact on Pricing Decision and Discounting Impact on Discounting Decision.

▶ Condition Pricing Impact and Discounting Decision as described, and change the data as follows:

Pricing Impact:	Discounting Impact:
Pricing Decision Lower: 1.03	Discounting Decision More: 1.02
Pricing Decision No Change: 1	Discounting Decision No Change: 1
Pricing Decision Higher: 0.98	Discounting Decision Less: 0.98

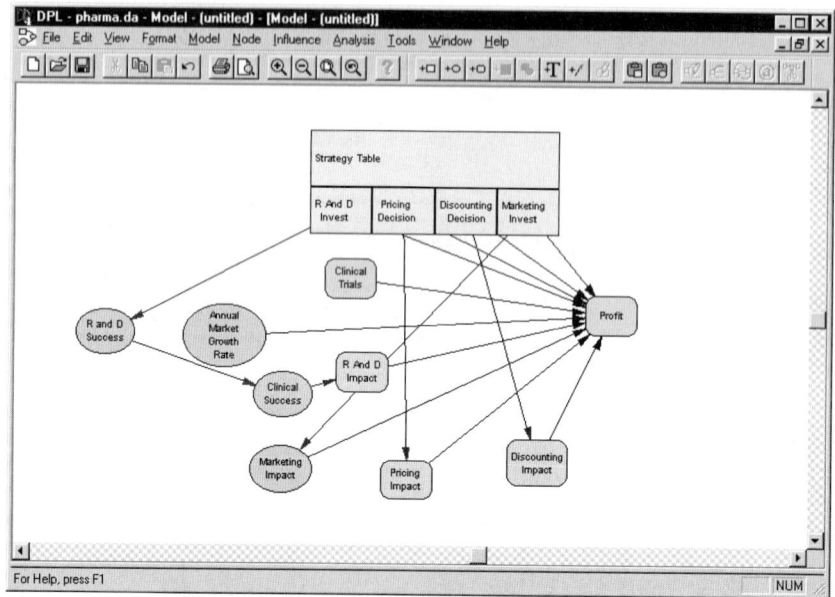

DPL has drawn an arrow from Discounting Decision to Discounting Impact, and another arrow from Pricing Decision to Pricing Impact.

At this point, the influence diagram is getting cluttered. Move nodes around to make the diagram more clear.

▶ Click-and-drag nodes to re-arrange the diagram as shown.

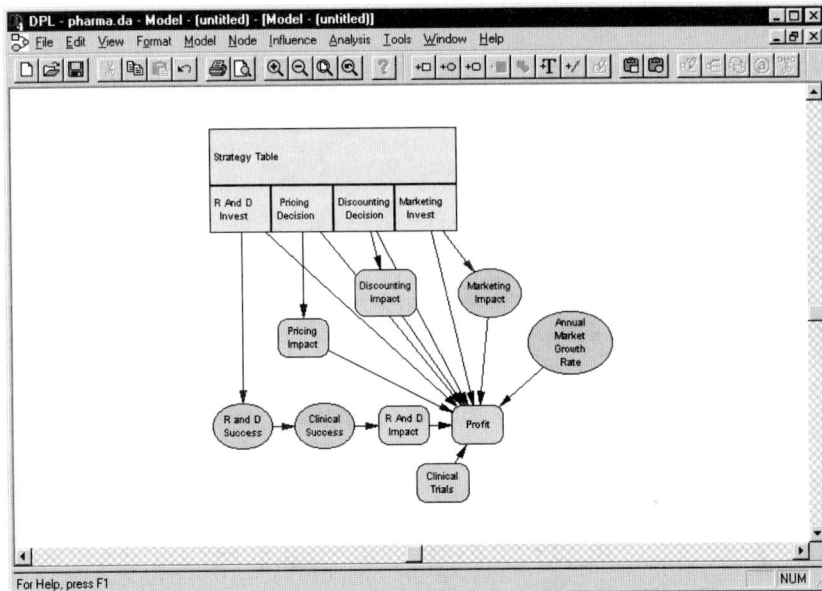

Now is a good time to save the model.

▶ Click the Save Project icon.

Analyzing the DPL Model

At this point you can run a decision analysis on the probabilistic model.

▶ Select Analysis Decision Analysis.

DPL displays the following warning dialog box:

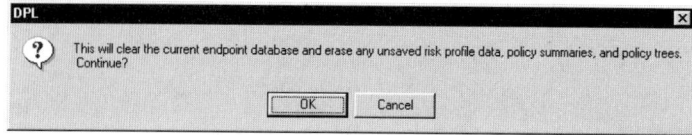

▶ Click OK.

▶ In the Decision Analysis Options dialog box, click the radio button labeled "Initial decision alternatives".

▶ Click OK.

Again, once the run is finished, DPL displays the Policy Tree window. Marketing Blitz is now the preferred strategy with an expected value for Profit of over $1.3 billion. While the expected values returned by the Technology Leader and Status Quo strategies are not far behind, Price War is by far the least attractive strategy.

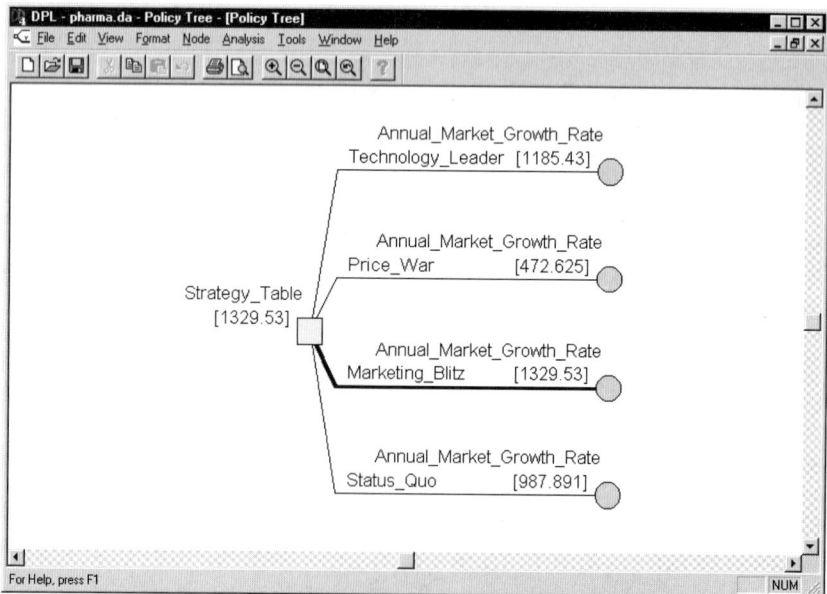

The blue triangles at the end of the branches of the Strategy Table have been replaced by green circles. The green circles indicate that a chance node follows the Strategy Table in the decision tree. You can expand the Policy Tree to see the full path of each scenario in the tree.

▸ Double-click on the green circle at the end of the Marketing Blitz branch.
▸ Click the Zoom Full icon.

The Policy Tree is now expanded to show the branches of Annual Market Growth Rate. You could continue to expand the tree until you reached the endpoints. This would be useful if you were interested in a particular scenario.

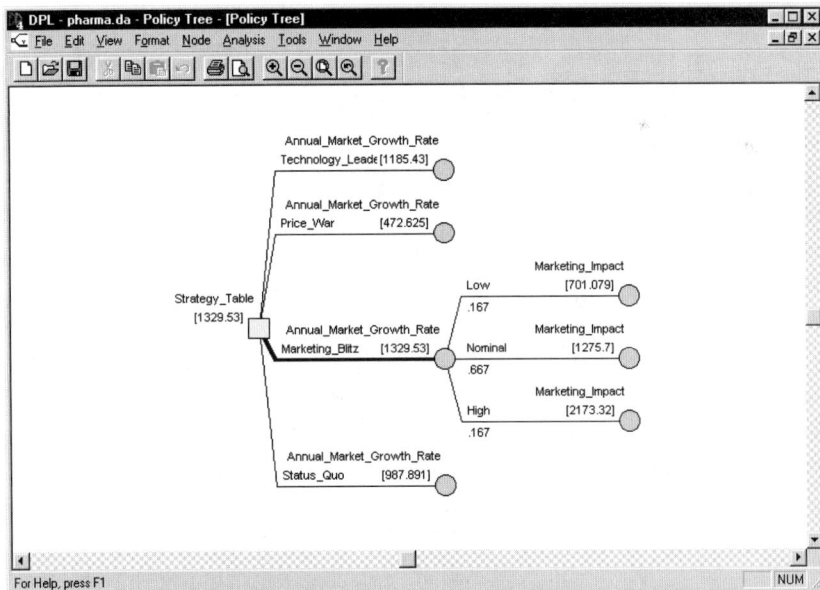

▸ Select Window Risk Profile Chart.

The Risk Profile Chart displays cumulative probability distributions for the four strategies.

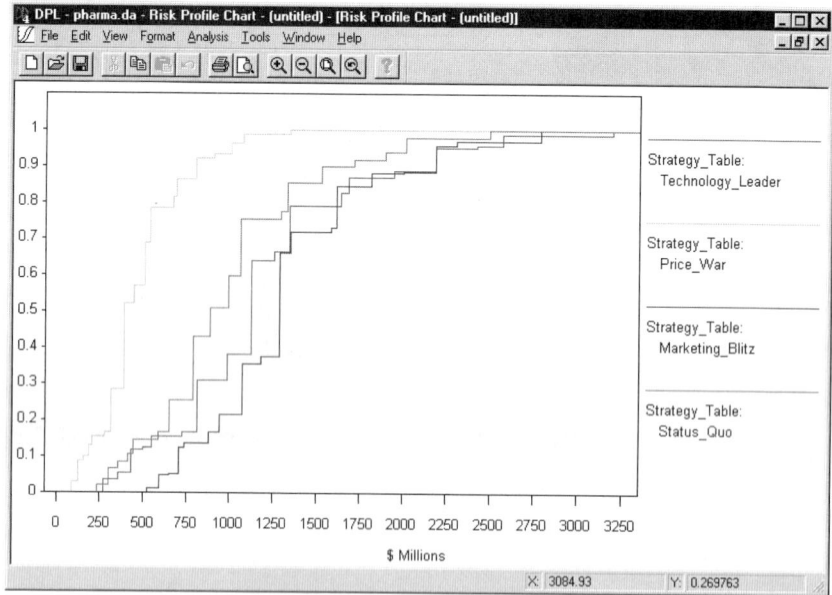

The line representing Price War is above and to the left of the others, indicating that it is certainly the worst of the four strategies. The line for Marketing Blitz is generally below and to the right of the others, implying it is generally less risky than the others. From the Policy Tree, you know that Marketing Blitz also has the highest expected value of the four strategies. Since it has the highest expected value and holds the least amount of risk of the four, the conclusion is that Marketing Blitz appears to be the best strategy.

Further Analysis

At this point you might wonder if one of the variables you chose not to model as an uncertainty has increased in sensitivity due to the modifications you have made. For example, the cost of Clinical Trials is set to a value of $200 million. You can run a Rainbow Diagram to determine if it is sensitive

▶ Select Analysis Rainbow Diagram.

DPL displays the Select Value for Sensitivity Analysis dialog box.

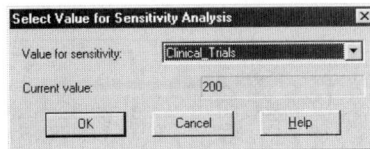

Because Clinical Trials is the first variable alphabetically, DPL displays it in the "Value for sensitivity" drop-down menu box. If you want to analyze a different variable, simply select it from the drop-down menu. DPL also shows that the current value of Clinical Trials is 200.

▶ Click OK.

DPL displays the Run Sensitivity Analysis dialog box.

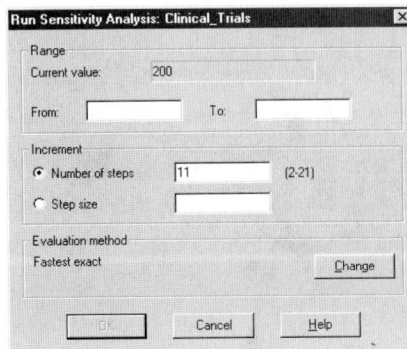

First you have to specify the range over which you want to analyze Clinical Trials. You think it may range from $150 to $250 million.

▸ In the "From" box, type "150".
▸ In the "To" box, type "250".

Next tell DPL the number of values within the range you want it to analyze. A good first cut is to use three steps, which will run it at either end of the range and at the midpoint.

▸ Enter "3" for the Number of Steps.

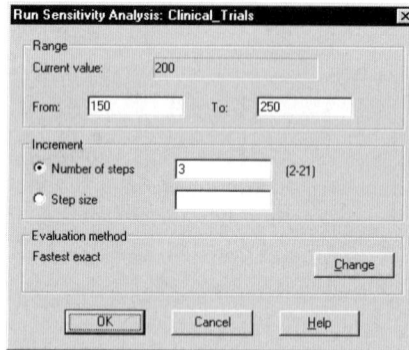

Run Sensitivity Analysis: Clinical_Trials	☒

Range
Current value: 200

From: 150   To: 250

Increment
◉ Number of steps   3   (2-21)
○ Step size

Evaluation method
Fastest exact   [Change]

[OK]   [Cancel]   [Help]

▸ Click OK.

The Rainbow diagram shows that the expected value of the model ranges from just under $1.38 billion to just over $1.28 billion as Clinical Trials varies over the given range. The graph shows no color changes, indicating that you would not change your decision policy based on this variance. Because the variable is not decision-sensitive and the variance it causes in Profit is not that great, you will likely continue to model Clinical Trials as a value node.

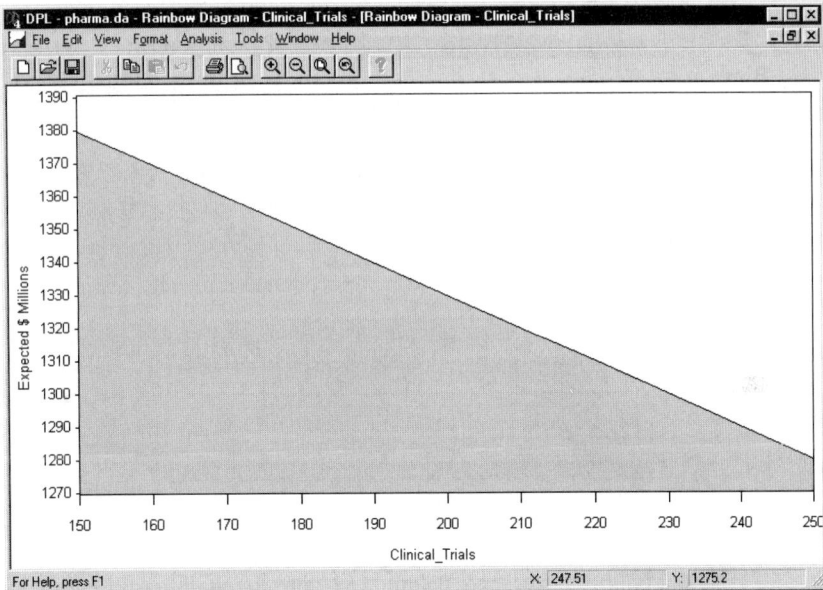

Wrapping Up

If you go back to the Project Manager, you can see all of the windows you have created in your analysis.

▸ Select Window Pharma.da.

You can rename windows at this time if you wish. If you do not rename each of the four Risk Profile Data, they will be overwritten the next time you generate the same type of output.

Finally, save the project.

▸ Select File Save Project.

A decision in a DPL model is a representation of a choice the decision maker faces. A strategy table is a way to model several decisions as a group. Together, decisions and strategy tables enable you to model a wide range of decision situations, capturing complex interactions and timing issues.

DECISIONS AND STRATEGY TABLES

Chapter 4: Decisions and Strategy Tables

What is a Decision?

A decision is an opportunity for you (or your organization) to choose between alternative states of the world. In DPL, a decision is a node, or event, with two or more alternatives.

In the influence diagram, a decision is represented by a yellow rectangle:

R&D
Invest

In the decision tree, a decision is represented by a yellow square with branches.

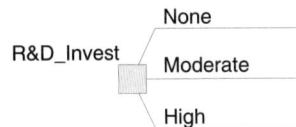

```
                          None
                    _____
        R&D_Invest /  Moderate
                   |  _____
                    \  High
                      _____
```

When you run a decision analysis, DPL will calculate the endpoint values of all the branches following from the decision. The alternative with the best expected value will be selected as the optimal alternative for this instance of the decision. "Best" depends on the objective function you've defined. If you've specified that the objective function should be maximized, the best alternative for the decision is the one with the highest expected value, and if the objective function should be minimized, the best alternative has the lowest expected value.

How Many Decisions Should You Model?

The number of decisions your model needs depends on the situation, your experience with decision analysis, and the amount of time you have for the analysis.

No decisions. DPL doesn't require that a model include any decisions. A model with no decisions is usually called a risk analysis, and may be used to evaluate the riskiness of a plan that has already been committed to.

One decision. The simplest structure for the decisions in your model is to have a single decision followed by as many chance events as necessary. This structure is good for smaller decisions, decisions with short time frames, or just getting started.

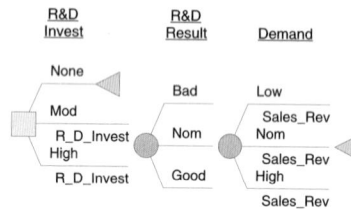

One decision now, one later. Many times people find that a model with one all-or-nothing decision has a surprising amount of risk. Upon further investigation, it turns out that much of the downside risk comes from sticking with a plan long after it becomes clear that the results are bad. It may also turn out that it misses the really good outcomes because the model assumes you can't expand to capture new opportunities. If you add a simple decision that occurs some time in the future, whose alternatives include a bail-out option, a continue alternative and maybe an expand option, you can often get a much more realistic and insightful result.

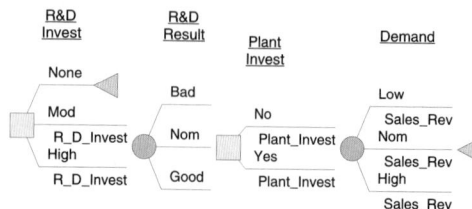

A few now, a few later. Some projects involve deciding about more than one issue. For example, you may have one decision about production, another about distribution and a third about marketing. Your model probably needs one decision for each of these at the beginning of the tree. You may also have decisions about these issues later in the tree.

Many Decisions. Although the software doesn't have a limit on the number of decisions you can usefully model, your brain does. The more decisions you have, the more data you will need to collect and the more complex the results will be. When you have more than three decisions, it may be more useful to start with a strategy table, in which you define all the relevant decisions, but then choose a few selected strategies (or combinations of decision alternatives) to evaluate. Once you have the model debugged and have a sense of which alternatives deserve a closer look, you can break the decisions apart and model them in more detail.

How to Model a Decision

First, give it a name. This is the only name that will appear on the influence diagram. It will also appear on the decision tree (along with the states). This name may also be used as a variable name in mathematical expressions. Some rules:

- The name may contain up to 255 characters. We recommend that you keep it under 32 characters to keep the outputs easier to read.
- The name may include spaces and special characters such as & - ?. These characters will be converted to underscores if you use the name in a formula (this prevents DPL from becoming confused about which characters belong in variable names and which provide the "punctuation" for the formula).
- The first character should be a letter. If the first character is a number, DPL will replace the number with the character "N" when the name gets used in formulas.

Second, define the alternatives. The alternatives of a decision are the states it can be in. The same naming rules apply. Some things to consider:

- **How Many States?** The decision must have at least two states. Technically, the maximum it can have is 64 (32,000 if you switch to DPL code). In practice, data entry, validation, and comprehension all start to get pretty hard at about six alternatives.

- **Remember the Status Quo!** Most decision analysis models should include a "Do Nothing" or "Business as Usual" alternative. Sometimes the Do Nothing alternative will be the best, which can be hard to accept if you really believe in a proposal — but wouldn't you rather find out now rather than after your effort and reputation are on the line? Sometimes the Do Nothing alternative is almost as risky, or even riskier than your proposed alternative, which can make your proposal look even better.

- **Don't Overlap.** During evaluation, DPL will only look at one alternative at a time and identify which one is optimal — it can say A is better than B, but it can't say do A **and** B. So if doing both A and B is a possibility, define another alternative called A and B.

```
                                    Sell Retail
       Distribution_Strategy
                                    Sell Wholesale

                                    Retail and Wholesale
```

Third, assign values if appropriate. If your decision's alternatives are naturally associated with values to be used in calculating the objective function values, you can assign them directly to the decision's states.

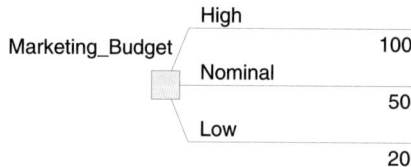

```
                            High
       Marketing_Budget                      100
                            Nominal
                                             50
                            Low
                                             20
```

Tutorial: Decisions

You have created a deterministic influence diagram from a spreadsheet, and are ready to add decisions.

▶ Open "Decisions.da".

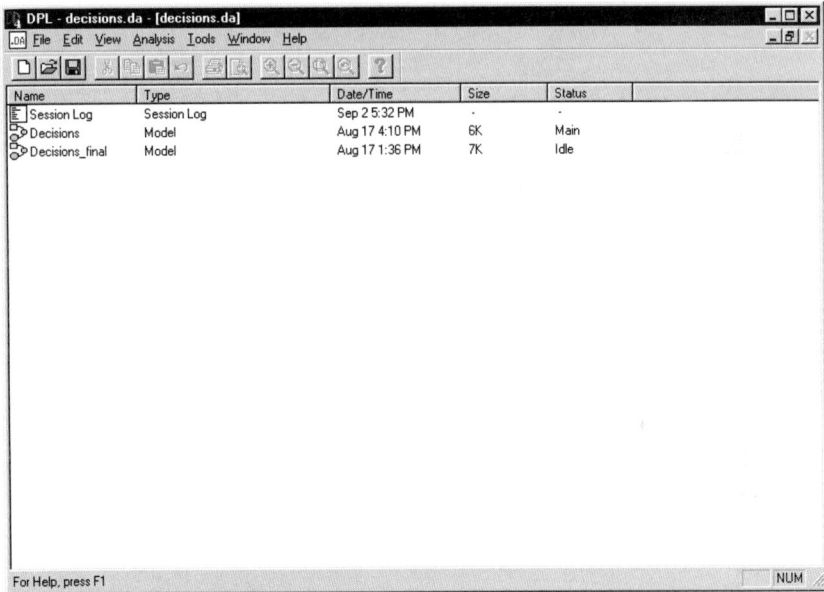

▶ Select Window Model-Decisions.

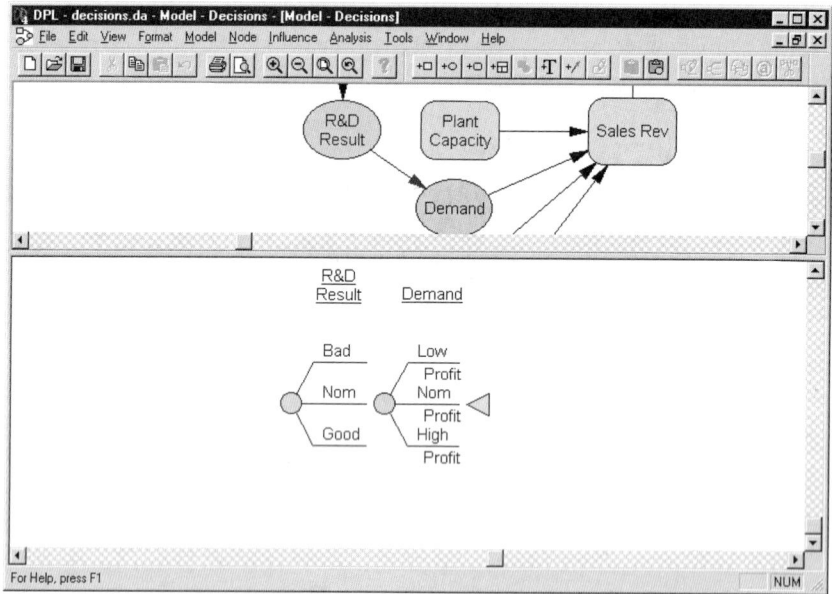

▸ Drag the splitter bar down to maximize the influence diagram pane.

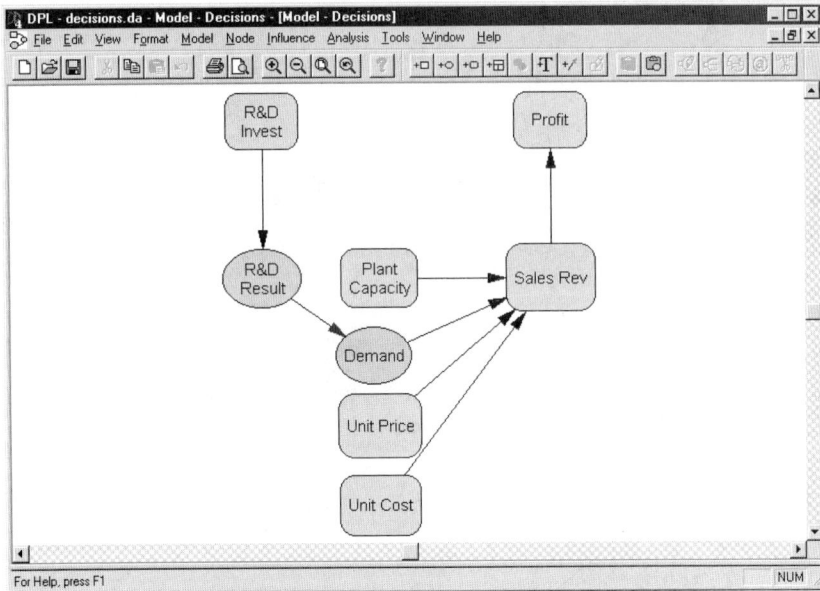

Creating New Decision Nodes

The current plant has a capacity of 5 units. You have the option of constructing a new plant to increase your capacity. Model this as a decision in DPL.

▸ Click the Add Decision icon.

+☐

▸ Place the new decision node above the Plant Capacity node.

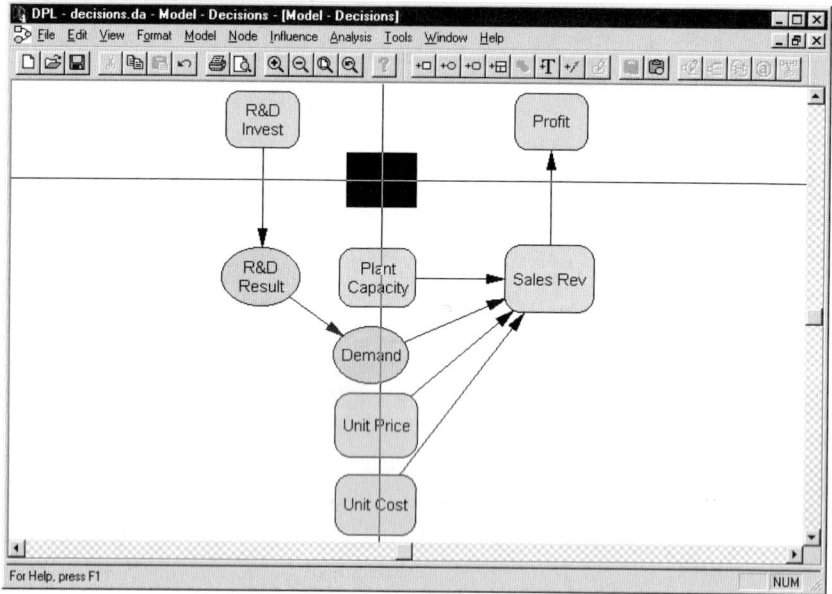

DPL displays the Node Definition dialog box.

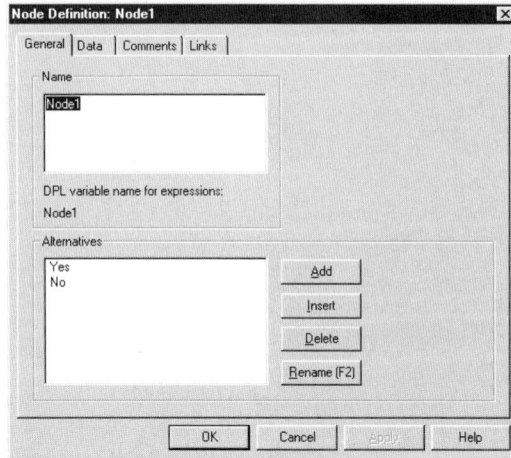

The General tab of the Node Definition dialog box contains the node name and alternative names. DPL assigns the new node the default name of "Node1." Change this to "Plant Invest."

▶ Type "Plant Invest", pressing the Enter key after "Plant" to put the name on two lines.

```
Node Definition: Node1                                    ☒
 General | Data | Comments | Links |
 ┌ Name ──────────────────────────────┐
 │ Plant                               │
 │ Invest                              │
 │                                     │
 │                                     │
 │                                     │
 │ DPL variable name for expressions:  │
 │ Plant_Invest                        │
 └─────────────────────────────────────┘
 ┌ Alternatives ──────────────────────┐
 │ Yes                     [  Add   ]  │
 │ No                                  │
 │                         [ Insert ]  │
 │                                     │
 │                         [ Delete ]  │
 │                                     │
 │                         [Rename (F2)]│
 └─────────────────────────────────────┘
              [   OK   ]  [ Cancel ]  [ Help ]
```

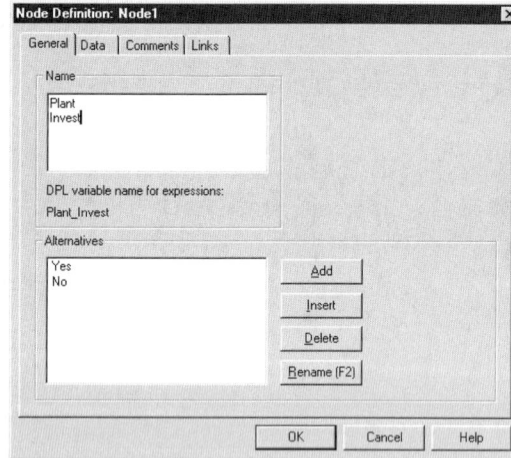

In the Alternatives section of the dialog box, DPL allows you to change the names of alternatives as well as the number of alternatives. By default, DPL assigns new decisions two alternatives named "Yes" and "No." (You can change the default alternative names by selecting Tools Options General.)

The two alternatives for the Plant Invest decision are Yes and No, so accept the default alternatives.

▶ Click OK.

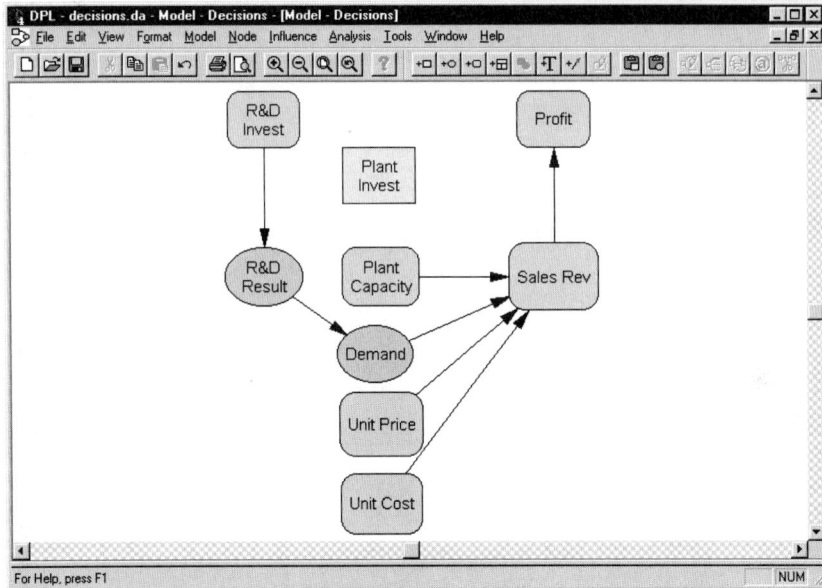

DPL displays decisions as yellow rectangles in the Influence Diagram.

Converting Value Nodes into Decision Nodes

R&D Invest is currently a value node, indicating that there is only one amount you
will invest in R&D. In reality, you choose the amount of money to invest in R&D, so
it makes more sense to model R&D Invest as a decision. You can change R&D
Invest from a value node to a decision node.

- Click on R&D Invest.
- Click on the Change Node Type icon.

DPL displays the Node Type dialog.

> **Node Type**
>
> ○ Decision
> ○ Chance   [OK]
> ◉ Value   [Cancel]
> ○ Controlled

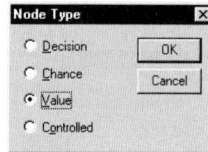

▸ Click the radio button labeled "Decision".
▸ Click OK.

DPL displays the General tab of the Node Definition dialog box.

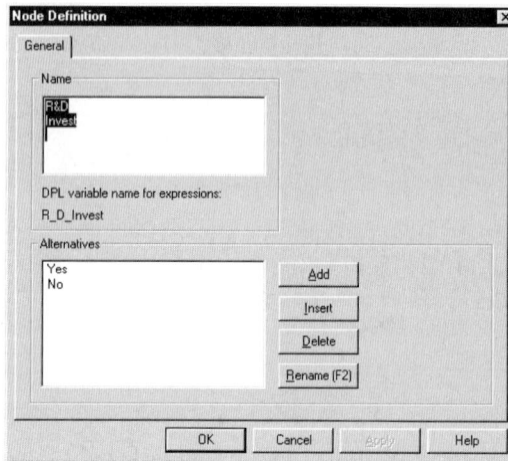

> **Node Definition**
>
> [General]
>
> **Name**
>
> R&D
> Invest
>
> DPL variable name for expressions:
> R_D_Invest
>
> **Alternatives**
>
> Yes   [Add]
> No
>   [Insert]
>   [Delete]
>   [Rename (F2)]
>
> [OK] [Cancel] [Apply] [Help]

You do not need to change the node's name. In the Alternatives section, DPL displays the default alternatives Yes and No. In reality, the R&D Invest decision has three alternatives: "None," "Moderate," and "High." Add a third alternative and rename the alternatives correctly.

▸ Click on the Yes alternative.
▸ Click the Insert button.
▸ Type "None," then press Enter.

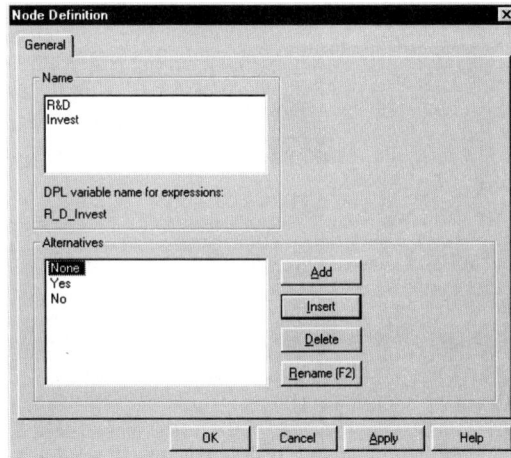

Next, rename the other two alternatives.

▶ Click the Yes alternative.

▶ Click Rename.

▶ Type "Moderate," then press Enter.

▶ Repeat the previous steps to rename the No alternative as "High."

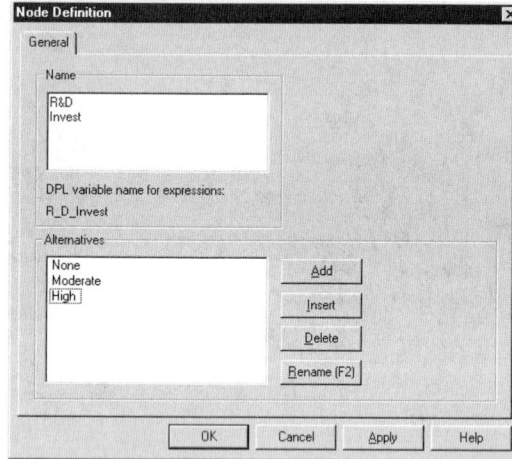

▶ Click OK.

DPL now displays R&D Invest as a decision node.

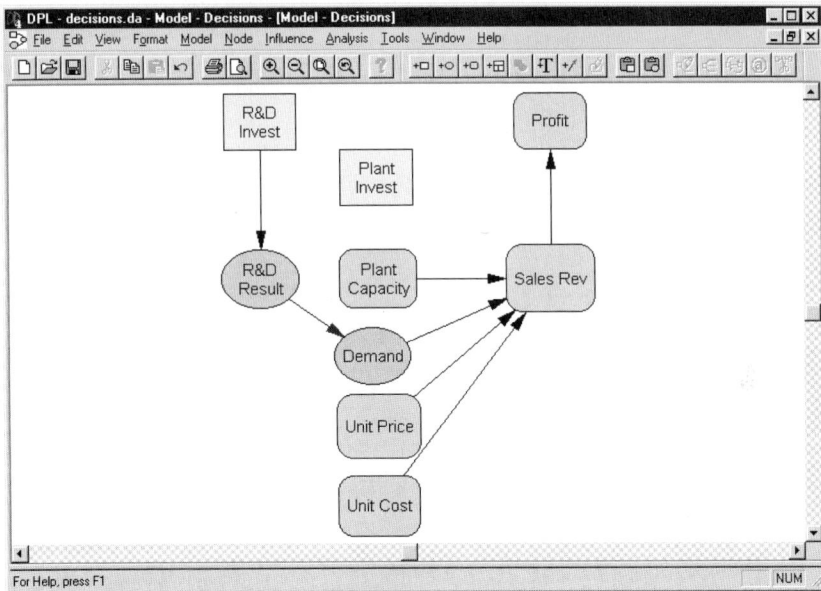

Assigning Values to Decision Alternatives

Sometimes you want to assign values to the different alternatives of a decision. For example, constructing the plant will cost $100 million, while doing nothing will cost nothing. Model these two costs as values associated with the alternatives of Plant Invest.

▸ Double-click on Plant Invest.

DPL displays the Data tab of the Node Definition dialog box.

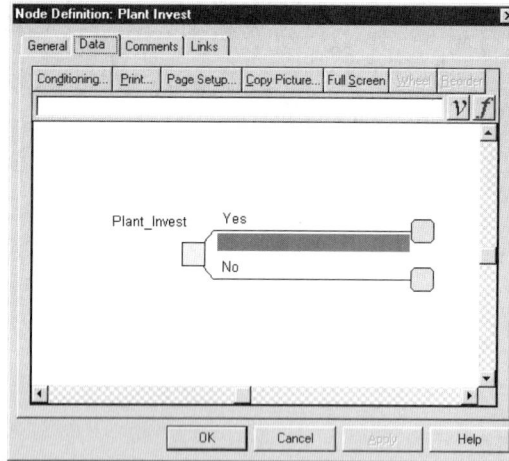

▸ Click the highlighted "Yes" branch of Plant Invest.

▸ Type "100," then press Enter.

▸ Type "0", then press Enter.

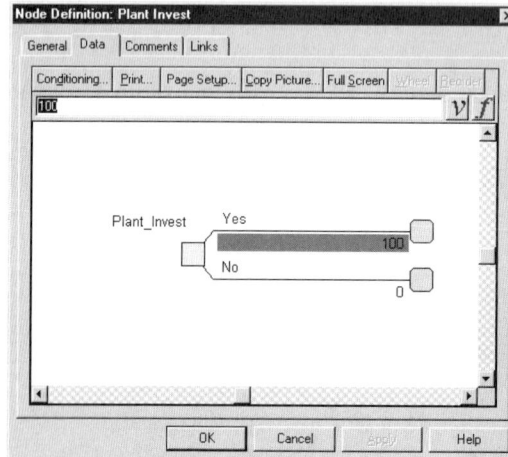

▸ Click OK.

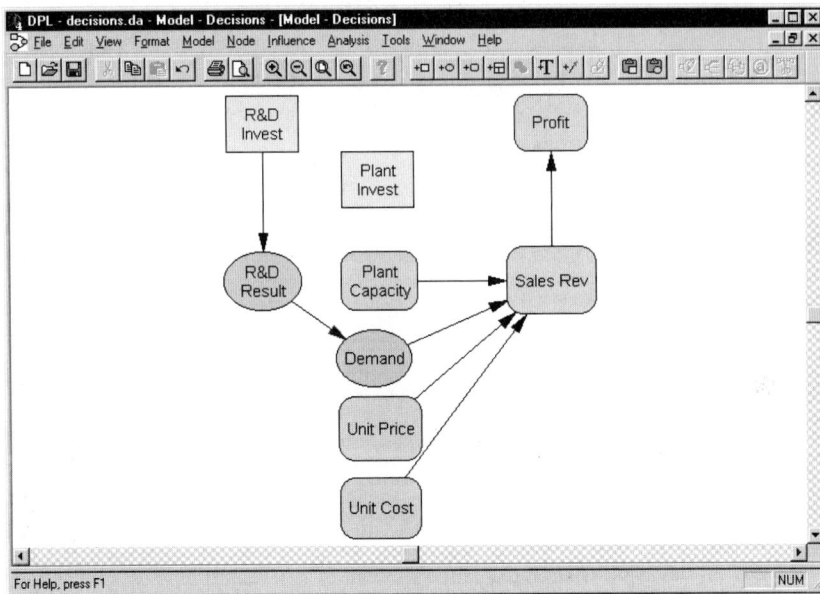

Next, make the values of **R&D Invest** reflect the costs associated with different levels of investment.

▸ Double-click on **R&D Invest**.

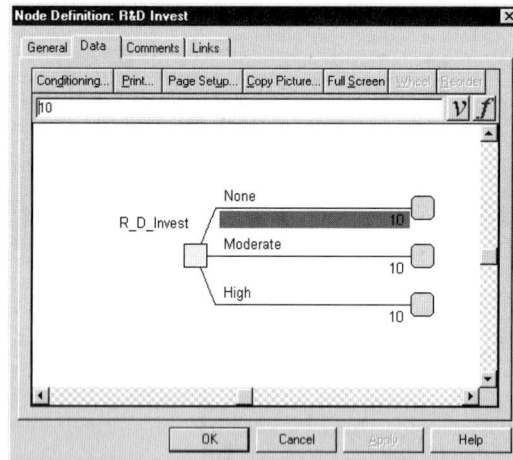

As a value node, R&D Invest had a single value of 10. DPL has copied this value to all three alternatives. Change the values associated with the None and High levels of investment, which are 0 and $20 million, respectively.

- ▶ Click on the highlighted "None" branch.
- ▶ Type "0", then press Enter.
- ▶ Press Enter again to highlight the "High" branch.
- ▶ Type "20", then press Enter.

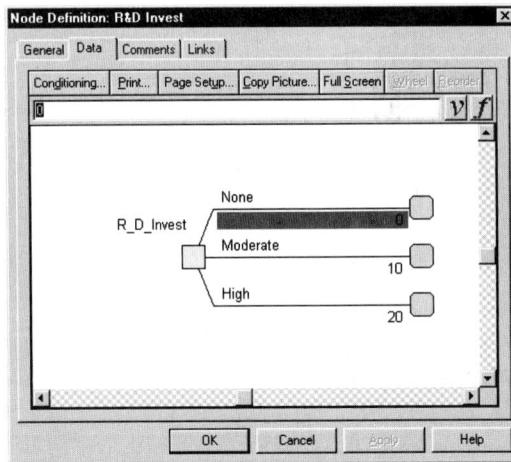

▶ Click OK.

You can now use the values associated with R&D Invest and Plant Invest in calculations elsewhere in the model.

Conditioning Other Variables on Decision Nodes

You can have your decision nodes condition other nodes in the model. For example, the plant capacity will depend on whether or not you decide to invest in a new plant. Therefore, you can condition Plant Capacity on Plant Invest.

▶ Double-click on Plant Capacity.

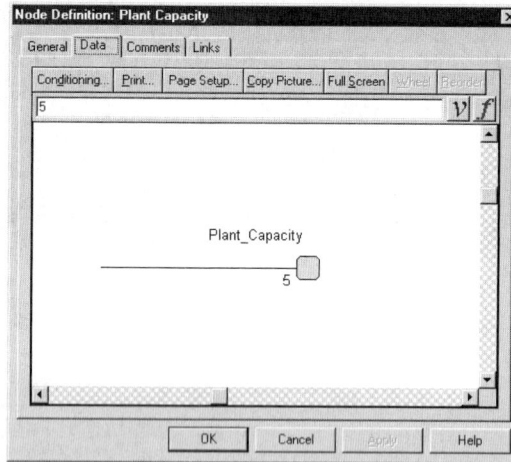

▸ Click Conditioning.

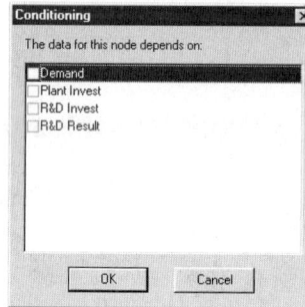

▸ Click the box labeled "Plant Invest".
▸ Click OK.

You can now enter different values for Plant Capacity based on the alternative you select for Plant Invest. If you choose to invest in a new plant, capacity will increase to 10. If you do not build a new plant, the capacity will remain at 5.

▸ Click on the highlighted "Yes" branch.
▸ Type "10," then press Enter.

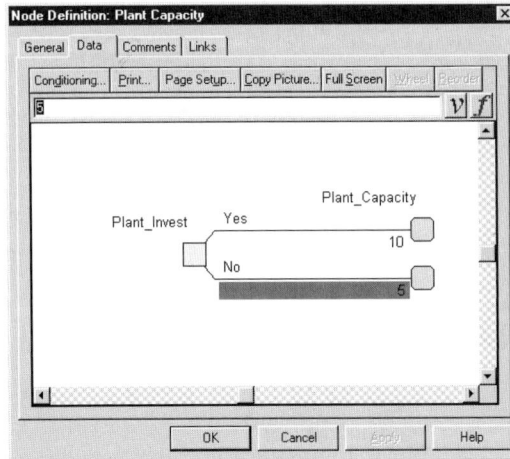

Node Definition: Plant Capacity

General | Data | Comments | Links

Conditioning... | Print... | Page Setup... | Copy Picture... | Full Screen | Wheel | Reorder

5

Plant_Invest   Plant_Capacity

Yes   10

No   5

OK | Cancel | Apply | Help

▸ Click OK.

Similarly, the amount you invest in R&D will affect your chances of having a positive R&D result. Condition R&D Result on R&D Invest.

▸ Double-click on R&D Result.

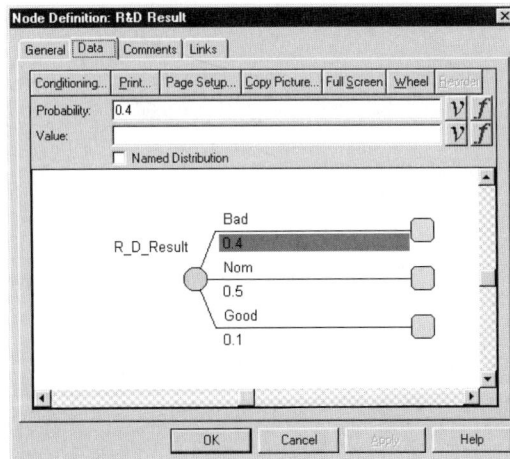

Node Definition: R&D Result

General | Data | Comments | Links

Conditioning... | Print... | Page Setup... | Copy Picture... | Full Screen | Wheel | Reorder

Probability: 0.4

Value:

☐ Named Distribution

R_D_Result

Bad
0.4

Nom
0.5

Good
0.1

OK | Cancel | Apply | Help

▶ Click Conditioning.

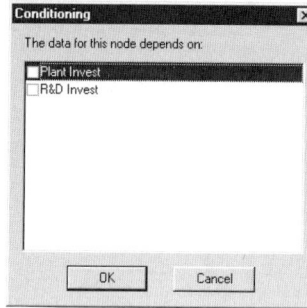

▶ Click the box labeled "R&D Invest".
▶ Click OK.
▶ Click Full Screen.

DPL has replicated the original probability distribution for all three alternatives of R&D Invest.

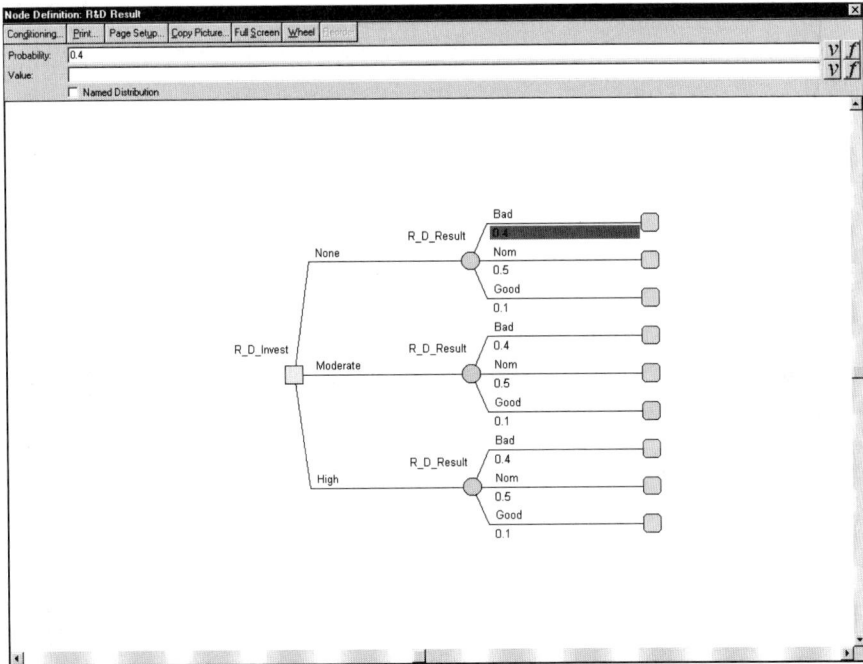

▶ Click on the Bad alternative of R&D Result associated with the None alternative of R&D Invest (it is highlighted).

You are guaranteed a "Bad" R&D Result if you invest nothing in R&D, so assign this outcome a probability of 1.

▶ Type "1", and press Enter twice (we are not entering any values for this node).
▶ Assign both the Nominal and Good outcomes probability 0.

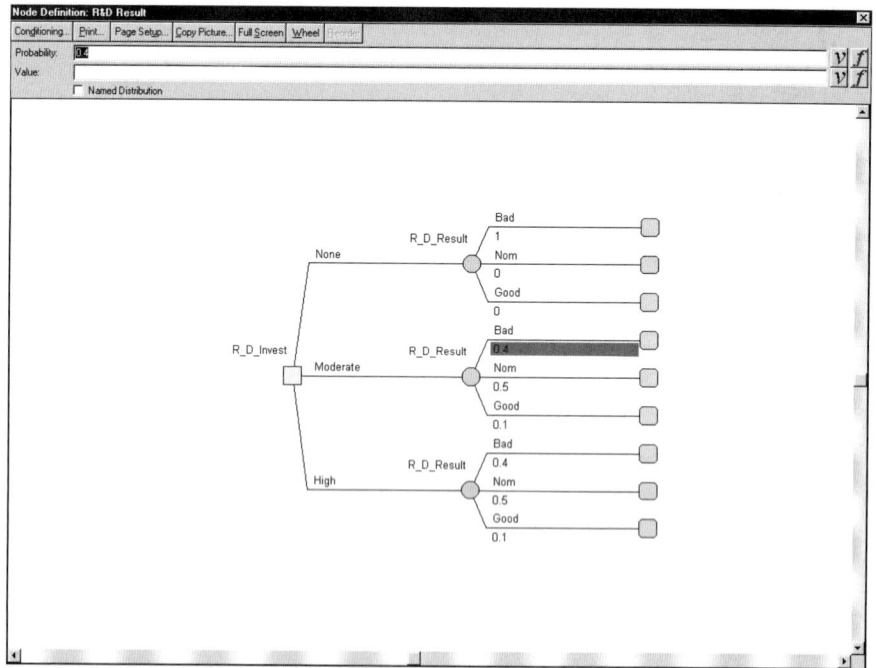

The original probability distribution accurately reflects a moderate amount of R&D investment. Therefore, you only need to change the distribution associated with the High alternative of R&D Invest.

▸ Enter the following probability distribution corresponding to the High alternative of R&D Invest:

R&D Result Bad: 0.2

R&D Result Nominal: 0.5

R&D Result Good: 0.3

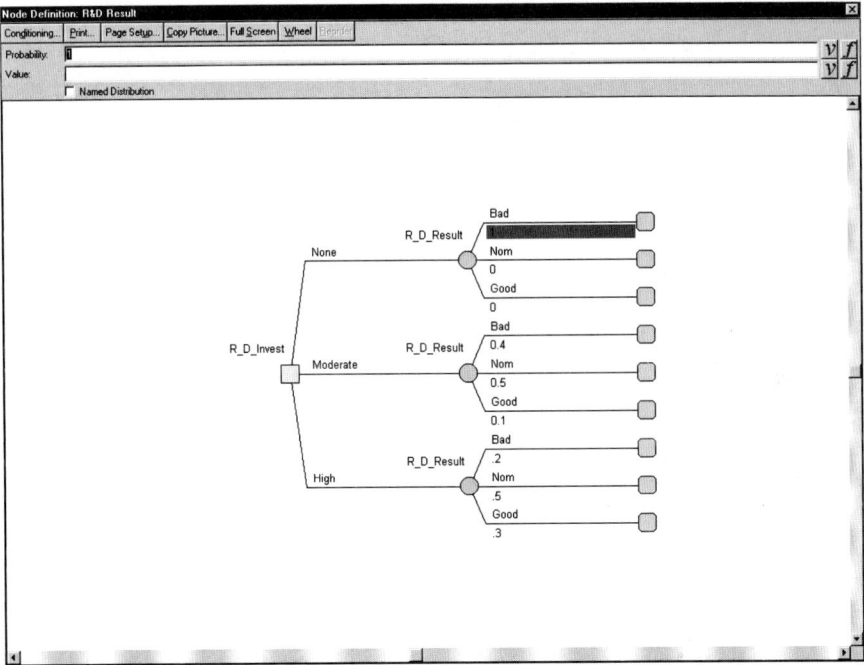

▶ Click Full Screen.
▶ Click OK.

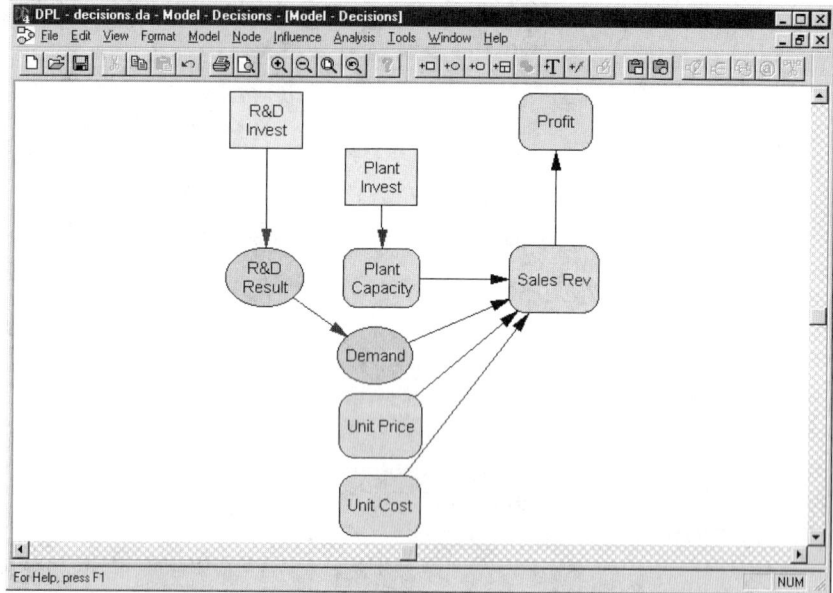

DPL has changed the arrowhead from R&D Invest to R&D Result from black (timing only) to brown (timing and conditioning).

Modeling Downstream Decisions

You need to tell DPL the order in which events occur in the Decision Tree.

▸ Select View Split Vertically.
▸ Drag the splitter bar to the middle of the screen.
▸ Right-click inside the Decision Tree pane.
▸ Select Zoom Full from the context menu.

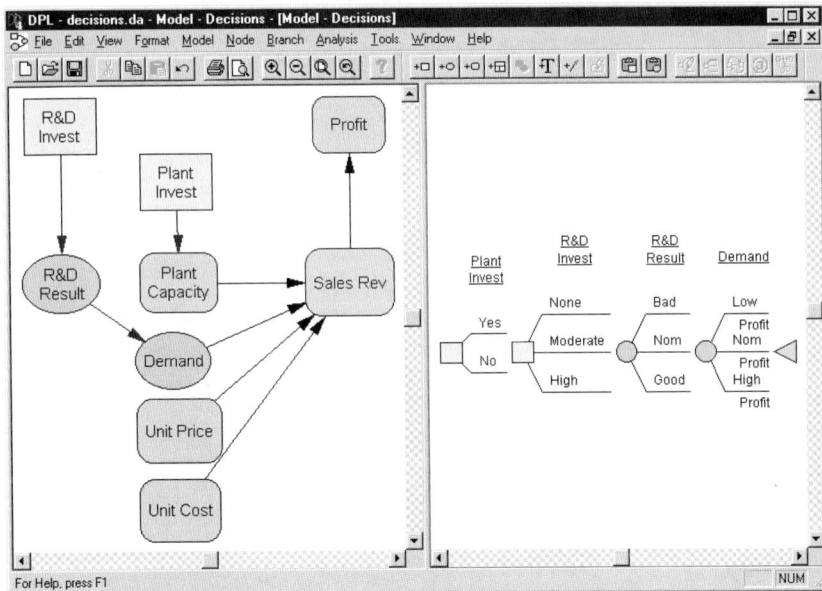

DPL places the two decisions first in the decision tree, indicating that you make your decisions before the uncertainties are resolved. In this case, you will actually know the result of the R&D effort before you decide whether or not to invest in a new plant. You can specify this by drawing a timing arrow from R&D Result to Plant Invest.

▶ Click inside the Influence Diagram pane.
▶ Click on the Create Influence icon.

▶ Click on R&D Result.
▶ Click on Plant Invest.

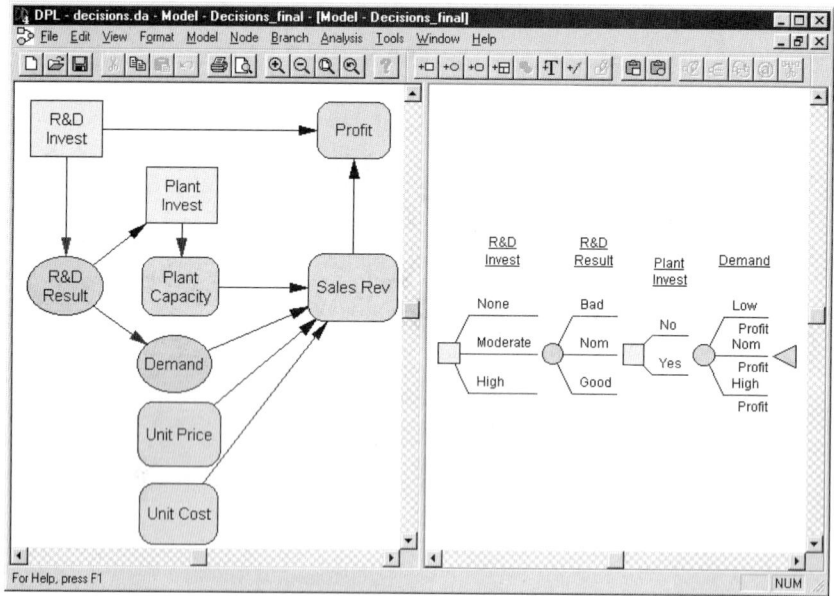

In the Influence Diagram, DPL has drawn a black arrow from R&D Result to Plant Invest. In the Decision Tree, DPL has switched the order of R&D Result and Plant Invest. You will now know the outcome of R&D Result before making the Plant Invest decision. Decisions made after one or more uncertainties are resolved are known as "downstream decisions."

Strategy Tables

How to Model with a Strategy Table

A strategy table contains several decisions that will be treated as a single decision in the influence diagram and decision tree. The alternatives are strategies, where a strategy is a representative setting for each of the decisions in the table. A strategy table is useful when you have many decisions — rather than trying to model, validate and understand hundreds of possible combinations of alternatives (if you have six decisions with three alternatives each, you have 3^6 or 729 combinations), you can focus on four or five representative ones. Once you've started to understand the issues, and have gained some insight into how the decisions and risks relate, you can trim the possibilities down to the most promising ones and model them in more detail.

Strategy tables are especially good group framing tools.

First, name the decisions. If you have only two or three decisions, you should probably use individual decisions rather than a strategy table. If you have more than eight or nine decisions, ask yourself if they are all of the same strategic level. For instance, if you started with "should we build a new manufacturing plant in China?" and your ninth decision is "what color should we paint the conference room in the new plant?" perhaps you should stop.

Second, list the alternatives for each decision. Be creative, be complete. Include a "Do Nothing" or "Business as Usual" alternative.

Third, define strategies. A strategy is one path through the table. Your goal is to define three to five strategies that represent the full span of the possibilities. One strategy should be "Do Nothing" or "Business as Usual." It may help to give the strategies names, such as "Very Aggressive," "Outsource Everything," or "Minimize Environmental Impact." At first, keep the strategies simple and "pure" — once you have run the model and have a better understanding of the risks and opportunities, you can come back and create hybrid or compromise strategies. Or, you can break some decisions out of the strategy table and let DPL choose the optimal combinations (especially useful for evaluating the benefit of delaying a decision until after some uncertainties have been resolved).

Decisions

	R And D Invest	Pricing Decision	Discounting Decision	Marketing Invest
▦ Technology Leader	Lower ▪▪	Lower ▫	More ▫	Lower ▪
▨ Price War				
▪ Marketing Blitz				
▫ Status Quo	No Change ▫	No Change ▪▫	No Change ▪▫	No Change ▪▪▫
	Higher ▪	Higher ▪	Less ▪	Higher ▪

Strategies } Alternatives

If you find that a decision isn't sensitive to the strategy — you always pick the same alternative — ask yourself (or your team) these questions:

- Is this a no brainer? It may be that, although you can think of other alternatives, there is really only one thing you would ever do. If you believe that this is really the case, declare a winner, for that decision at least, and remove the other alternatives from consideration.
- Is this a tactical decision? If this decision is much less strategic than some of the other decisions, the impact of its alternatives is hard to gauge. Pick a representative alternative and shelve this decision until the bigger issues have been resolved.
- Are the other alternatives experimental or risky? Perhaps you should add another strategy. Or, you should plan to revisit these alternatives when you create hybrid/ compromise strategies.

Tutorial: Strategy Tables

Imagine that we are in charge of developing a business strategy for a product of HealthDrug, a large pharmaceutical company. Our business strategy involves making decisions about four key aspects of our business: Marketing, pricing, research and development, and discounts for large customers. We have identified four strategies, each encompassing these four decisions, which we think might be successful for HealthDrug, and have named them Technology Leader, Price War, Marketing Blitz, and Status Quo. Technology Leader positions us as a premium pharmaceutical company, with emphasis on leading edge R&D and high-priced products. Price War attempts to increase our market share by selling our products for less than our competitors', placing little emphasis on R&D. Marketing Blitz relies on strong advertising, holding steady on pricing, and sacrificing R&D. Status Quo has us continuing to do what we have been doing.

Creating Strategy Tables

You can create new decisions within a Strategy Table, incorporate existing decisions into a Strategy Table, or do a combination of both.

▸ Open "Strategies.da".

Name	Type	Date/Time	Size	Status
Session Log	Session Log	Sep 2 6:13 PM	-	-
Strategies_final	Model	Aug 18 1:27 PM	7K	Idle
Strategies	Model	Aug 18 9:17 AM	8K	Idle
Downstream_final	Model	Aug 18 10:29 AM	10K	Main
Downstream	Model	Aug 18 1:27 PM	10K	Idle

▸ Select Window Model-Strategies.

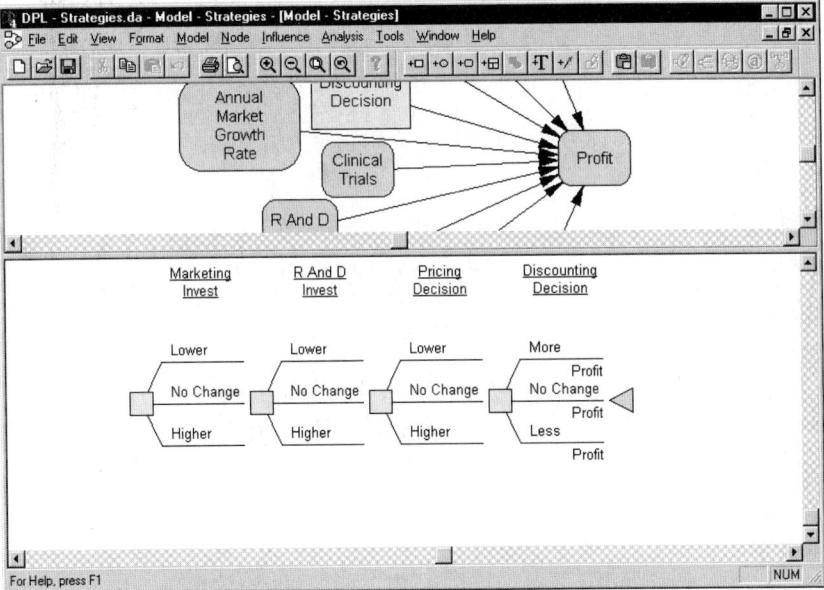

▸ Drag the splitter bar down to maximize the influence diagram pane.

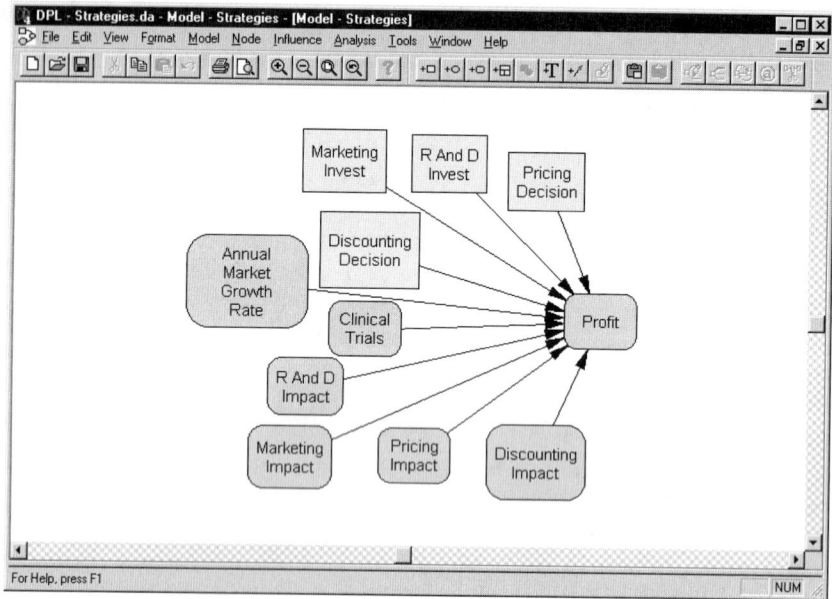

You have four decisions in this model. Rather than evaluate each decision independently, you can create a strategy table from these four decisions.

▶ Click on the Create Strategy Table icon.

▶ Place the strategy table above Annual Market Growth Rate.
▶ Click OK to accept the default name for the Strategy Table.

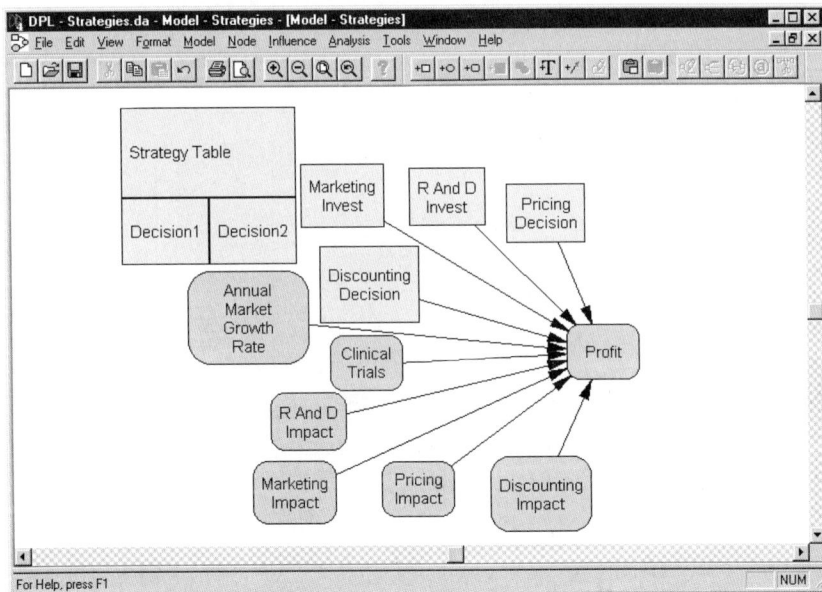

The strategy table node contains one section denoting the strategy table, and additional sections for each of the component decisions. New strategy nodes are created with two new component decisions.

▶ Double-click on the Strategy Table section of the Strategy Node.

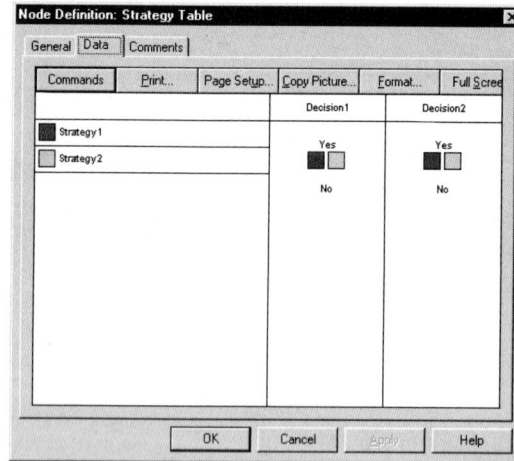

You can make the strategy table easier to see by changing the view to full screen.

▶ Click Full Screen.

You want to incorporate the four decisions into the Strategy Table, and remove the two decisions DPL has created automatically.

Strategy Table commands are available from the Commands button or from the context (right click) menu. Use these commands to add the existing four decisions to the strategy table.

▶ Click the right mouse button.
▶ Select Add Decision from the context menu.

DPL displays a list of the decisions.

Select Event

Discounting Decision
Marketing Invest
Pricing Decision
R And D Invest

[OK] [Cancel]

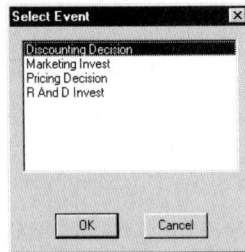

▶ Click OK to select Discounting Decision.
▶ Repeat the previous three steps to add the remaining three decisions to the strategy table.

Node Definition: Strategy Table

Commands | Print... | Page Setup... | Copy Picture... | Format... | Full Screen

	Decision1	Decision2	Discounting Decision	Marketing Invest	Pricing Decision	R And D Invest
■ Strategy1	Yes ■□	Yes ■□	More ■□	Lower ■□	Lower ■□	Lower ■□
□ Strategy2	No	No	No Change	No Change	No Change	No Change
			Less	Higher	Higher	Higher

Next, delete the two decisions which were created with the strategy table.

▶ Click on Decision1.

▶ Click on the right mouse button.

▶ Select Delete from the context menu.

▶ Repeat the previous three steps to delete Decision2.

Node Definition: Strategy Table				⊠
Commands Print... Page Setup... Copy Picture... Format... Full Screen				
	R And D Invest	Pricing Decision	Discounting Decision	Marketing Invest
▣ Strategy1	Lower ■ ■ □	Lower ■ ■ □	More ■ ■ □	Lower ■ ■ □
□ Strategy2	No Change	No Change	No Change	No Change
	Higher	Higher	Less	Higher

Defining Strategies

The four decisions each have three alternatives, combining into 81 possible strategies. Some of these strategies might not be feasible. For example, you can't afford to invest heavily in R&D and marketing while cutting your prices. Other strategies may be very similar. For example, the difference between choosing "No Change" and "Less" for the Discounting Decision when you choose the "Higher" alternative for the other three decisions is probably trivial.

To eliminate having to investigate 81 possible combinations of decision alternatives, you have identified four themes which can be turned into strategies. These themes are Technology Leader, Price War, Marketing Blitz, and Status Quo. Technology Leader positions you as a premium pharmaceutical company, with emphasis on leading-edge R&D and high-priced products. Price War attempts to increase market share by selling products for less than your competitors', placing little emphasis on R&D. Marketing Blitz relies on strong advertising, holding steady on pricing, and sacrificing R&D. Status Quo has you continuing to do what you have been doing.

Now, turn these themes into four strategies in the strategy table.

- ▸ Click on the right mouse button.
- ▸ Select Create Strategy from the context menu.
- ▸ Repeat the two previous steps.
- ▸ Click on Strategy1.
- ▸ Click the right mouse button.
- ▸ Select Rename from the context menu.
- ▸ Type "Technology Leader".
- ▸ Repeat the previous four steps to rename the remaining three strategies "Price War", "Marketing Blitz", and "Status Quo".

The icons in the strategy table indicate which decision alternatives are chosen in each strategy. You need to assign each strategy an alternative to choose for each decision. By default, all strategies are assigned the first alternative of each decision. Change the assignments to reflect your strategies. For example, in the strategy Technology Leader, you'll set R And D Investment to be Higher, Pricing Decision to be Higher, Discounting Decision to be Less, and Marketing Invest to be Lower.

▸ Click-and-drag the icon next to Technology Leader (a red square).
▸ Place the icon under the Higher alternative of R And D Invest.

This tells DPL that the strategy Technology Leader will choose the "Higher" alternative of R And D Invest.

▶ Repeat the previous two steps for all decisions in Technology Leader and then for all strategies until the strategy table has the following structure:

	R and D Invest	Pricing Decision	Discounting Decision	Marketing Invest
Technology Leader	Higher	Higher	Less	Lower
Price War	Lower	Lower	More	No Change
Marketing Blitz	Lower	No Change	No Change	Higher
Status Quo	No Change	No Change	No Change	No Change

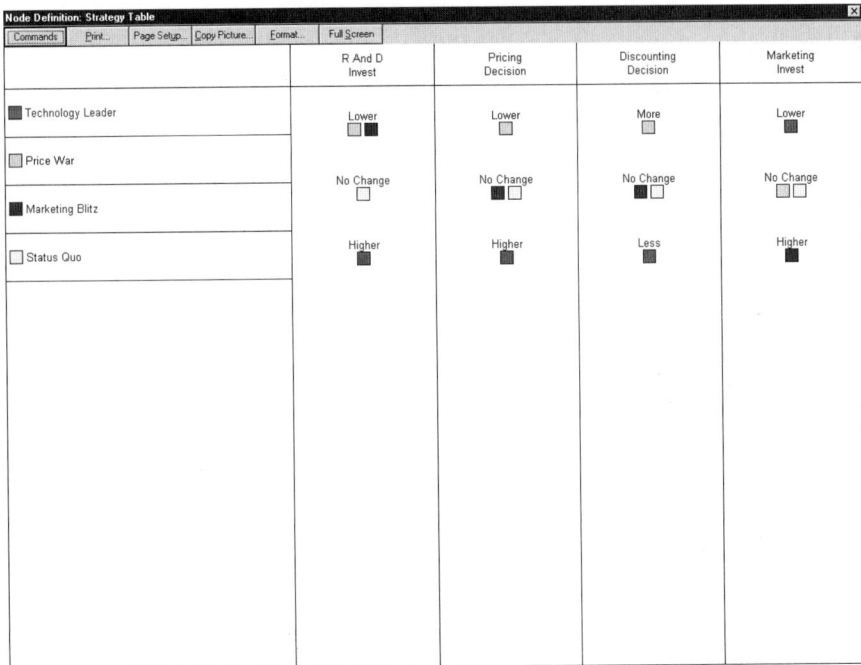

▶ Click Full Screen.
▶ Click OK.

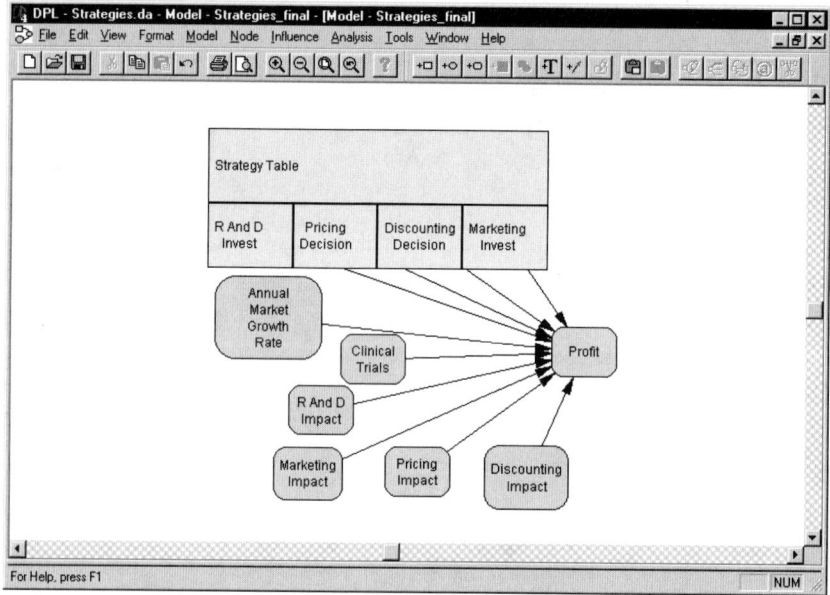

The strategy table appears in the decision tree.

▸ Select View Decision Tree (or press the Tab key).

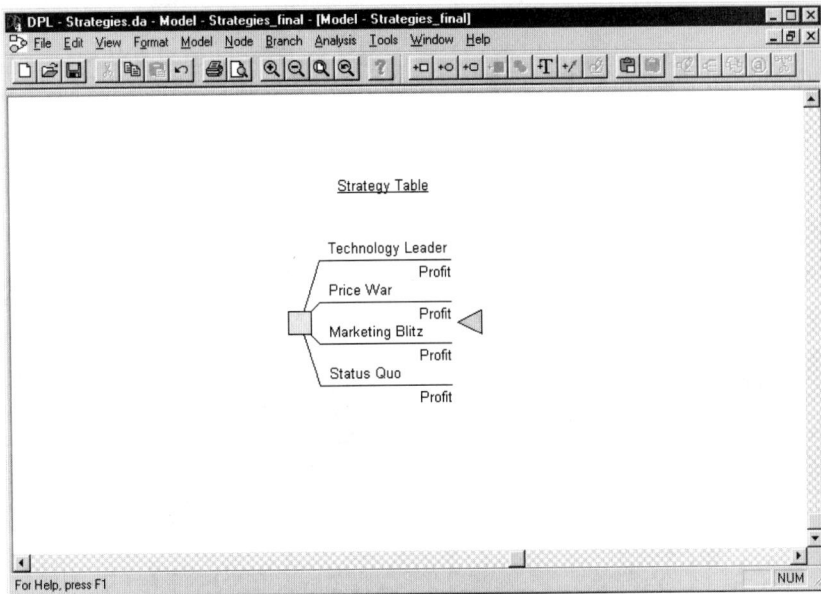

In the decision tree, DPL displays the strategy table as a single decision node with the strategies as the decision's "alternatives".

▸ Select View Influence Diagram (or press the Tab key again).

Conditioning Nodes on Strategy Table Decisions

You can have the decisions in a strategy table condition other nodes. You can also have the Strategy Table itself condition other nodes.

The value for Pricing Impact depends on the alternative you select for Pricing Decision. You can indicate this by conditioning Pricing Impact on Pricing Decision. This is done from the Data tab of the Node Definition dialog box.

▸ Double-click on Pricing Impact.

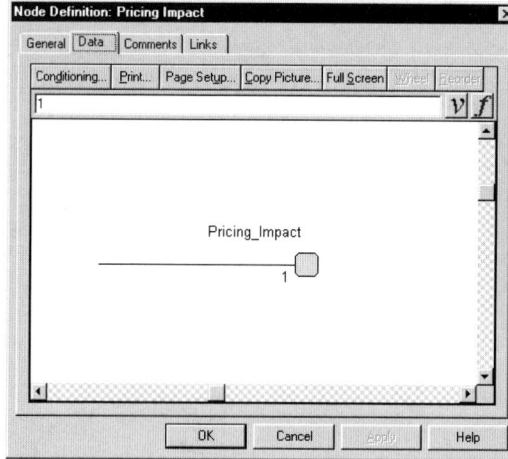

▶ Click Conditioning.

DPL displays the Conditioning dialog box

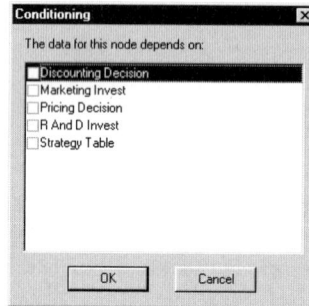

▶ Click the box labeled Pricing Decision.
▶ Click OK.

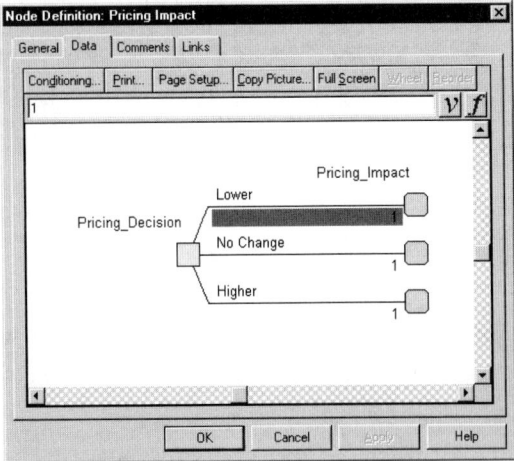

There are now three values for Pricing Impact, one corresponding to each alternative of Pricing Decision.

▶ Change the data for Pricing Impact as follows:

Pricing Decision Lower: 1.03

Pricing Decision No Change: 1

Pricing Decision Higher: 0.98

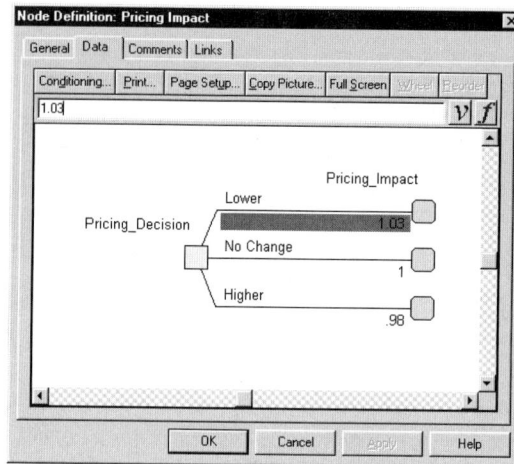

▶ Click OK.

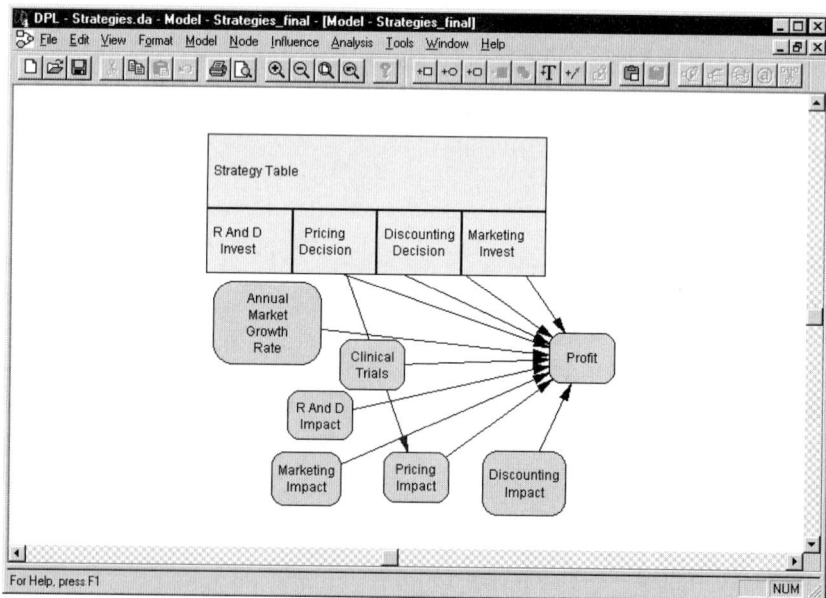

DPL has drawn an arrow from Pricing Decision to Pricing Impact. The arrow has a blue arrowhead, indicating that the values of Pricing Impact depend on the alternatives of Pricing Decision.

Removing Decisions From Strategy Tables

You should not feel that once you assign certain decision alternatives to a strategy you have to keep those forever. One way to investigate whether you have optimized your strategy is to remove a decision from the strategy table and model it independently.

For this exercise, you will use another model in the project "Strategies.da". You need to make it the Main model in the Project Manager before you can work on it.

▸ Select Window Strategies.da.

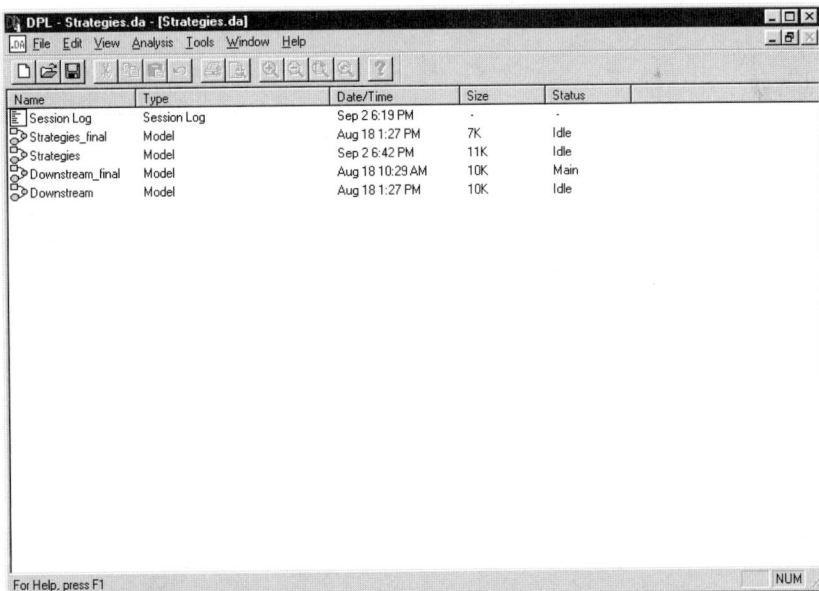

▸ Right-click on the model "Downstream."

▸ Select Make Main from the context menu.

▸ Double-click on the model Downstream.

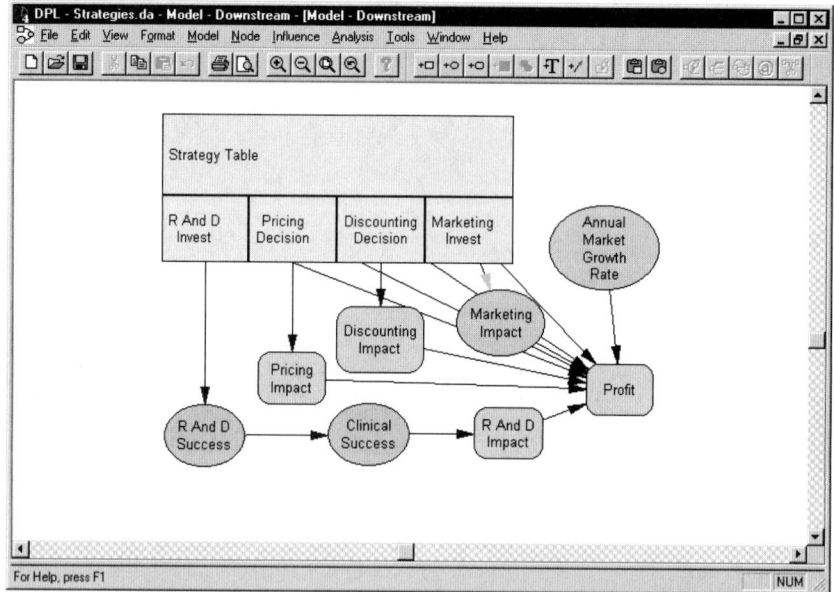

This is the same model, with some uncertainties added. Run a decision analysis to see which is the optimal strategy.

▸ Select Analysis Decision Analysis.

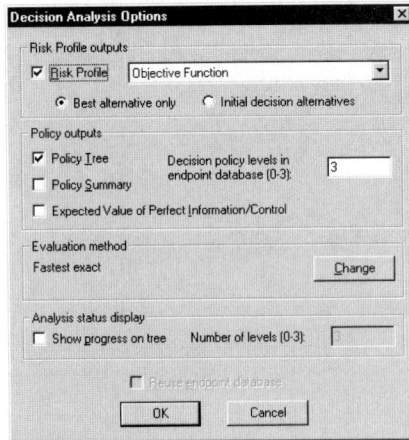

▶ Click OK.

DPL displays the Policy Tree.

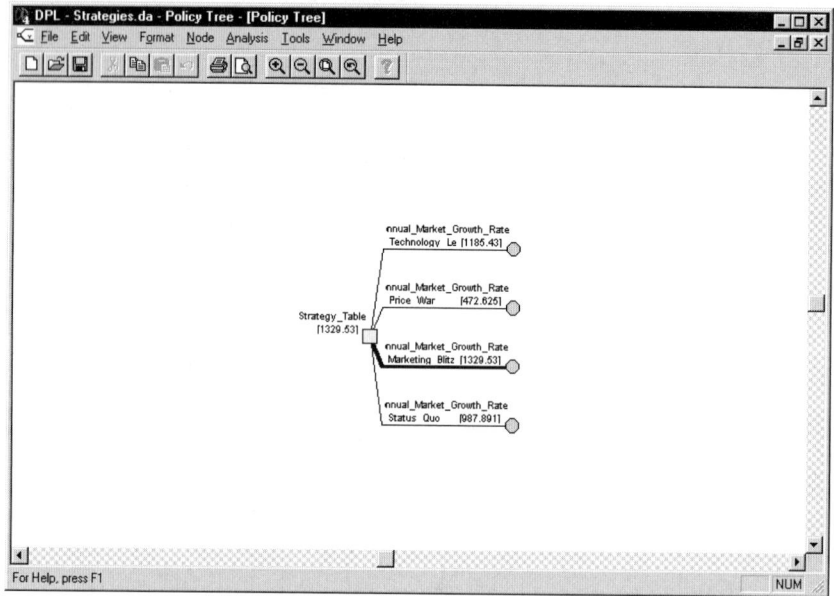

You can see that Marketing Blitz is the optimal strategy, and returns an expected value of $1.3 billion.

Look at the strategy Marketing Blitz in the Strategy Table.

▶ Select Window Model-Downstream.
▶ Double-click on the Strategy Table section of the Strategy Node.

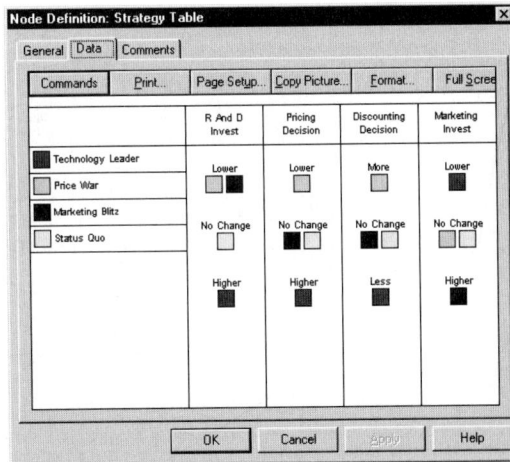

The strategy Marketing Blitz selects the No Change alternative for Pricing Decision. Is this the optimal setting for the Marketing Blitz strategy? You could run the model two more times, once each with Pricing Decision set to its other two alternatives for Marketing Blitz. Instead, you will remove Pricing Decision from the Strategy Table and model it independently.

▸ Right-click on Pricing Decision.
▸ Select Remove Decision from the context menu.

DPL has removed Pricing Decision from the Strategy Table.

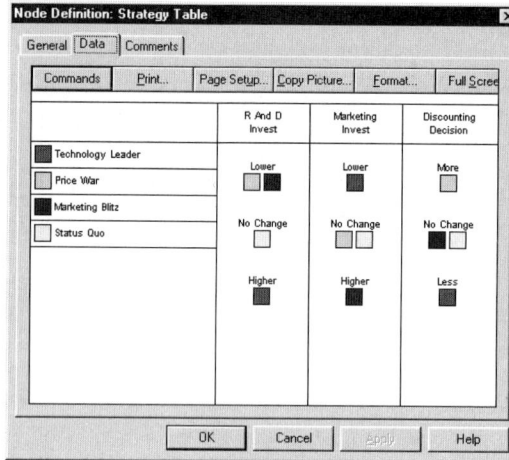

▸ Click OK.

You can see that Pricing Decision is now an independent decision.

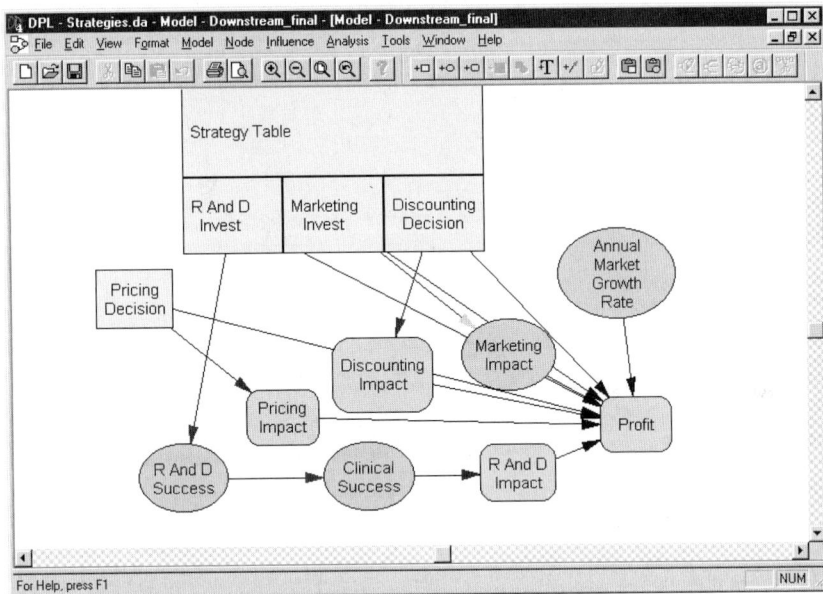

You can re-run the decision analysis to see if this changes the results.

- Select Analysis Decision Analysis.
- Click OK to remove the current decision structure from the project.
- In the Decision Analysis Options dialog box, click OK to run the analysis.

The model takes longer to run because you have increased the number of paths by a factor of three (from 216 to 648).

DPL displays the new Policy Tree.

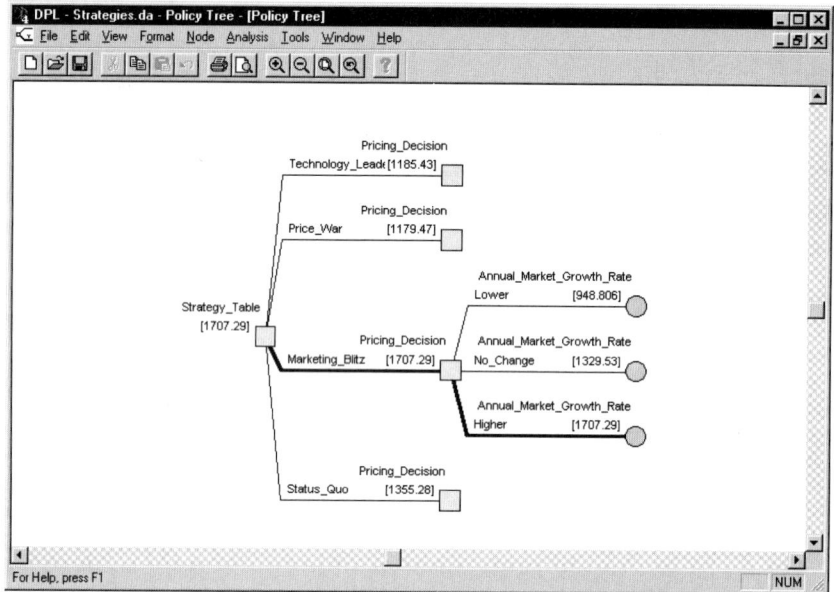

The Policy Tree reports that the optimal setting of Pricing Decision for the strategy Marketing Blitz is Higher, rather than the previous setting of No Change. The new expected value is $1.7 billion-an increase of $400 million.

At this point, you can continue to model Pricing Decision independently, or you could add it back into the Strategy Table and simply change its setting for the Marketing Blitz strategy from No Change to Higher. This would reduce the total number of paths in the model, which may save you time if you are planning on performing further analyses.

Chance events in DPL models represent variables about which there is uncertainty. DPL enables you to define chance events that are independent or conditioned by other decision or chance events. You can also define chance events using named distributions such as the normal or exponential distributions.

Chapter 5

Chapter 5: Chance Events

What is a Chance Event?

A chance event is an element in your decision environment about which you are uncertain. It can be a number, such as production cost, or a state of the world, such as the competitiveness of the market. A chance node reflects the state-of-information that you or your team has. Others may have different states of information, and may even know with virtual certainty something that is uncertain to you.

In DPL, a chance event is a node with two or more outcomes.

In the influence diagram, a chance event is represented by a green oval:

In the decision tree, a chance event is represented by a green circle with branches:

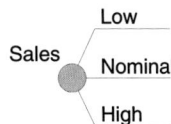

How to Model an Uncertainty

First, give it a name. This is the only name that will appear on the influence diagram. It will also appear on the decision tree (along with the outcomes). This name may also be used as a variable name in mathematical expressions. Some rules:

- The name may contain up to 265 characters. We recommend that you keep it under 32 characters to make the outputs easier to read.
- The name may include spaces and special characters such as ? _ ( ). These characters will be converted to underscores if you use the name in a formula (this prevents DPL from becoming confused about which characters belong in variable names and which provide the "punctuation" for the formula).
- The first character should be a letter. If the first character is a number, DPL will replace the number with the character "N" when the name gets used in formulas.

Second, define the outcomes. The outcomes of an uncertainty are the states it can be in. The same naming rules apply. Some things to consider:

- **How many outcomes?** An uncertainty must have at least two states. Technically, the maximum number of states it can have is 64 (32,000 if you switch to DPL code). If you plan to use a named distribution (discussed later in this chapter), the maximum number of outcomes is 6. In practice, data entry, validation and comprehension all start to get pretty hard at about 6 outcomes.
- **Cover all the possibilities.** The outcomes of an uncertainty must be "collectively exhaustive," which means that, taken together, they include all the possible states of the world. For example, an uncertainty about an EPA regulation expected in the year 2000 might have these outcomes:

But what happens if the EPA doesn't publish its ruling in 2000? (It has been known to miss a date or two.) If your model includes a decision that must be made in 2000, you need to model the possibility that you will have to make the decision before the regulation is known. In this case, the uncertainty would have these outcomes:

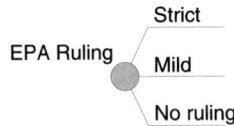

```
                            Strict
                           _____
          EPA Ruling    ⦿   Mild
                           _____
                            No ruling
```

- **Don't overlap.** The outcomes of an uncertainty must be "mutually exclusive," which means that you can be in only one state once the uncertainty is resolved. For example, you wouldn't want a chance node with these outcomes:

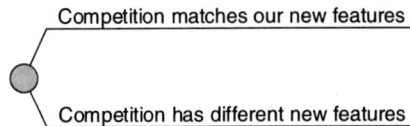

```
          Competition matches our new features
          _____
    ⦿
          Competition has different new features
          _____
```

These outcomes overlap if it's possible that a competitive product might include all the same features as your new product **and** add some of their own.

- **Consider low-probability, high-consequence outcomes.** When people go back and look at decision models they built several years ago, it is fairly common to find that, in at least one case, the future that really happened was not included in the modeled range of outcomes. If the error is just a few percentage points off, the consequences aren't terribly dire. But what if you model a range of nuclear fuel disposal costs, only to find in the future that all nuclear power plants get closed down pending a final decision on a disposal site? What if you model a range of production costs for your product, but do not consider the possibility that a new technology is introduced that cuts costs in half?

Third, assign probabilities to the outcomes. The probabilities of the outcomes must sum to 1.0. If you are having trouble with this, check to see that your outcomes really are mutually exclusive (if your probabilities add to more than 1.0) and collectively exhaustive (if your probabilities add to less than 1.0).

```
                                 Strict
                                 .2
                    EPA Ruling   Mild
                         ●       .6
                                 No ruling
                                 .2
```

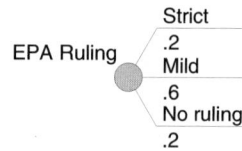

These probabilities represent your best information about the relative likelihood of the outcomes. When you are assessing probabilities, be sure to take advantage of all available sources of information — historical data, model runs, company experts, etc. For all sources of information, ask yourself (or your experts) these questions:

- Is this the best information available?
- What are its known limitations?
- What would it cost (time and money) to get better information? How much better would that be? (Later in the analysis, you can evaluate whether the advantages of getting this better information outweigh the costs.)
- Upon what assumptions or historical experiences is this information based? Is there anything we know we need to adjust to make this apply to the future?

Fourth, assign values, if appropriate. If your chance event's outcomes are naturally associated with values to be used in calculating the objective function values, you can assign them directly to the chance event's states.

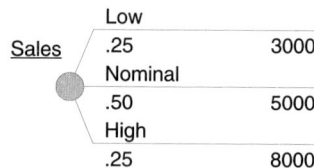

```
                              Low
                              .25            3000
                    Sales     Nominal
                       ●      .50            5000
                              High
                              .25            8000
```

"Turbo" Decision Analysis

If you have limited time for analysis, or want to do a quick once-over-lightly pass at the beginning of a more intensive project, you can use the "10-50-90" approach to probability assessment. This approach works best on chance events that represent ranges of numbers (such as construction costs) rather than discrete states (such as winning or losing a lawsuit).

First, define three outcomes, called Low, Nominal (or Basecase) and High.

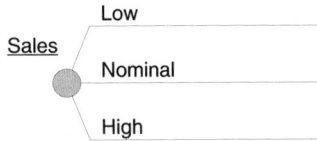

```
              Low
  Sales  ╱
        ╭──── Nominal
        ●
         ╲
              High
```

Second, assign probabilities of .25, .5 and .25, respectively, to these outcomes.

```
              Low
  Sales  ╱   .25
        ╭──── Nominal
        ●    .50
         ╲   High
             .25
```

Third, assign values to the outcomes. The Low outcome should be the value which you think represents the 10th percentile point for the distribution on this value — in other words, you think there is only a 10% chance that the actual value will fall below your Low value. The Nominal value should be the 50th percentile point — the actual value could be above or below with equal probability. The High value should be the 90th percentile point — there is a 10% chance the actual value will be higher.

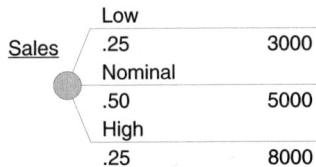

```
              Low
  Sales  ╱   .25          3000
        ╭──── Nominal
        ●    .50          5000
         ╲   High
             .25          8000
```

You will notice that these are the same rules we recommend for setting the ranges for a deterministic expected value tornado diagram. DPL's defaults are set to facilitate these "turbo" models — if you build a deterministic model and do an expected value tornado diagram, when you make deterministic variables into chance events, DPL's defaults assume you want three states, .25/.5/.25 probabilities, and the values from the tornado — you don't have to edit anything!

This approach requires two assumptions:

- The variables are normally distributed
- You can directly assess the 10th, 50th and 90th percentile points on the distribution

Decision analysis practitioners have found that these assumptions are acceptable in a surprising number of cases.

Conditional Chance Events

One of the most powerful things about decision analysis is its ability to model relationships among variables — the ability to capture the "our sales will depend on the state of the economy, what our competitors do, whether we get to market before the Christmas season, and whether we can 'hire' a famous cartoon character as our mascot" relationships that make predicting the future so difficult.

<u>One Conditioning Event</u>

If one chance event depends upon another, the influence diagram will have an arrow between them. The dependent, or conditional, chance event requires a complete probability distribution for each of the conditioning event's outcomes.

Sometimes, the effect of the conditioning event is shown as a shift in the probability distribution for the conditional event:

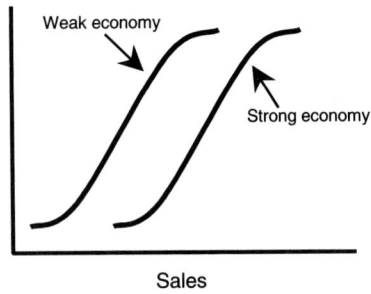

This is modeled by changing the values, but not the probabilities, for the conditioned event's outcomes.

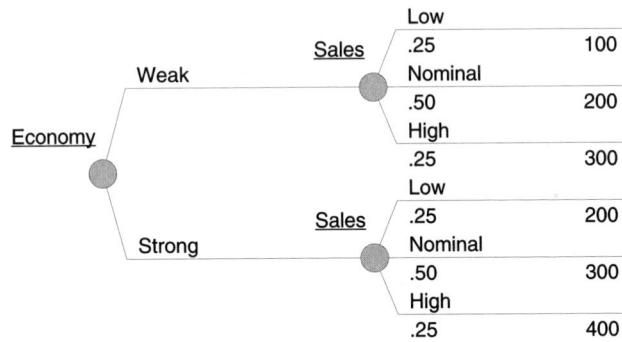

Other times, the effect is shown in the probabilities:

```
                                                  Low
                                     Sales        .2              200
                          Yes                     Nominal
                                            ●     .3              300
         Before                                   High
         Christmas  ●                             .5              400
                                                  Low
                                     Sales        .5              200
                          No                      Nominal
                                            ●     .3              300
                                                  High
                                                  .2              400
```

Of course, the conditioning event may affect both the probabilities and the values for the conditional event.

Multiple Conditioning Events

If a chance event depends on more than one other event, it requires a probability distribution for each combination of the conditioning events' states. This means that an event that depends on two three-state events requires nine separate probability distributions. Three conditioning events with three states each mean 27 distributions! The DPL software can handle this with ease, but your experts and analysts will have a very hard time keeping everything straight.

Some suggestions:

- Try to have only one, two or three arrows leading into any chance event.
- Try to turn a chance event with three or more conditioning events into a value node. For example, instead of:

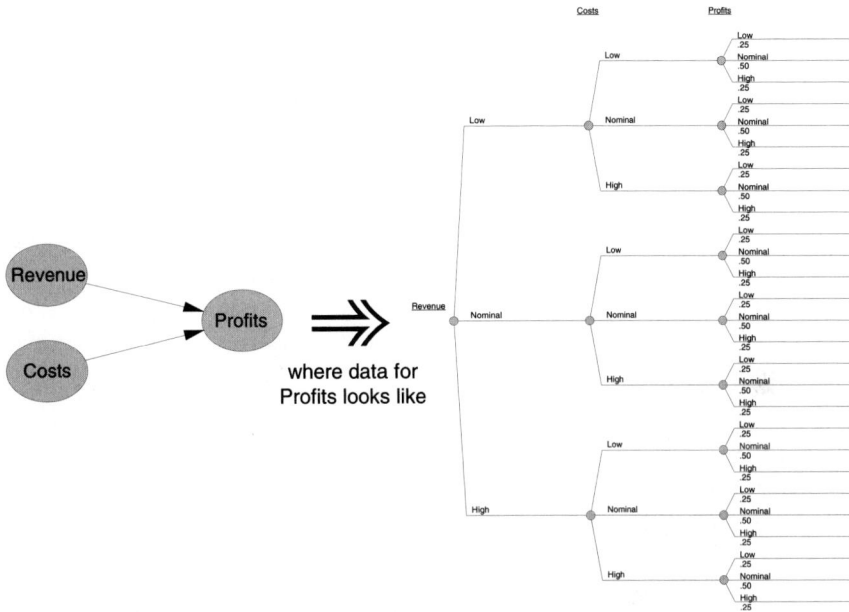

where data for
Profits looks like

Try:

where Profit = Revenues - Costs

- Try writing a formula to describe a conditioning event's impact. Instead of:

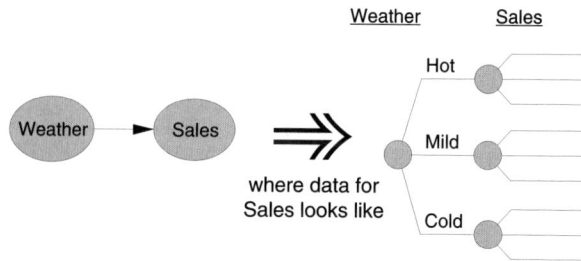

Try:

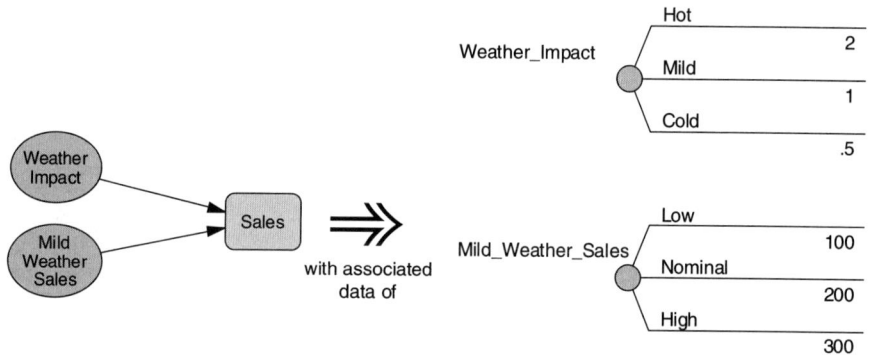

Sales = Mild Weather Sales × Weather Impact

Chance Events Depending on a Decision

It is common to have chance events in a model that depend on the decisions in addition to other chance events. There is nothing wrong with this, and DPL will allow you to enter these dependencies easily. However, you will not be able to calculate the value of information for such chance events.

To see why not, consider a model in which the sales of a product depend on whether you do your own marketing or form a partnership with a larger, more established company. It might seem natural to draw the influence diagram like the following:

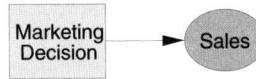

and to enter the data for sales as:

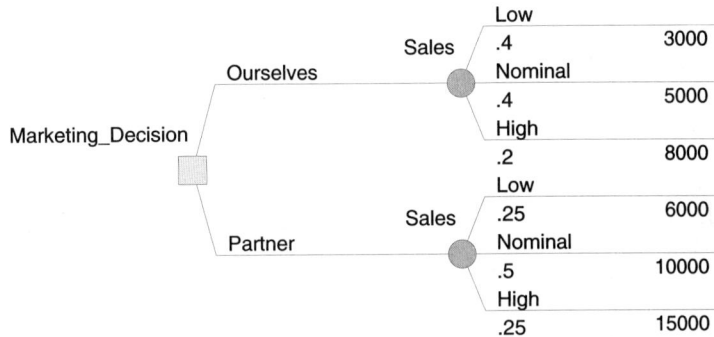

You will be able to enter data and run decision and sensitivity analyses with no trouble. But the value of information calculations will return a value of zero for Sales. This is because, to calculate the value of information, DPL must rearrange the decision tree to put Sales before the decision:

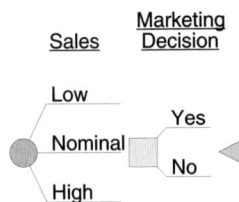

But what probability distribution should it use for Sales? DPL needs to pick one, but it won't know which one to pick until after the decision node appears, which is too late.

There are three possible solutions to this quandary.

- **Two chance nodes.** Revise your model to include one chance node for each decision alternative.

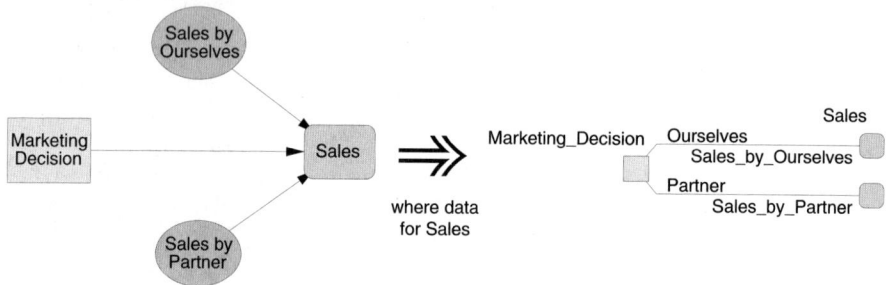

Now you can get the value of information on each sales chance event.

- **Decision affects values, not probabilities.** Turn on the advanced chance node features using the Tools Options Advanced dialog box. Then, define the chance node using separate probability and value data input trees. Make sure the probability data tree does **not** depend on the decision.

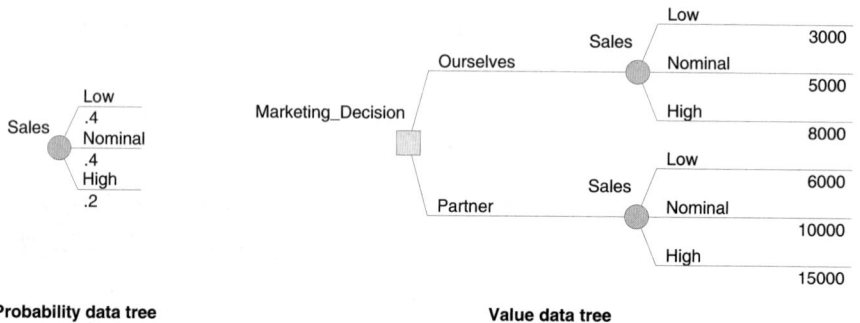

Probability data tree   **Value data tree**

Of course, this approach won't work if the probabilities really do depend on the decision!

- **Decision effect modeled with formula.** Revise your model so that the effect of the decision can be modeled in a formula:

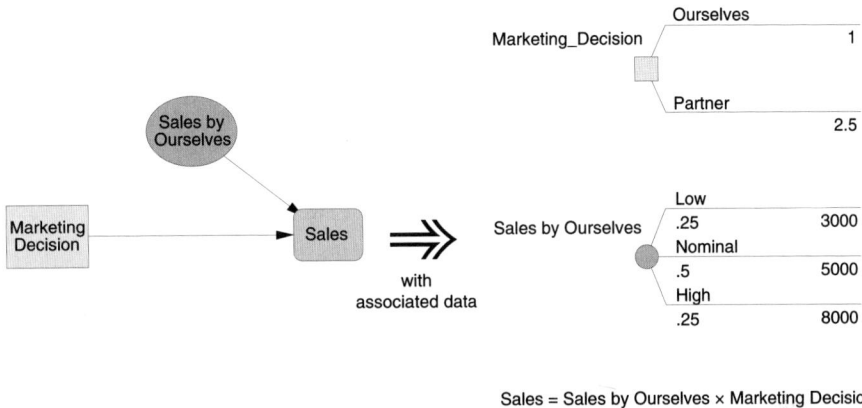

Sales = Sales by Ourselves × Marketing Decision

Chance Events Modeled with Named Distributions

The probability distribution(s) for a chance event can come from a wide variety of sources, and you may use a number of techniques to develop these distributions. For uncertainties such as the probability of winning a lawsuit or the relative likelihood of alternative regulatory proposals, you are likely to speak to the best-informed person you can find on the subject, and they will give you direct estimates that can go straight into the DPL model.

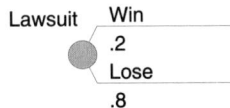

For other uncertainties, such as sales for a new product, you are still likely to be speaking with an expert, but they probably can't give you exactly what you need to put into DPL. Instead, your assessment meeting is more likely to produce a set of data points like this:

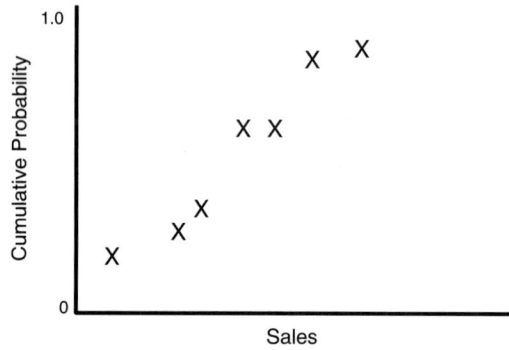

To put this into DPL, you need to fit a curve to these points (whether you do this with a pencil or a computer depends on your preferences) to create a continuous probability distribution for the uncertainty.

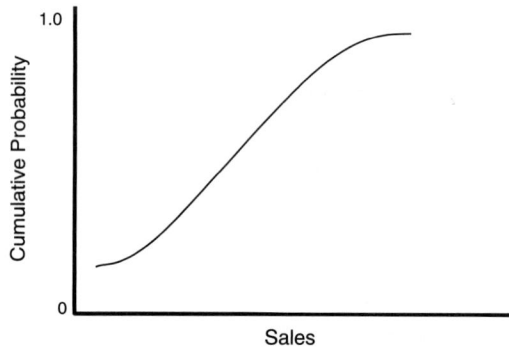

DPL uses discrete outcomes rather than continuous distributions, so you need to create a discrete approximation of this continuous distribution. Again, there are a variety of techniques for this.

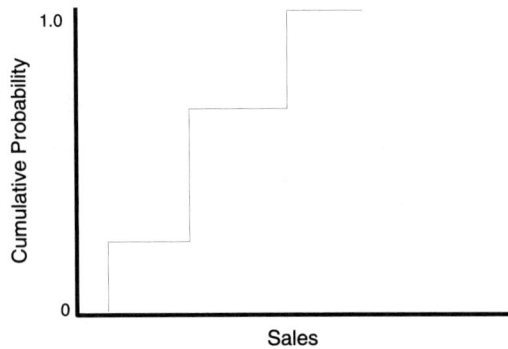

This discrete approximation can now be entered into DPL.

Some uncertainties are easier to describe as continuous distributions from the start, rather than starting with directly assessed points and fitting a curve. For these uncertainties, your expert is able to say, for example, that engine failures can be modeled with a Poisson distribution, or that the amount of oil in a new field can be represented by a lognormal distribution.

DPL offers 21 of these so-called "named distributions." You can assign one of these distributions to any chance event with two to six outcomes. To use a named distribution:

- First, name the chance event as usual.
- Second, specify the outcomes. Remember, the chance event can have a maximum of six outcomes. Arrange the outcomes in order from lowest to highest.
- Third, assign a named distribution.
- Fourth, enter the parameters for the distribution.

Whenever the model is run, DPL will create a discrete approximation for the chance event, assigning probabilities and values to the outcomes. It uses an analytic technique called Gaussian Quadrature, which creates an approximation that matches the moments of the original named distribution. (The moments of a distribution are descriptive statistics such as the mean, the variance, and so forth.) For example,

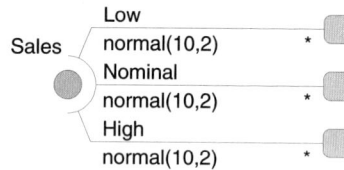

```
                          Low
            Sales       normal(10,2)          *
                         Nominal
                        normal(10,2)          *
                          High
                        normal(10,2)          *
```

will be approximated by DPL as

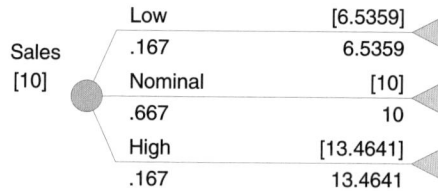

```
                          Low              [6.5359]
            Sales         .167              6.5359
            [10]         Nominal             [10]
                          .667               10
                          High            [13.4641]
                          .167             13.4641
```

If you calculate the mean of the approximation, .167×6.5359+.667×10+.167×13.4641, the value is 10, just as in the original normal distribution. The standard deviation,

$$= \sqrt{0.167(6.5359 - 10)^2 + 0.667(10 - 10)^2 + 0.167(13.4641 - 10)^2}$$

$$= \sqrt{0.167(-3.4641)^2 + 0.667(0)^2 + 0.167(3.4641)^2}$$

$$= \sqrt{0.333(3.4641)^2} = \sqrt{4} = 2$$

is also an exact match.

For information on how this technique compares to Monte Carlo simulation, you can download a white paper from our website, "DPL and Monte Carlo Simulation."

Tutorial: Chance Events

Consider a simple capital investment decision in which you are the owner of a manufacturing company. You need to decide how much to invest in R&D and in additional manufacturing capacity.

You have created a deterministic influence diagram with a strategy table and run an Expected Value Tornado Diagram to determine which variables should be modeled as uncertainties. Now you're ready to add uncertainty to the model.

▸ Open "chance.da".

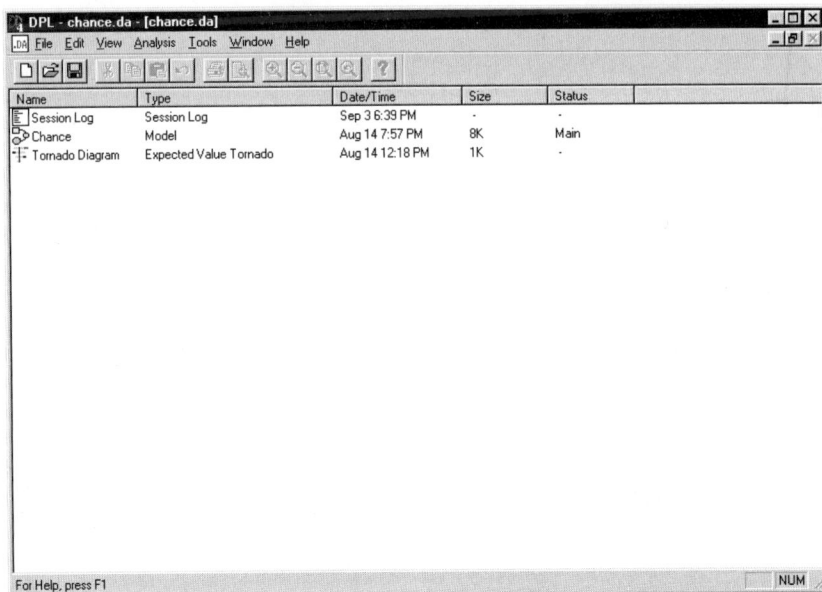

▸ Select Window Model-Chance.

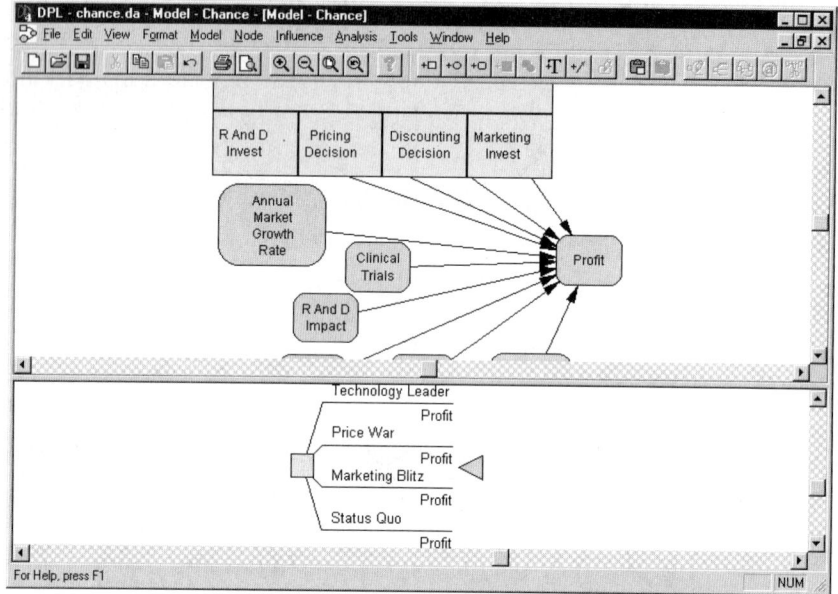

▶ Drag the splitter bar down to maximize the influence diagram pane.

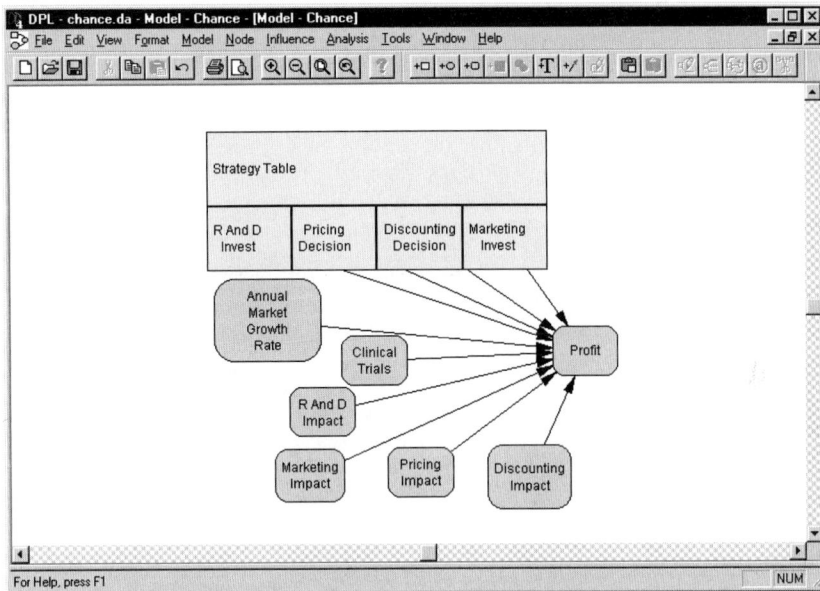

Creating New Chance Nodes

First, incorporate uncertainty into R And D Impact. The success of the R & D effort is dependent on the amount invested in R&D. Model this dependency by creating a chance node.

▶ Click on the Create Chance icon.

▶ Place the node to the left of R And D Impact.

DPL displays the General tab of the Node Definition dialog box. The General tab has two sections: Name and Outcomes.

Next, rename the node R And D Success. DPL has highlighted the default name "Node1", so you simply have to type to replace it.

▸ Type "R And D Success", pressing Enter after "R And D" to put the name on two lines.

Underneath the name entry box DPL displays the variable name as it will appear in expressions. The variable name for expressions is simply the variable name with characters other than numbers or letters replaced by underscores.

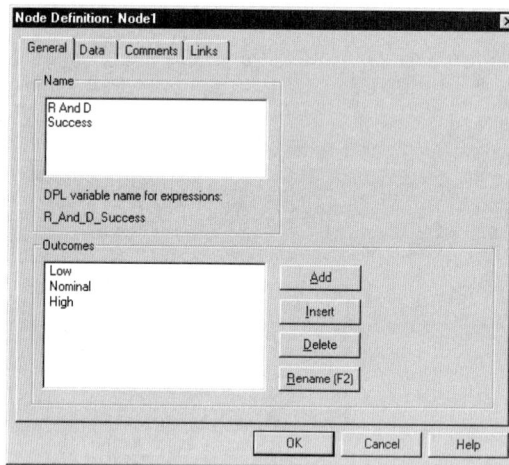

R And D Success has four possible outcomes: Bad, Acceptable, Good, and Excellent. By default, chance nodes have three outcomes, named Low, Nominal, and High. (You could change the default outcome names by selecting Tools Options General.) Rename the default outcomes to match the desired outcome names.

▸ In the Outcomes section, click the Low outcome.
▸ Click Rename (or press F2).
▸ Type "Bad", then press Enter.
▸ Repeat the previous three steps to rename Nominal as "Good" and High as "Excellent."

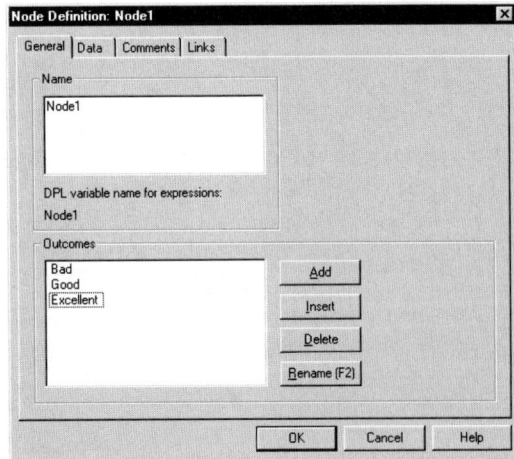

Because R And D Success has four outcomes, you need to insert another outcome for Acceptable between Good and Excellent.

- ▶ Click on Good.
- ▶ Click Insert.
- ▶ Type "Acceptable", then press Enter.

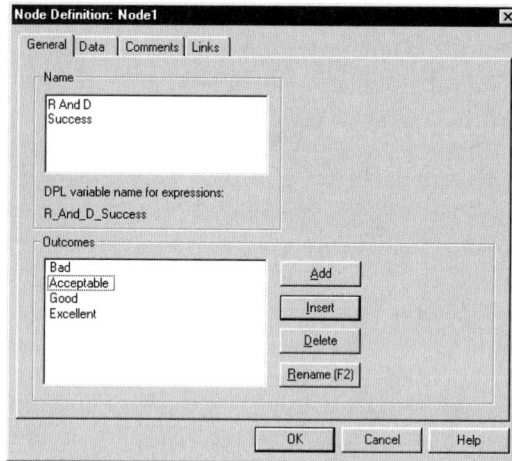

▶ Click OK.

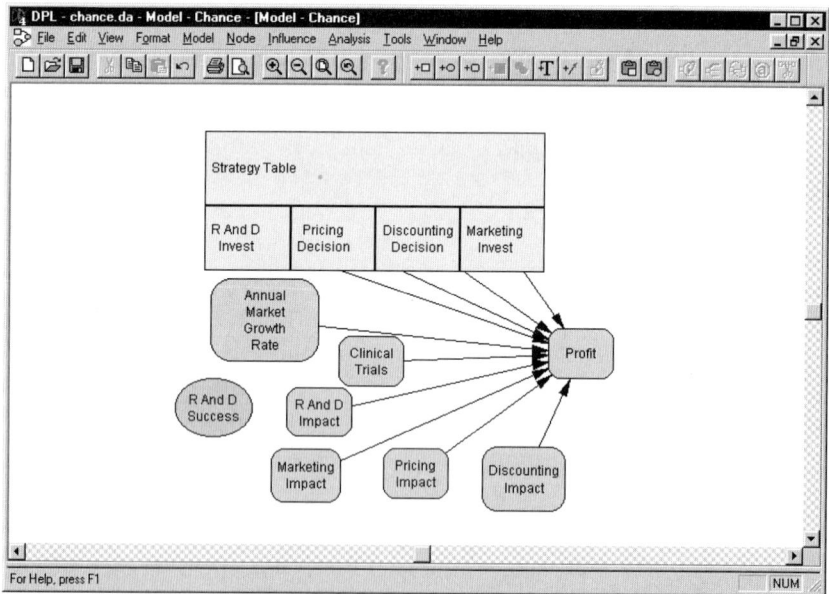

Now edit the data for R And D Success.

▸ Double-click on R And D Success.

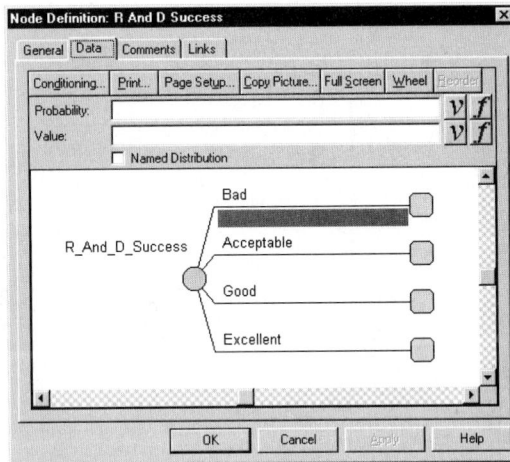

The Data tab is where you enter the probabilities and the values associated with the outcomes of R And D Success.

There is a 30% chance that the outcome of R And D Success will be Bad, yielding a value of 0. Enter this information into the node data.

▸ Click in the entry bar labeled Probability.
▸ Type "0.3". Press Enter.
▸ Type "0". Press Enter.

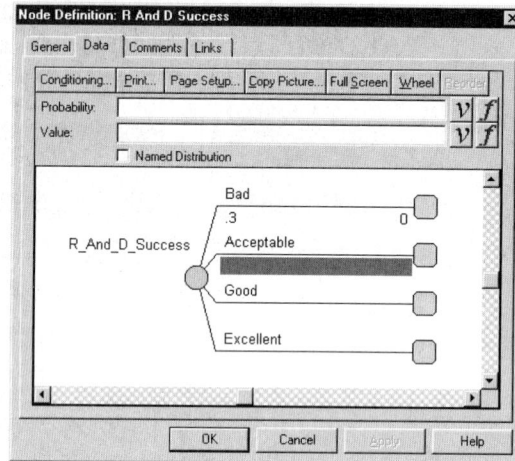

DPL displays the data you have entered for the outcome Bad; the probability is on the left end of the branch and the value is on the right. The magenta bar has moved down to the Acceptable outcome, indicating that data entered next will be placed on this branch.

▸ Repeat the previous two steps to enter the following probability-value pairs for the remaining three branches:

      Acceptable: 0.4, 100

      Good: 0.2, 120

      Excellent: 0.1, 150

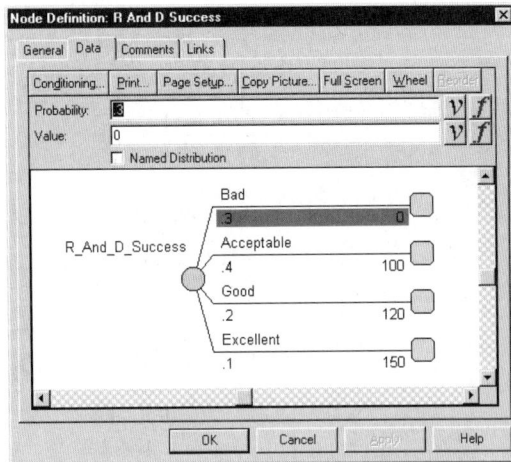

▸ Click OK.

DPL displays chance nodes as green ovals in the Influence Diagram.

Converting Value Nodes into Chance Nodes

Now change Annual Market Growth Rate into an uncertainty.

▸ Click on Annual Market Growth Rate, then click on the Change Type icon.

▸ Select Chance from the Node Type dialog box.
▸ Click OK.

DPL displays the General tab of the Node Definition dialog box.

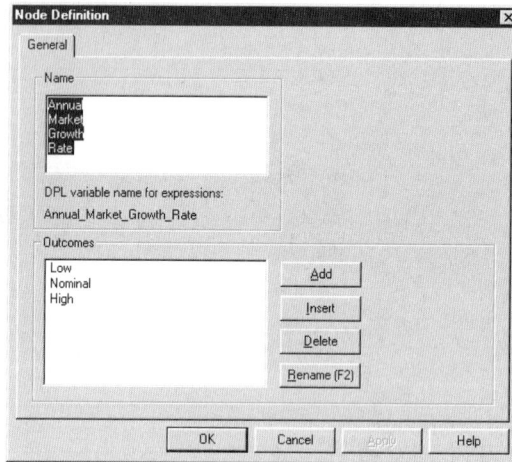

You do not need to change the node's name, and can accept the default outcome names of Low, Nominal, and High.

▶ Click OK.

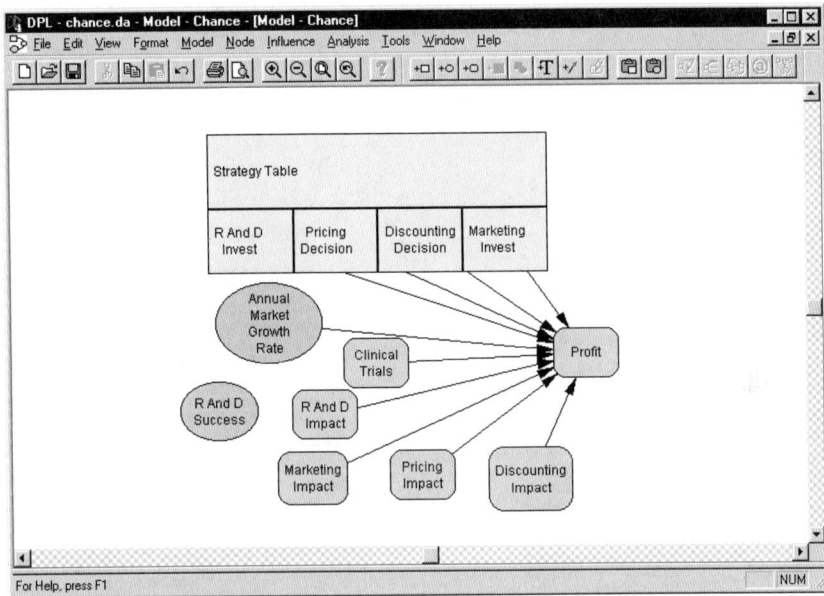

DPL has changed the node for Annual Market Growth Rate into a green oval to indicate that it is an uncertainty. Now edit the data for Annual Market Growth Rate.

▶ Double-click on Annual Market Growth Rate.

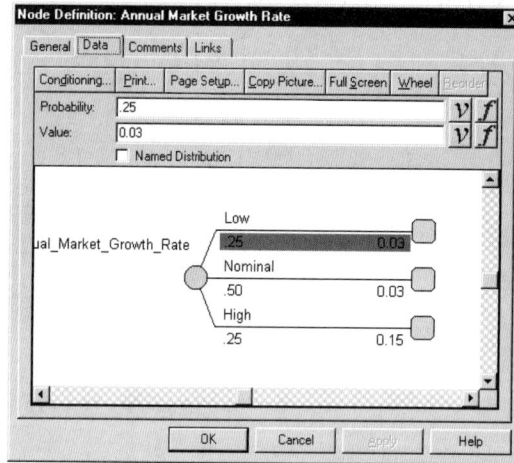

DPL has taken the Low and High values used for Annual Market Growth Rate when running the Expected Value Tornado Diagram and inserted them in the Low and High outcomes of the node data. It has assigned the outcomes the default probability distribution of 0.25, 0.5, 0.25. (You can change the default probabilities by selecting Tools Options General.)

▶ Click OK.

Conditioning a Chance Event's Probabilities and Values

You can have the values and probabilities of a chance node depend on the state of another chance or decision node. The success of the R&D effort depends on the amount invested in R&D, so condition R And D Success on R&D effort.

▶ Double-click on R And D Success.
▶ Click Conditioning.

DPL displays the Conditioning dialog box.

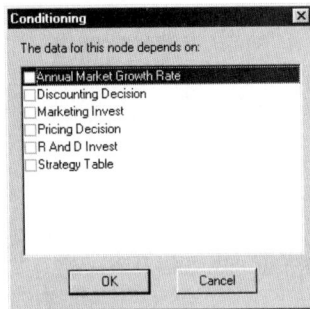

▸ Check the box next to R And D Invest.

▸ Click OK.

The new data tree is difficult to read, so make the view full-screen.

▸ Click Full Screen.

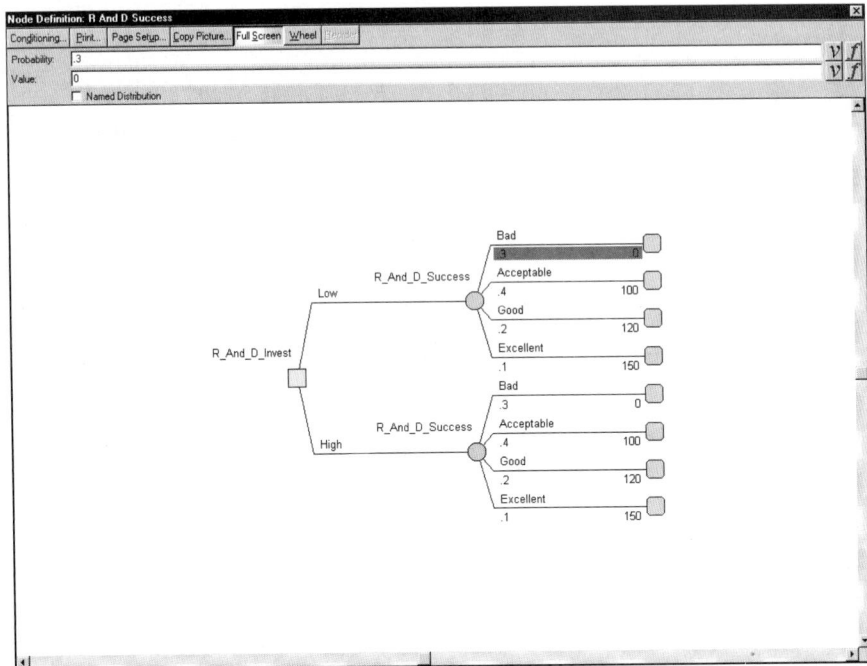

DPL shows a separate set of outcomes for R And D Success for each alternative of R And D Invest. The probabilities and values entered for R And D Success represent its distribution when you select the Low alternative of R And D Invest. Change the distribution for R And D Success corresponding to the High alternative of R And D Invest.

▸ Click on the Bad outcome of R And D Success corresponding to the High alternative of R And D Invest.

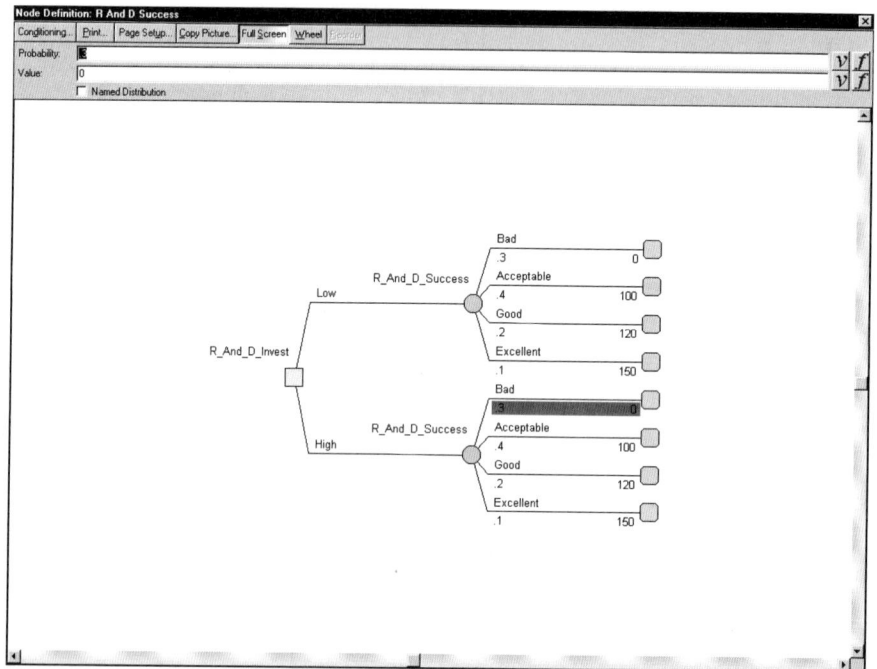

▸ Enter the following probabilities and values for the outcomes of R And D Success:

    Bad: 0.1, 0

    Acceptable: 0.4, 110

Good: 0.3, 140

Excellent: 0.2, 200

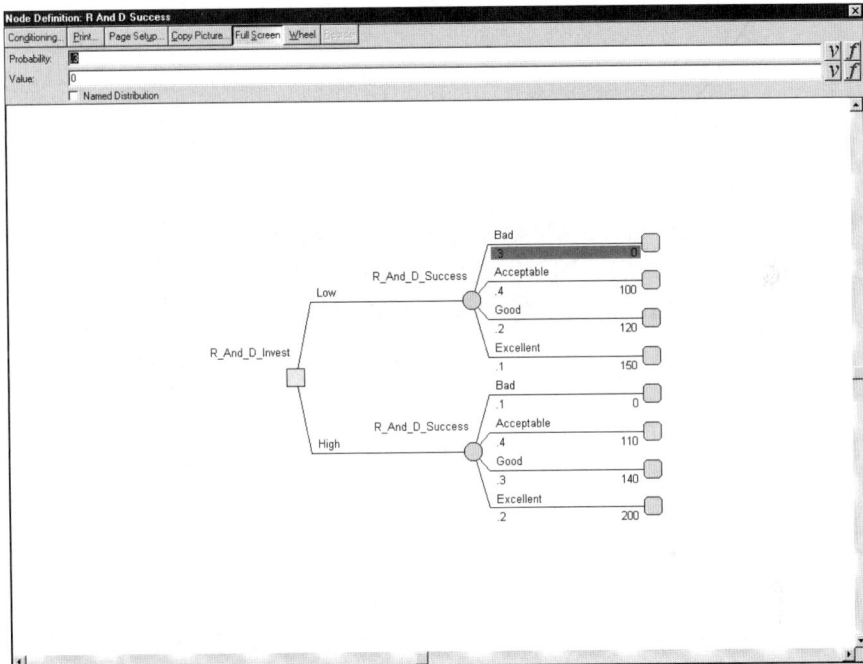

▸ Click Full Screen.

▸ Click OK.

DPL has drawn an arrow from R And D Success to R And D Invest to indicate the conditioning you specified. The burgundy arrowhead indicates that both the probabilities and values of R And D Success depend on the state of R And D Invest.

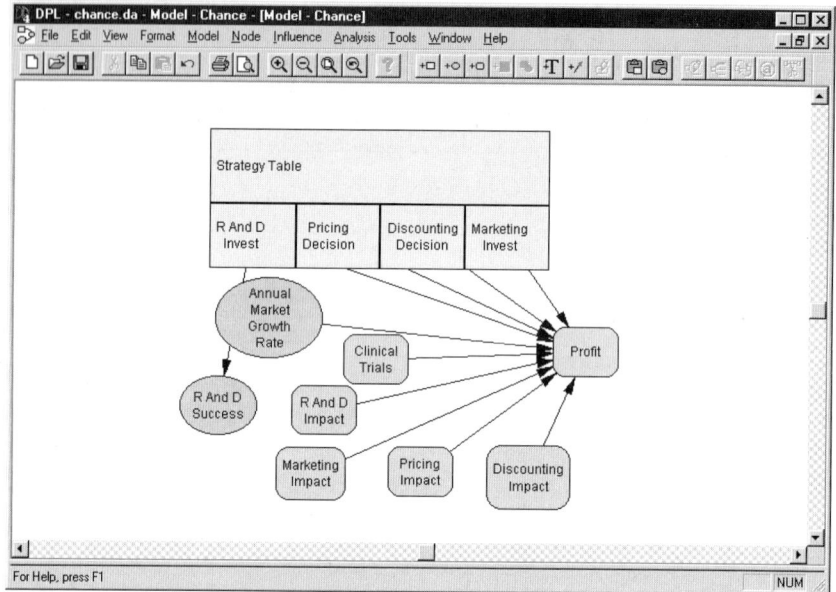

Using Named Distributions

You have already defined Annual Market Growth Rate as a chance node with a discrete probability distribution. However, a more accurate representation would be to model it as a Normal Distribution with mean of 0.03 and standard deviation of 0.1.

▸ Double-click on Annual Market Growth Rate.

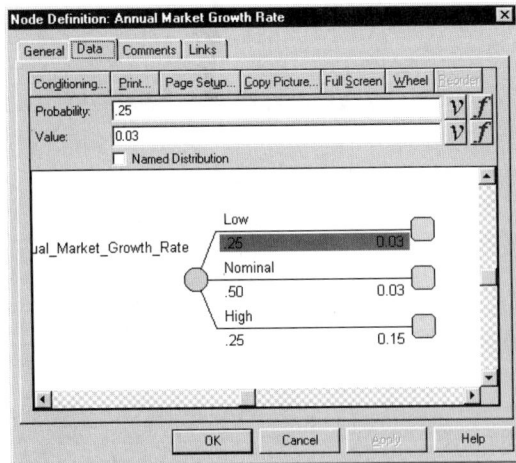

▶ Click the check box labeled Named Distribution.

DPL displays a list of Named Distributions.

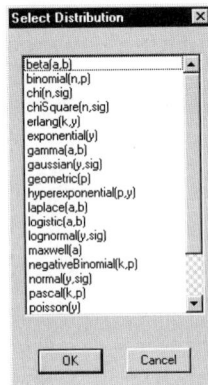

▶ Click on "normal(y,sig)"
▶ Click OK.

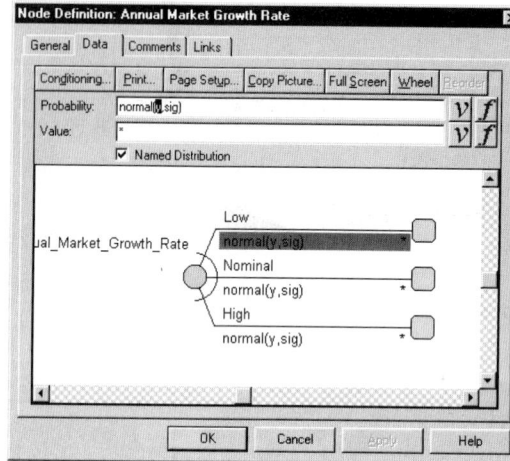

Notice that each branch of the data tree has "normal(y, sig)" as the probability expression and an asterisk as the value expression. This means that DPL is using a normal distribution with the parameters y (mean) and sig (standard deviation). The asterisk indicates that the value for each branch is defined by the Named Distribution defined in the probability section of the branch.

Enter the correct parameters.

▶ Change the probability expression to "normal(0.03,0.1)"
▶ Press Enter.

DPL updates all three branches to reflect the new distribution.

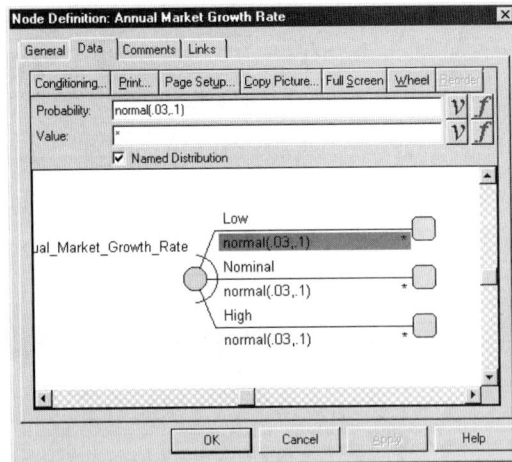

DPL will now base the probabilities and values for the three branches on the Normal distribution.

▸ Click OK.

Separating Probability and Value Distributions (Advanced Feature)

The next variable you want to change into a chance node is Marketing Impact.

▸ Change Marketing Impact into a chance node, accepting the default state names.

The probability distribution for Marketing Impact will change based on the amount you choose to spend on marketing in Marketing Invest. Therefore, condition Marketing Impact on Marketing Invest. However, while the probability distribution for Marketing Impact will change depending on the state of Marketing Invest, the multiplier values will not. In order to minimize the amount of data entry you have to do, you will separate the probability and value data in Marketing Impact.

Note: This is considered an advanced feature of DPL. It can be confusing, especially to new users, and it is recommended that you refrain from implementing it until you are very comfortable with DPL. Separating probability and value distributions exists purely as a time-saver, and does not provide additional functionality.

Because the ability to separate probability and value data is considered an advanced feature, it is turned off by default. You can turn it on in the Options dialog box.

▸ Select Tools Options.

▸ Click on the Advanced tab.

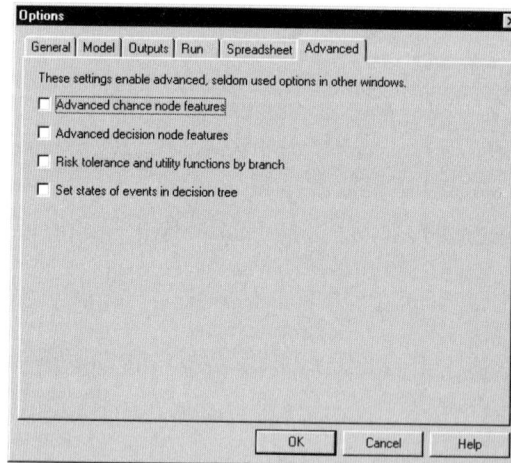

▸ Check the box labeled "Advanced chance node features" by clicking it.

▸ Click OK.

Next, condition Marketing Impact on Marketing Invest.

▸ Double-click on Marketing Impact.

▸ Click on Conditioning.

DPL displays the Conditioning dialog box.

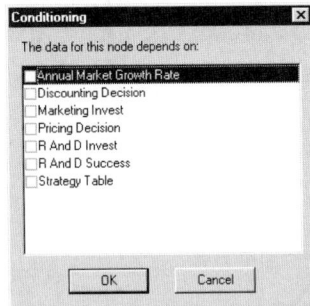

▶ Check the box labeled Marketing Invest.

▶ Click OK.

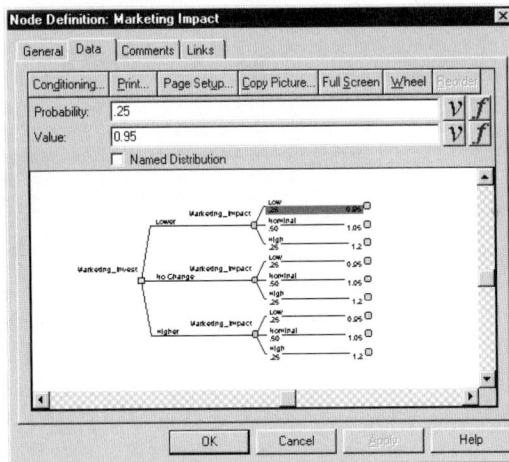

Next the probability and value distributions need to be separated. This is done in the General tab of the dialog box.

▶ Click on the General tab.

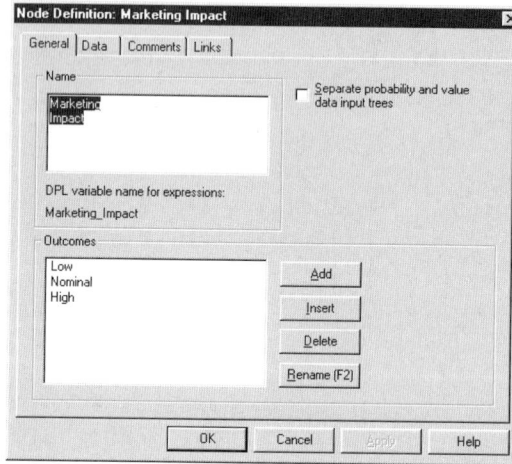

▶ Check the box labeled "Separate probability and value data input trees."

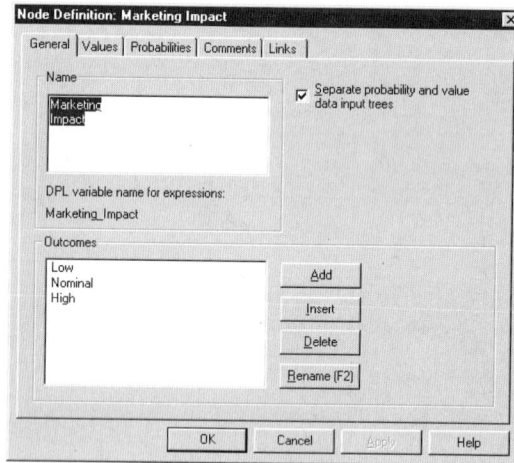

Instead of having a single tab for data, there are now two tabs labeled "Values" and "Probabilities." Remove conditioning from the Values tab, then enter the conditioned probability distributions in the Probabilities tab.

▶ Click on the Values tab.

▶ Click Conditioning.

▶ Uncheck the box labeled Marketing Invest.

▶ Click OK.

▶ Click on the Probabilities tab.

▶ Click Full Screen.

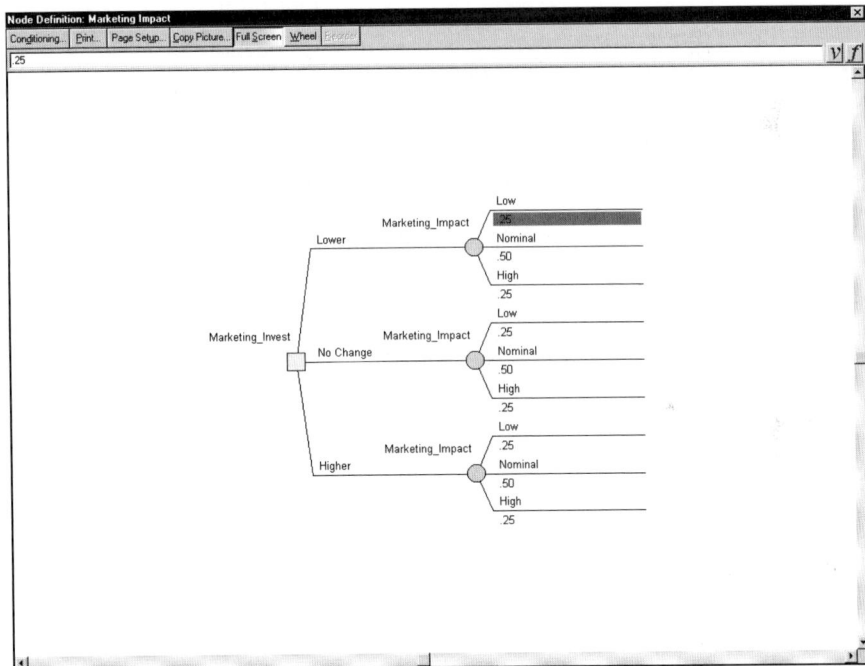

▶ Enter the following probabilities for the Marketing Impact outcomes corresponding to the Lower and Higher states of Marketing Invest:

Lower:	Higher:
Low: 0.6	Low: 0.1
Nominal: 0.3	Nominal: 0.3
High: 0.1	High: 0.6

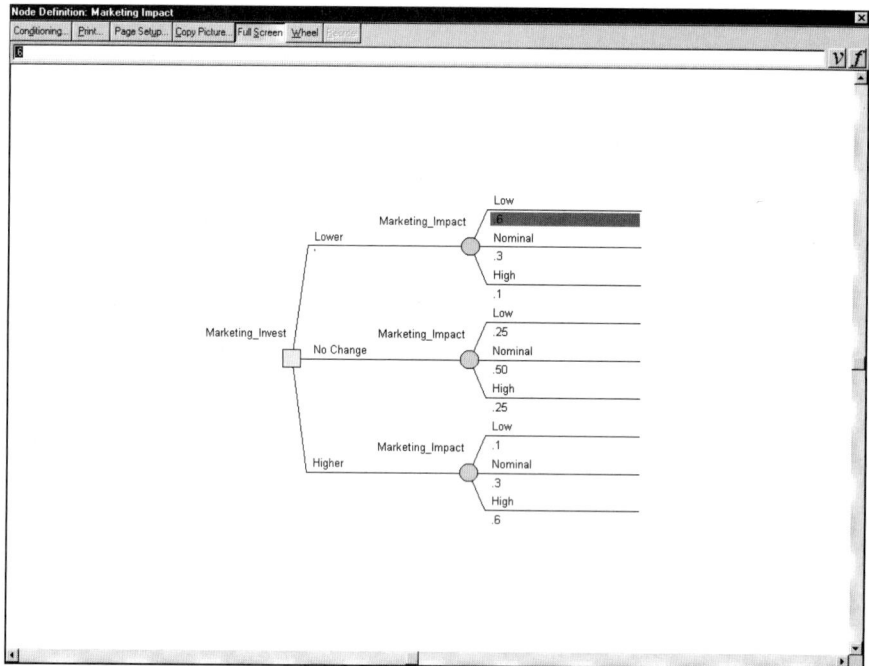

▶ Click Full Screen.
▶ Click OK.

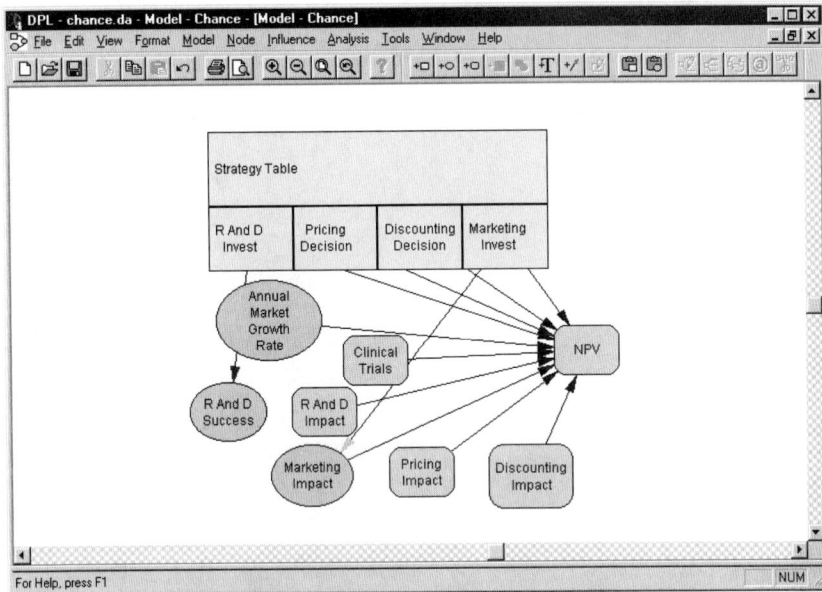

There is now an arrow with a green arrowhead from Marketing Invest to Marketing Impact. A green arrowhead indicates that the probabilities for the outcomes of Marketing Impact depend on the state of Marketing Invest, but the values do not.

D ecision analysis requires a model that assigns a value to each unique scenario defined by the influence diagram and decision tree. DPL offers several features for building these value models, including named variables, calculated values, get/pay expressions on the decision tree, and objective and constraint functions. This chapter discusses these features in the context of models with a single objective, or attribute. The next chapter discusses models with multiple objectives.

Chapter 6

Chapter 6: Values, Get/Pay Expressions, and Objective Functions

In every decision model there are two main components. The first component is the set of scenarios, which are possible future states of the world; we use decisions and chance events to model them. The second component is the objective, or value model, which is our way of keeping score, to allow comparisons between scenarios.

The objective of your model is what you want to maximize or minimize. It is sometimes also called the model's criteria or measure of value. It is a rule you give to DPL to tell it how to rank any two scenarios. For example, in a simple business decision, your objective might be to maximize the profit from a new business venture. If you were presented with two scenarios and asked to rank them, what information would you ask for? Well, profit. And you would prefer the scenario with the higher profit number to the one with the lower profit number.

DPL uses several tools for modeling your objectives. Although they are to some degree interchangeable, they each contribute some unique features to your modeling arsenal.

Values. Values are named, deterministic variables. They are either numbers or formulas, and are used for calculations.

Get/Pay Expressions. Get/Pay expressions are used to assign particular numbers or values to particular paths in the decision tree.

Objective Function. The objective function gets calculated at every endpoint. The number that gets calculated by the objective function is used in the expected value and certain equivalent rollback calculations.

In the simplest possible model, these three all do the same thing, which is to assign a single number to each scenario. For example, consider a simple two-event model with four scenarios. The influence diagram would have a single value node:

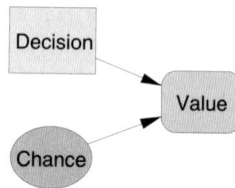

The data for the value node would be:

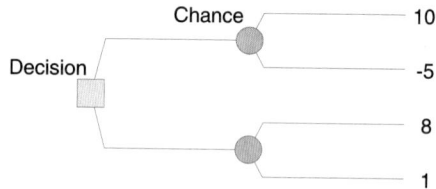

When you build the decision tree, the get/pay expression is simply the name of the value node:

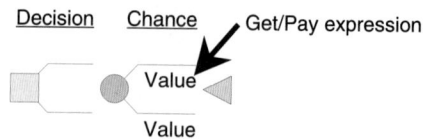

The objective function is simply the endpoint values, which equals the get/pays, which equal the value node:

Objective function value

```
                                        [10]
                          Chance          10
                                         [-5]
          Decision                        -5
                                          [8]
                                           8
                                          [1]
                                           1
```

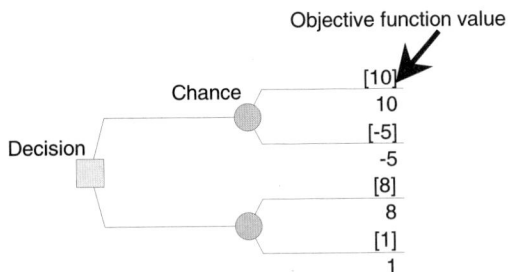

With a model this simple, using these tools seems a lot like cracking walnuts with a sledgehammer (or building a Rube Goldberg contraption). We could, in fact, eliminate one step by not defining a value node and using get/pay expressions to assign the value directly to the decision tree branches:

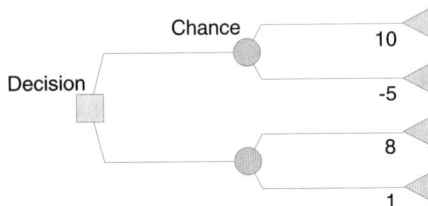

```
                        Chance
                                    10
          Decision
                                    -5

                                     8

                                     1
```

So why three tools? Each allows us to build models with more size, complexity, flexibility, and usefulness. In the rest of this chapter, we will discuss single attribute models, focusing mostly on the power of value nodes, less on get/pay expressions, and hardly mentioning objective functions. In the next chapter, we discuss multi-attribute models, looking at the power of get/pay expressions, objective functions, and constraints.

Modeling with Value Nodes

A value is a named, deterministic variable defined with constants, formulas, or both. In DPL, a value is a node that has no states of its own (so it isn't an event) but which can be conditioned by decision or chance events, so that it has a number or formula for each of the conditioning event's states. Value nodes are useful because they allow us to do calculations within the model. For example, consider a model where we are deciding whether to introduce a new product or continue selling our current one. We are uncertain about the number of units we will sell and our marketing and production costs. Our objective is to maximize profit. The initial influence diagram might look like this:

We could define profit directly for each scenario:

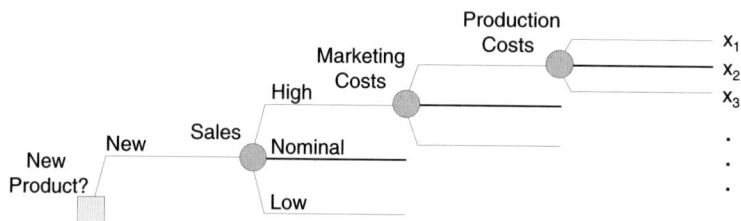

But we would have to enter 54 ($2\times3\times3\times3$) separate values for Profit. To get those values, we'd have to make 54 calculations. If, for instance, our estimates for marketing costs, changed, we'd have to track down each affected scenario and recalculate Profit. And, we could not do sensitivity analysis.

With multiple value nodes, on the other hand, we can make DPL do most of the work:

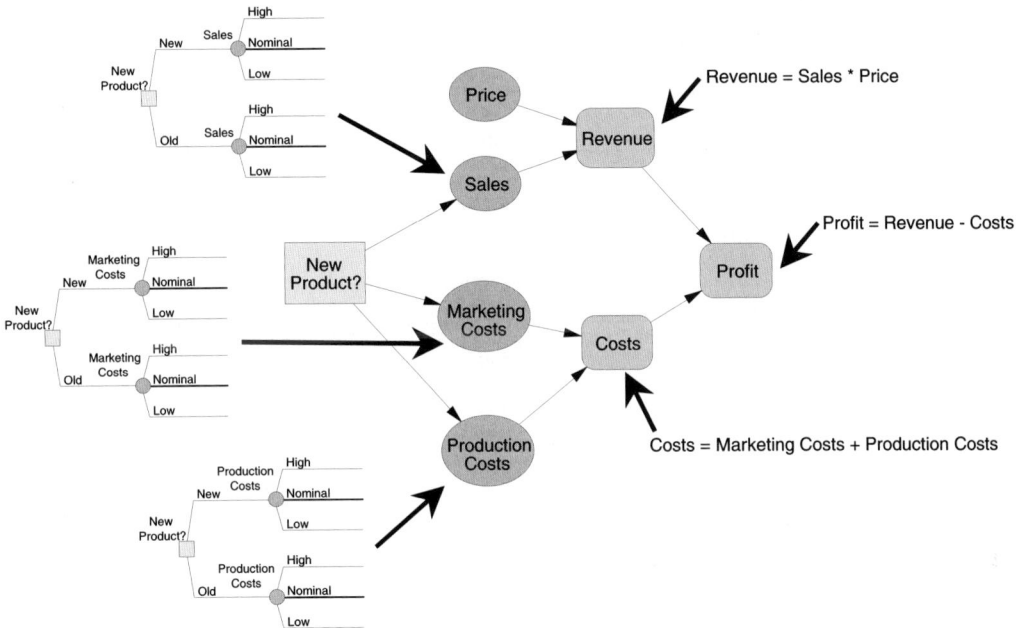

We enter price, six values each for sales, marketing costs and product cost, and three formulas. If marketing costs change, we change the data for the marketing cost node and all of the profit values will be correct the next time we run an analysis. And, we can do sensitivity analysis.

The decision tree get/pay expression doesn't change:

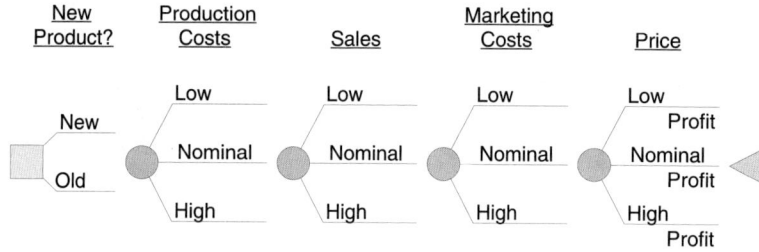

In short, multiple value nodes (seen only in the influence diagram, but defined from either the influence diagram or the decision tree) allow you to build models where the modeler only has to enter the basic elements and the formulas required to calculate the objective, and DPL does all the math.

In the influence diagram, a value is represented by a blue rounded rectangle:

Because they have no states, and therefore do not define scenarios, value nodes do not appear in the structure of a decision tree. Instead, they appear in the get/pay expressions and objective functions that assign values to the scenarios of the tree.

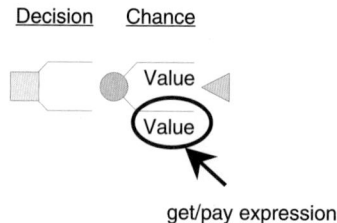

get/pay expression

When you run a decision analysis, DPL calculates the value of each formula and uses these to assign an endpoint value to each path, or scenario, in the tree. The value calculations are smart, or dynamic — they are done in the context of a specific location in the decision tree and therefore the same value calculation may have different results on different paths.

How to Model a Value Node

First, give it a name. This is the name that will appear on the influence diagram. It may appear on the decision tree in get/pay expressions. This name will very likely be used as a variable name in mathematical expressions. Some guidelines:

- The name may contain up to 256 characters. We recommend that you keep it under 32 characters to keep the formulas and outputs easier to read.
- The name may include spaces and special characters such as ? & - ( ). These characters will get converted to underscores if you use the name in a get/pay expression or formula. (This prevents DPL from becoming confused about which characters belong in variable names and which provide the "punctuation" for a formula.)
- The first character should be a letter. If the first character is a number, DPL will add an "N" to the front of the name.

Second, assign values or formulas. DPL offers a great deal of flexibility when it comes to defining values and formulas (we use the term "expression" to refer to a constant or formula). The options include:

- **Constants.** Constants are numbers. You can enter them as integers (5) or decimal numbers (9.8). Negative numbers use a minus sign (-10). You may use floating-point notation (1E+6) or even octal (037375) or hexadecimal (OX3EFD). Don't use formatting such as commas (enter 1000000 not 1,000,000).
- **Formulas with DPL Operators.** DPL offers a basic set of operators for calculation. They work the same as they do in the C programming language and most spreadsheet applications. The operators are listed in the table on page 206. They are described in more detail in the On-Line Help system.
- **Formulas with DPL Functions.** DPL offers a wide range of functions for calculation. Most are based on the mathematical and financial functions used in

Microsoft Excel and Lotus 1-2-3. DPL also supports the use of array formulas similar to those in Excel. The functions are listed in the table below. They are described in detail in the on-line help system, including any differences between DPL functions and those in Excel or 1-2-3. C programmers can write their own function libraries; again, the details are in the On-Line Help system.

- **Formulas with Variable Names.** You may use the names of other variables in formulas. This includes both value nodes and decision and chance events that have values associated with their branches. When entering these names, remember that DPL is case sensitive and converts spaces and other special characters to underscores (e.g., Marketing Costs becomes Marketing_Costs). It is usually easier to use the variable button and select the variable name from the list box than to type it in yourself.

DPL Operators

Arithmetic Operators
Unary plus (+)
Negation (-)
Addition (+)
Subtraction (-)
Multiplication (*)
Division (/)
Exponentiation (^)

Relational Operators
Equality (==)
Inequality (!=)
Greater than (>)
Less than (<)
Greater than or equal to (>=)
Less than or equal to (<=)

Logical Operators
AND (&&)
OR (||)
NOT (!)

Conditional Operator (? :)

DPL Functions

Arithmetic Functions
abs(x)
exp(x)
log(x)
log10(x)
max(x,...)
max(x,...)
mod(x,y)
pow(x,y)
sqrt(x)

Mathematical Spreadsheet Functions
@abs(x)
@exp(x)
@int(x)
@ln(x)
@log(x)
@mod(x,y)
@pi
@round(x,y)
@sqrt(x)
xltranspose(s)

Error Spreadsheet Functions
@na

Selection Spreadsheet Functions
@choose (z,x,y,...)
@cols(s)
@hlookup(x,s,y)
@index(s,x,y)
xlindex (s,x,y)
xlmatch(x,s,y)
@rows(s)
@vlookup(x,s,y)

Financial Spreadsheet Functions
@cterm(x,y,z)
@ddb(x,y,z,w)
@fv(x,y,z)
@irr(x,s)
@npv(x,s)
@pmt(x,y,z)
@pv(x,y,z)
@rate(x,y,z)
@sln(x,y,z)
@syd(x,y,z,w)
@term(x,y,z)

Logical Spreadsheet Functions
xland (x,y,x,...)
@false
@if(z,x,y)
xlor(x,y,z,...)
@true

Decision Tree Functions
display( )
prob( )
state( )
statename( )

Statistical Spreadsheet Functions
@avg(x,y,z,...)
@count(x,y,z,...)
@max(x,y,z,...)
@min(x,y,z,...)
@std(x,y,z,...)
@sum(x,y,z,...)
@var(x,y,z,...)

Conditioning Value Nodes

A value node does not have states like a decision's alternatives or a chance event's outcomes. This means that a plain value node with no arrows leading into it can be defined with a single number or formula:

where the data entry
tree for Price looks like

However, once a value node is conditioned by a decision or chance event, it can have a different number or formula for each of that decision's alternatives or that chance event's outcomes.

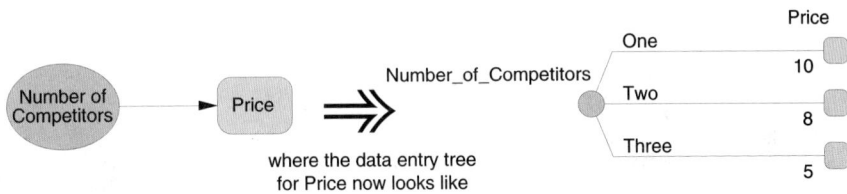

where the data entry tree
for Price now looks like

This is called direct, or full, conditioning and is specified using the Conditioning button in the Data tab of the Node Definition dialog box. In fact, when you assign values to the alternatives of a decision or the outcomes of a chance event, to DPL you are defining a value with the same name as and conditioned by that decision or chance. It's just such a common modeling tool that we combined the steps for defining them to make life easier.

In some cases you don't want, or need, direct conditioning. For example, if you were defining the following revenue value node, full conditioning would lead to typing the same formula three times, once for each state of the conditioning event Sales:

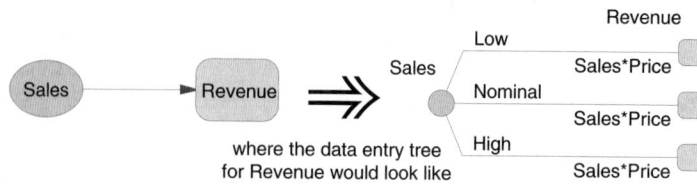

where the data entry tree
for Revenue would look like

It is easier to define, and maintain, this node with a single formula that refers to the conditioning event as a variable. This is called indirect conditioning. It is a very powerful technique for building complicated models with a minimum of typing. It is also extremely useful when assessing data for a variable with many conditioning events. To define indirect conditioning, simply use the name of the conditioning variable in a formula, without using the Conditioning button.

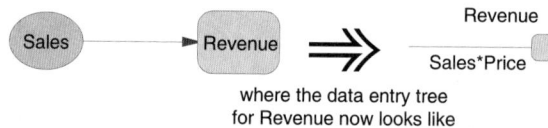

where the data entry tree
for Revenue now looks like

Value Nodes and Sensitivity Analysis

In DPL, you can perform a sensitivity analysis on any number that has been defined as a named variable.

Sensitivity Analysis on Constants

To perform a sensitivity analysis on a constant in your model, do not embed the constant in a formula. Instead, define the constant as a value node, then use the node's name in formulas. For example, if revenue is defined as sales multiplied by the price, which is 10, do not define Revenues as Sales*10. Instead, define a value node called Price, enter 10 as its value, and define Revenues as Sales*Price. This way, you can conduct sensitivity analysis on Price.

Sensitivity Analysis on Probabilities

To perform a sensitivity analysis on the probability of a chance event, do not enter the probabilities as constants in the chance events data tree. Define a value node, then enter the probabilities as functions of that value. For example, if the probabilities for an event are .2 and .8, define a value node called Prob, enter .2 as its value, and enter Prob as the probability for the first outcome of the chance event. DPL will automatically use 1–Prob as the probability of the 2nd outcome. You **must** leave the last probability expression blank; if you do not, DPL will give you the error message "Invalid Initializer."

Modeling with Get/Pay Expressions

A get/pay expression is a number, formula or variable name assigned to a particular branch or set of branches in the decision tree. A get expression adds its contents to the path value; a pay expression subtracts from it.

Most single attribute models have only a single get/pay expression, assigned to the endpoints of the decision tree. However, multiple get/pay expressions can be useful in single attribute models, especially when you have an asymmetric decision tree and wish to have more control over which values get assigned to which paths. Multiple get/pay expressions can also significantly reduce model run times in some situations. (Get/pay expressions are more important in multi-attribute models, discussed in the next chapter.)

Get/pay expressions do not appear in the influence diagram, although they often correspond with the value node or nodes at the "end" of the influence diagram — the ones that have arrows leading into them, but none leading out.

For example, in the following influence diagram, the Profit node is likely to be the variable referred to in the get/pay expression in the decision tree.

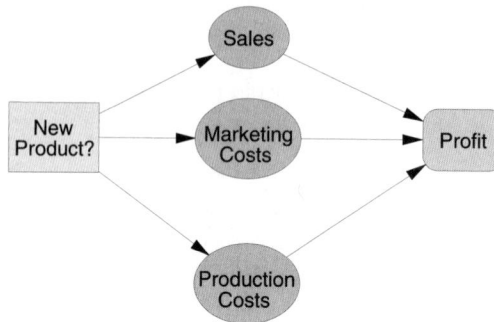

In the decision tree, get/pay expressions appear under, and to the left of, the branches. The get/pay expression for the model above would look like this:

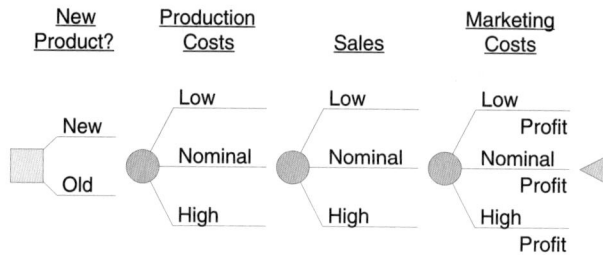

How to Define a Get/Pay Expression

First, select the branches or endpoints to which the expression will attach. Some things to consider:

- If you want the get/pay expression to be evaluated at the endpoints of the decision tree, select the branches that lead into the blue triangles:

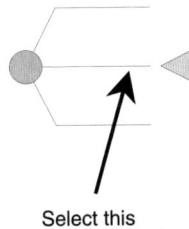

Select this

- If you want the get/pay expression to be evaluated at a particular node, select the branches of that node:

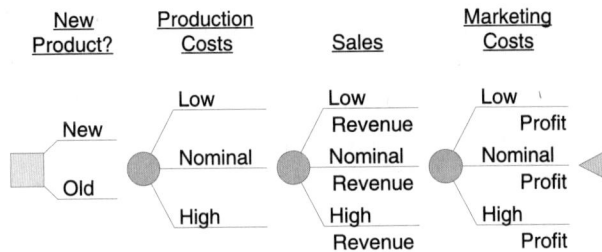

The get/pay expression will move with the branch if you restructure the decision tree.
- If you attach a get/pay expression anywhere except the endpoints, be sure that it appears after, or to the right of, all of the events the expression depends on, directly or indirectly.

Second, define the get/pay expression. This expression can be a constant or a formula using any of DPL's functions and variable names. In practice, life is much easier if you put all of your formulas in value definitions and limit get/pay expressions to only constants and variable names

How Many Get/Pay Expressions?

The number of get/pay expressions a model needs depends on the situation and your experience with DPL.

One get/pay expression. A model that is completely symmetric, meaning that all paths from root to endpoint go through the same events, in the same order, and has only one attribute, or measure of value, can usually be modeled with a single get/pay expression at the endpoints.

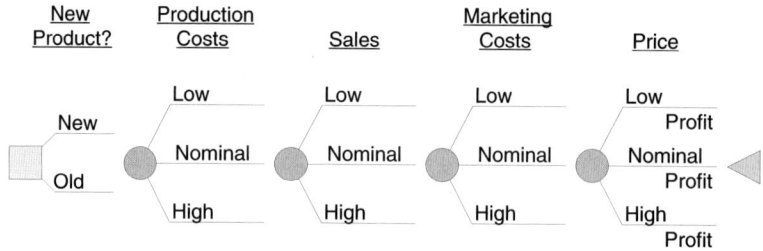

One get/pay expression per path. A decision tree that has a single attribute but is asymmetric, meaning not all paths have the same events in the same order, needs at least one get/pay expression for each path. They don't necessarily have to be different.

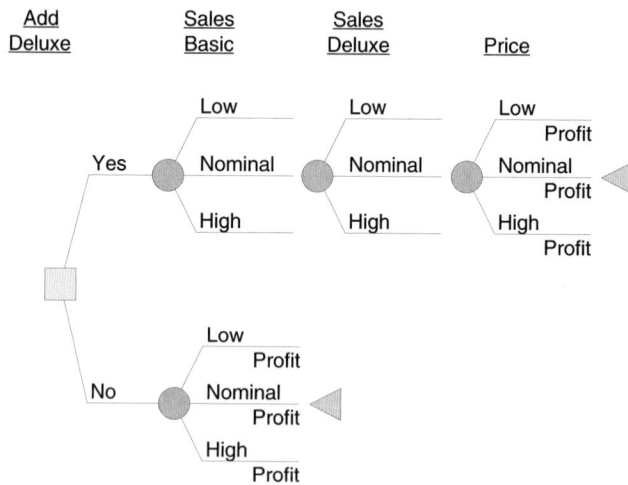

If you forget to put a get/pay expression on a path (such as forgetting the second "profit" expression in the model above), DPL will assign 0 to each endpoint without a get/pay expression. If you don't have any get/pay expressions in the tree at all, DPL will present the warning message "No get/pay expressions encountered" when you attempt to run an analysis. If you continue to run the analysis, all the paths will have 0 endpoint values.

More than one get/pay expression per path. If your model has a "separable value function," meaning that the calculation of the endpoint has pieces that are combined with only addition and subtraction, you can put those pieces into separate get/pay expressions. For example, if you have a model in which profit is calculated as Revenues less Costs, you could build a decision tree with Revenues and Costs in the get/pay expressions, rather than Profit.

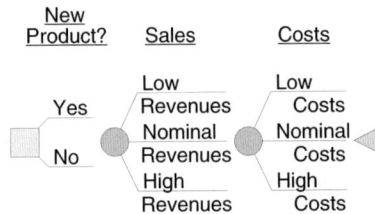

This is called "promoting" the terms of the value function.

Why bother? Some people prefer to break terms out for communication purposes, especially to show the timing of costs and revenues. More importantly, promoting terms can reduce runtimes by an astonishing amount for some models. (We've seen runtimes drop from 30 hours to 3 minutes!) This trick only works if the model has fairly independent pieces or groups of events — if any of the terms depend on all or most of the events in the decision tree, you won't see any performance improvement.

Working with multiple get/pay expressions requires a bit more care than working with only one. Some things to watch out for:

- Don't double count. For example, if you put Revenues and Costs in their own get/pay expressions, don't also put Profit in one.
- Be consistent with units. If you are working in dollars (or millions of dollars), all the get/pay expressions must be in dollars (or millions of dollars).
- Keep "gets" and "pays" straight. If you are calculating Profit, Revenue is a "get" and Cost is a "pay."

Because it is harder to work with multiple get/pay expressions, we recommend that you start building your model with a single get/pay (such as Profit). Once the model has stabilized, experiment with promoting terms to see if it improves communication or runtime.

"Conditioning Event Not in a Known State"

When DPL evaluates a decision tree, it starts at the root, or left-most node, and works toward the right. (For a complete description of the evaluation process, refer to Chapter 10.) As it works its way to the right, it attempts to replace all get/pay expressions containing variable names with the appropriate constants and calculations. Often, the variable names in the get/pay expressions refer to values that depend on decision or chance events. This means that the variable will have different values in different parts of the tree. To replace this variable name with a number, DPL needs to know which of the variable's possible values to use at this particular location in the tree. For it to do so, all of the events that the variable depends on must have already occurred — that is, they must appear to the left of the get/pay expression containing the variable name. For example, if you have the following influence diagram, Revenues depends on Sales only, but Costs depends on both Sales and Unit Production Costs.

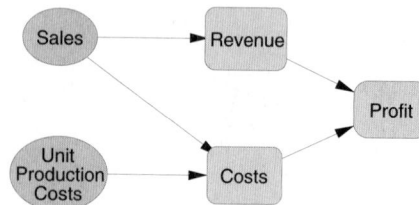

If you build a simple tree with Profit at the endpoint, DPL can calculate it with the events in either order:

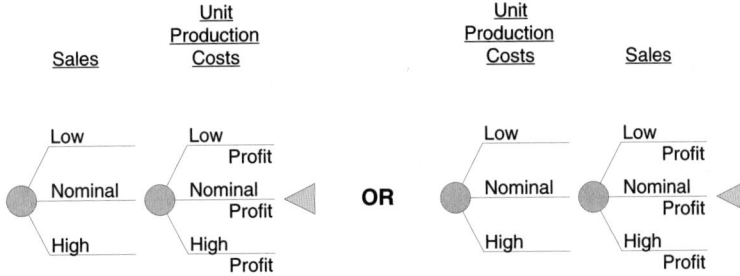

If you decided to promote terms, the first decision tree would work:

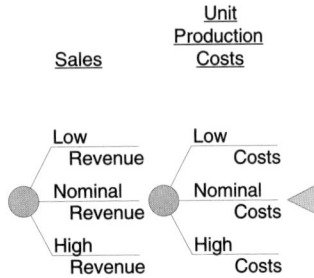

but the other order would not:

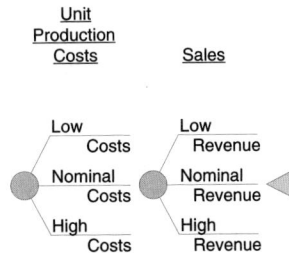

In this case, you would get the error message "Conditioning Event Sales Not in a Known State." When you clicked on OK, DPL would open the get/pay dialogue box for Costs, indicating that this is where the trouble is. If you look at the decision tree, you can see why DPL is unable to calculate costs — working from the left, at this point in the tree, DPL knows what state Unit Production Costs is in, and therefore which Unit Production Costs value to use when calculating Costs, but it has not yet come to Sales, and therefore does not know which Sales value to use in the Costs calculation. You can also run into this problem if you are building your own decision tree, rather than using DPL's default tree, and forget to put an event from the influence diagram on all the appropriate paths.

Does this mean you can only build symmetric decision trees if you use variable names? Not at all. There are several useful techniques for handling asymmetric decision trees. For example, consider a simple model where you are deciding whether to add a deluxe version of a popular item in your product line. Your first influence diagram looks like this:

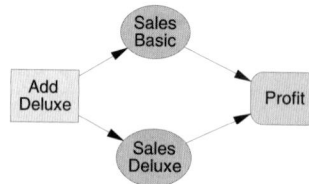

Your first decision tree looks like this:

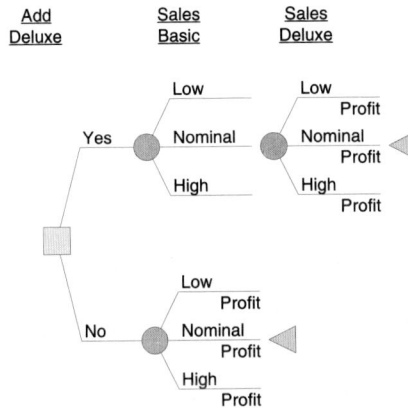

When you attempt to run a decision analysis, you get the error message "Conditioning Event Sales Deluxe Not in a Known State" and the get/pay dialogue box for the "No" half of the decision tree opens.

How can you fix this?

- **Build a symmetric tree.** This is probably the easiest approach, mechanically, but it does mean that the decision tree doesn't show the asymmetry inherent in this model.

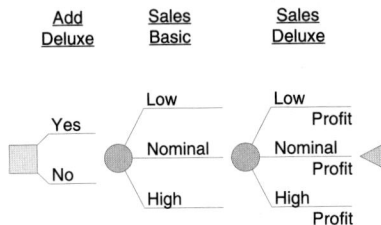

- **Condition formulas, not data.** If you modify your influence diagram, you can keep the asymmetry in the decision tree:

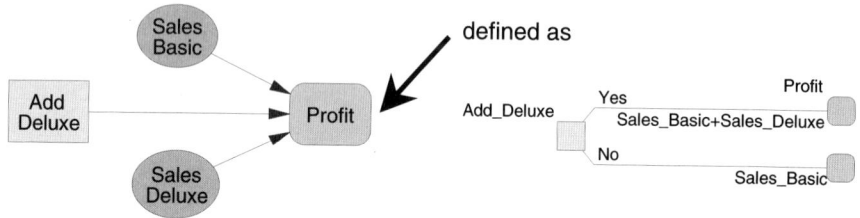

Since Profit is conditioned directly only by the decision, DPL goes into the definition for Profit and finds that Profit is indirectly conditioned by Sales Deluxe only when the decision is Yes. On the No part of the tree, Profit is not conditioned by Sales Deluxe, so the event is not necessary.

- **Define separate "final values."** You could define two profit nodes,

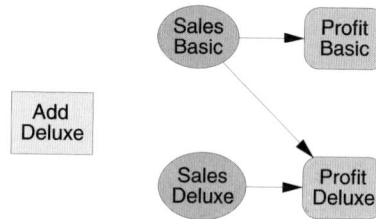

and refer to them in the decision tree:

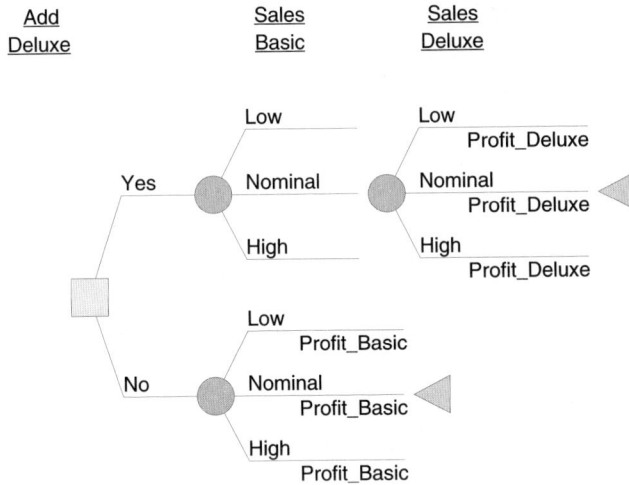

- **Separate the terms.** Because Profit is comprised of two additive terms, revenues from Basic Sales and revenues from Deluxe Sales, you can separate and promote them in the decision tree:

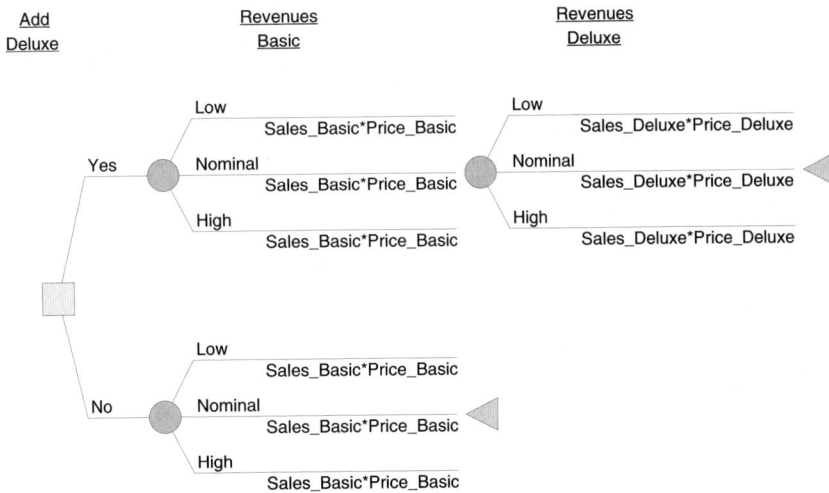

Modeling with an Objective Function

An objective function is used to process endpoint values between the roll-forward and the rollback steps of an evaluation. In most cases, and in almost all models where there is a single attribute, or measure of value (such as Profit), the objective function doesn't do anything — the endpoints are rolled back without change. One possible use for an objective function in a single attribute model might be to convert all endpoint values from one set of units ($) to another ($million or £). For example, if you had a model with these endpoints,

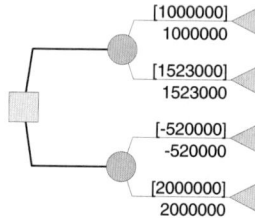

you might find all those 0's cumbersome. You could add an objective function,

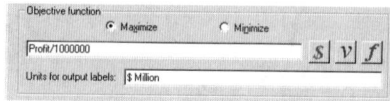

which would result in tidier expected values.

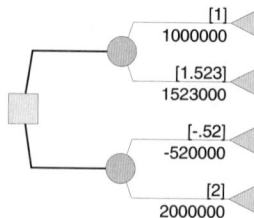

If you only want to divide by a factor of 10, you can use the output scaling option. If you want any other conversion, you must put it in the influence diagram or use an objective function.

Objective functions are much more important (and interesting) in multiple attribute models, discussed in Chapter 7.

Tutorial: Values, Get/Pay Expressions, and Objective Functions

‣ Open "Values.da".

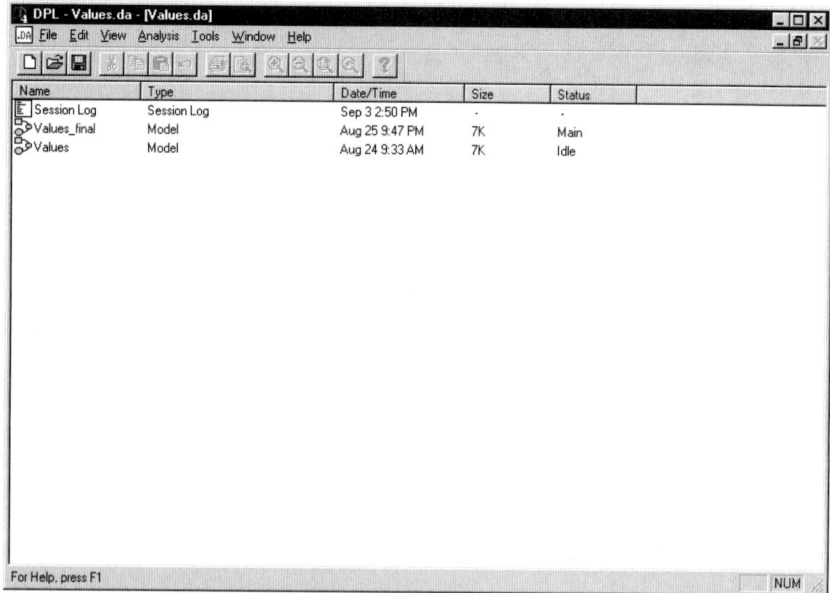

‣ Select Window Model-Values.
‣ Select View Split Vertically.

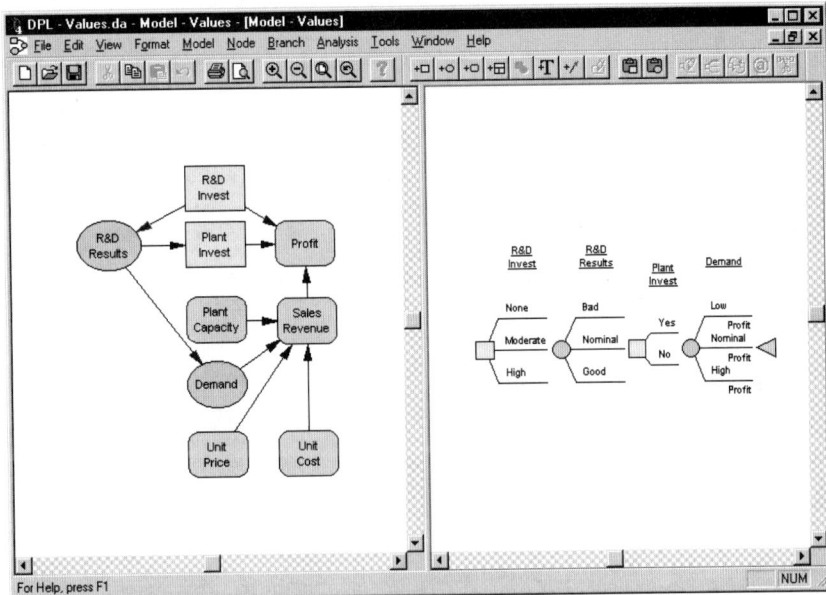

Value Nodes

This model represents a simple investment decision, the objective of which is to maximize Profit.

Value nodes appear as blue rounded rectangles in the Influence Diagram. They do not appear in the Decision Tree.

▸ In the Influence Diagram, double-click on the value node Profit.

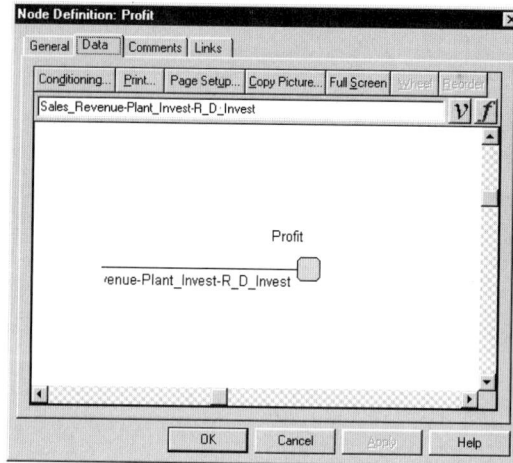

The data for the value node Profit contains an expression telling DPL to calculate a value for Profit by subtracting the values of R&D Invest and Plant Invest from the value of Sales Revenue. What is the value of Sales Revenue?

▶ Click OK.

Assigning Formulas to Value Nodes

▶ Double-click on Sales Revenue.

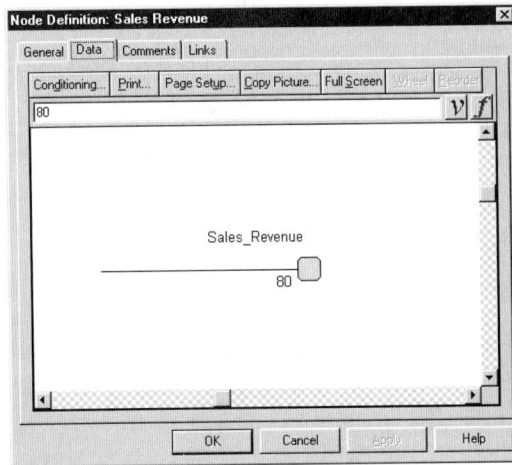

Sales Revenue currently has a value of 80. However, it is really a calculation based on how many items are sold and at what price. You can replace the value of 80 with the correct calculation.

▶ Click in the entry box. Delete "80".

The calculation uses values contained in other variables in the model. When you enter variable names, you should use the icon rather than typing the name to reduce data entry errors.

▶ Click the Variable icon.

DPL displays the Select Variable dialog box.

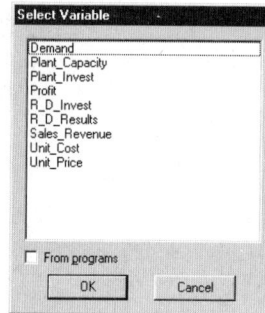

▸ Click on Demand, then click OK.

DPL places the variable name Demand in the data entry box. We want to multiply Demand by the difference between Unit Price and Unit Cost. DPL uses an asterisk to indicate multiplication.

▸ Type "*".
▸ Enter the rest of the expression, "(Unit Price-Unit Cost)", using the variable icon to enter the variable names.

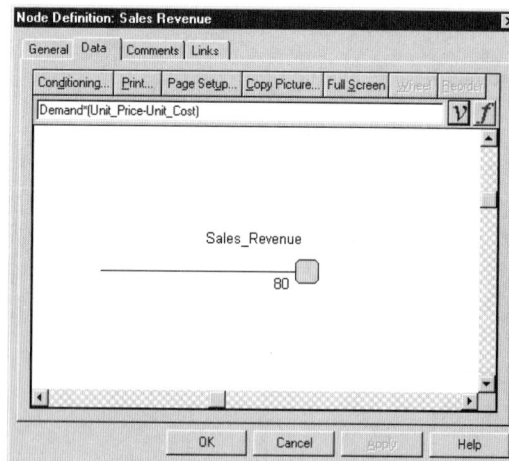

▸ Click OK.

Assigning Constants to Value Nodes

Value nodes can contain simple numbers rather than calculations. You can use value nodes to represent known quantities, then use the node name in formulas rather than numbers to make them easier to read. Using a value node to represent a number also allows you to perform a sensitivity analysis on it.

▶ Double-click on Unit Price.

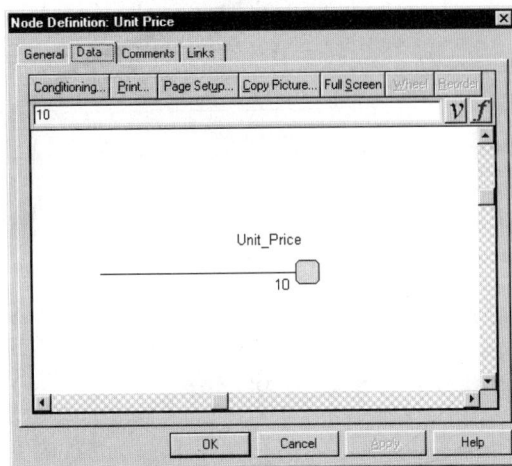

Unit Price has been assigned the known value of 10.

▶ Click OK.
▶ Double-click on Unit Cost.

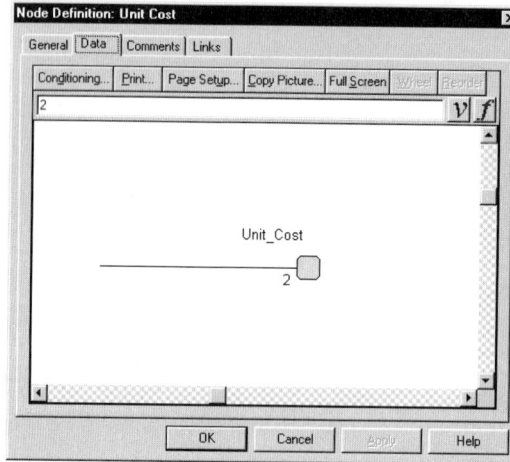

Unit Cost has been assigned the known value of 2.

▶ Click OK.

Conditioning Value Nodes

You can condition Value Nodes on decision or chance nodes. In this example, Plant Capacity depends on whether or not you build a new plant. If you don't build a new plant, capacity stays at its current level of 5. If you build a new plant, capacity increases to 50. You can indicate this by conditioning Plant Capacity on Plant Invest.

▶ Double-click on Plant Capacity.

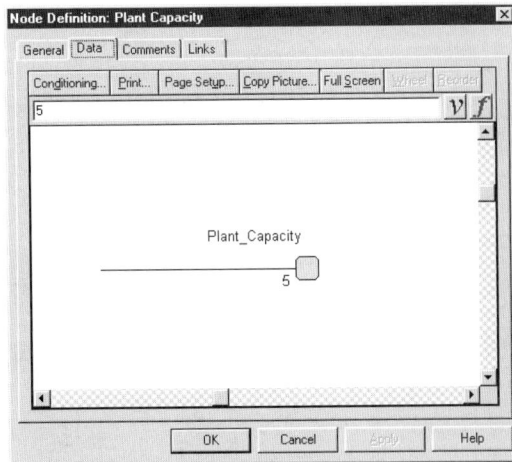

▸ Click Conditioning.

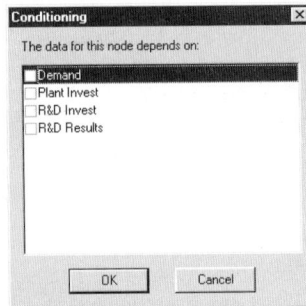

▸ Click the box next to Plant Invest.

DPL places a checkmark in the box next to Plant Invest, indicating that the value of Plant Capacity depends on the state of Plant Invest.

▸ Click OK.

There are now two values for Plant Capacity, one corresponding to each alternative of Plant Invest.

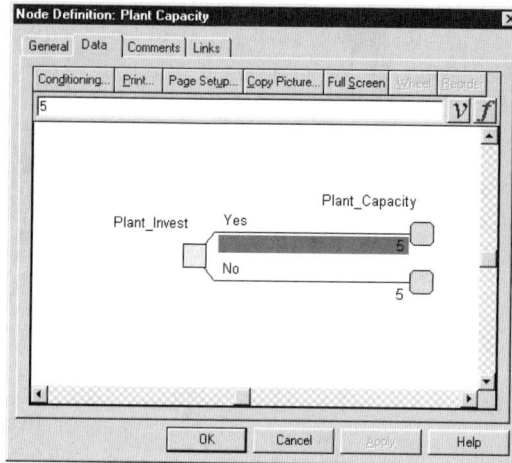

Next, change the value corresponding to the Yes alternative of Plant Invest.

▶ Click the value of 5 under the Yes branch of Plant Invest (it's highlighted in magenta).

▶ Type "50", then press Enter.

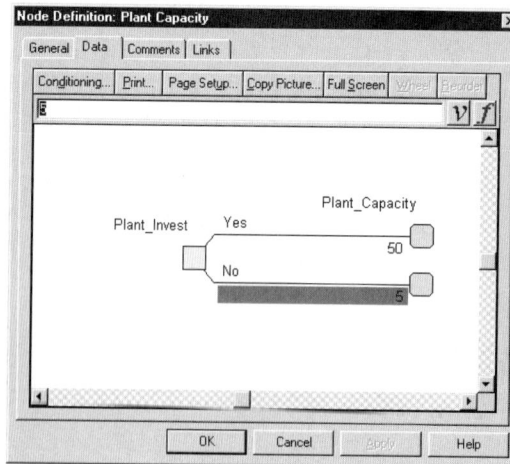

▶ Click OK.

DPL has drawn an arrow with a blue arrowhead from Plant Invest to Plant Capacity. The blue arrowhead tells you that the values of Plant Capacity depend on the alternative of Plant Invest you have selected.

Using Functions in Expressions

The demand for our product is uncertain, but may range from 0 to 40 units. However, if we do not construct the new plant, our production will be limited to 5 units. Accordingly, Sales Revenue will be limited by Plant Capacity, not Demand. We can indicate this by using the "min" function in the calculation of Sales Revenue.

▶ Double-click on Sales Revenue.

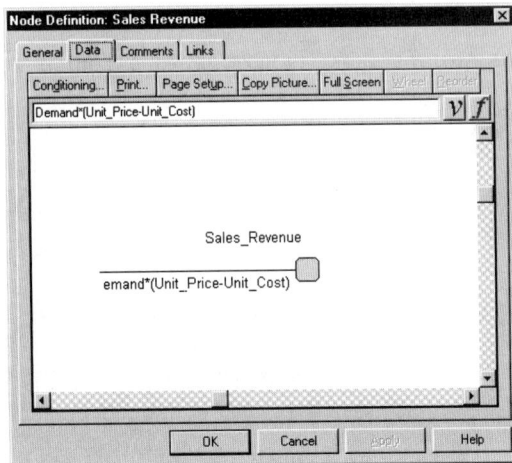

▶ Click at the beginning of the entry box.
▶ Click the Function icon.

DPL displays the Select Function dialog. The dialog shows a list of all of the functions available in DPL. The "@" character is used before function names to differentiate them from variable names in DPL expressions.

Select Function

```
@abs(x)
@avg(list)
@choose(index,list)
@cols(name)
@count(list)
@cterm(int,fv,pv)
@ddb(cost,salvage,life,period)
@exp(x)
@false
@fv(pmt,int,term)
@hlookup(x,name,row)
@if(cond,true,false)
@index(name,col,row)
@int(x)
@irr(guess,name)
```

OK   Cancel

▸ Use the scroll bar to scroll down until you see the function "@min(list)". Click on it, then click OK.

DPL inserts the function "@min(list)" into the expression. The "list" indicates that DPL selects the minimum value from a list of values.

▸ Edit the function to read "@min(Demand,Plant_Capacity)*(Unit_Price-Unit_Cost)", using the variable icon to enter the variable name Plant_Capacity.

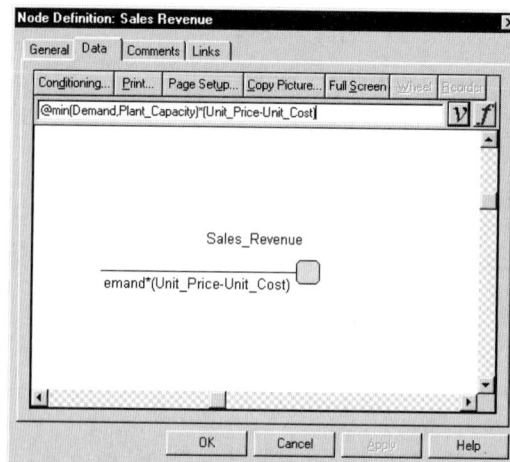

Node Definition: Sales Revenue

General | Data | Comments | Links

Conditioning... | Print... | Page Setup... | Copy Picture... | Full Screen | Wheel | Reorder

@min(Demand,Plant_Capacity)*(Unit_Price-Unit_Cost)   y f

Sales_Revenue

emand*(Unit_Price-Unit_Cost)

OK   Cancel   Apply   Help

▶ Click OK.

You have now told DPL how to calculate Sales Revenue. It will in turn use the calculated value of Sales Revenue to calculate values for Profit.

Single Get/Pay Expressions

In the Decision Tree, you can see that Profit is the only Get/Pay expression, and is located on the branches of Demand, the last node in the tree. This tells DPL to calculate the expression defined in the value node Profit at each endpoint of the tree.

Defining Your Model's Objective

While the Get/Pay expression tells DPL how to use the values of Profit as the endpoint values for this model, you have not yet told DPL whether to select the optimal policy based on maximizing or minimizing Profit. In this case, you obviously want DPL to select the policy which maximizes Profit. You can tell DPL to do this in the Objective dialog box.

▶ Select Model Objective.

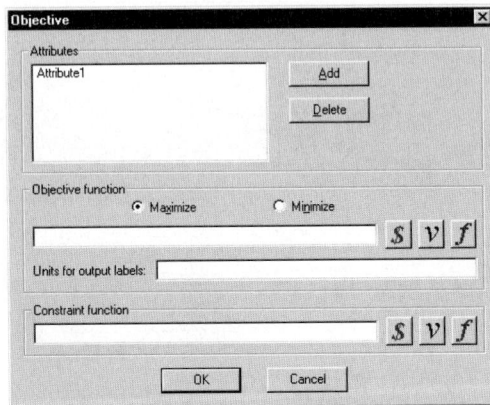

In the Objective Function section of the dialog, you can see that the "Maximize" radio button is selected by default. The entry bar underneath the radio buttons is where you can write an expression for your objective function. If you look at the Attributes section of the dialog box, you can see that there is only one attribute, "Attribute1." Because the model only has one attribute, DPL automatically sets the objective function equal to this attribute. (This is the equivalent of typing "Attribute1" in the objective function entry bar.) The result of this is that DPL will select the optimal policy based on maximizing Profit.

Because the model has only a single attribute which you would like to maximize, you do not have to modify the attributes or objective function.

Units for Output Labels

You can use the Objective dialog box to enter a units label to be used to label the output values of Profit.

▸ In the entry bar labeled "Units for output labels," type "$ Millions."

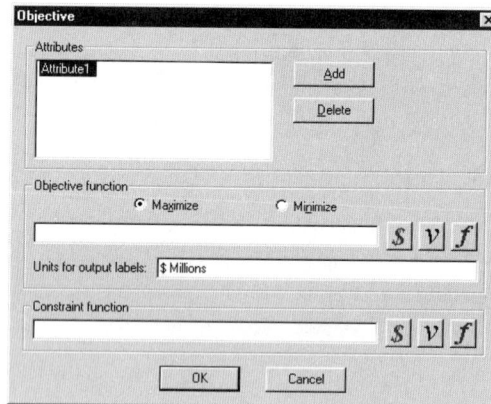

▸ Click OK.

Promoting Terms

The model will run in its current state, and produce an expected value of $24 million. Because the model is small, it runs very quickly. If the run-time were longer than you liked, you could re-arrange the get/pay expressions to reduce it.

▶ Double-click on Profit.

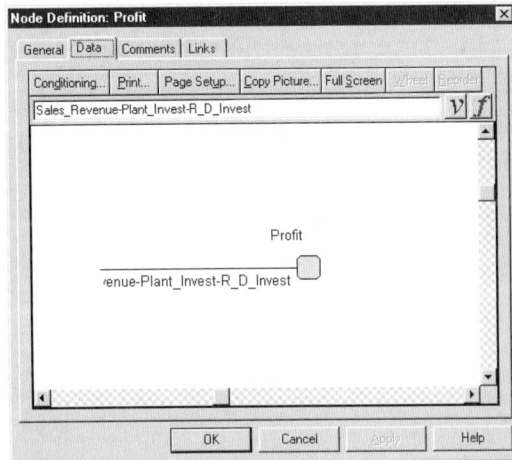

The expression for Profit is simply the sum (or difference) of three other variables in the model: Sales Revenue, Plant Invest, and R&D Invest. This makes it easy to promote terms, which means breaking our single get/pay expression into three get/pay expressions (one for each variable), then evaluating them earlier in the decision tree.

By promoting terms, there is no need for the calculation in Profit. You can leave it in the node data, but it will be ignored by DPL after you reconstruct the get/pay expressions.

▶ Click OK.

Promoting terms involves manipulating the decision tree, so it is useful to maximize the decision tree view.

▸ Slide the splitter bar to the left edge of the screen.

The decision tree currently has one get/pay expression, Profit, on the last node, Demand.

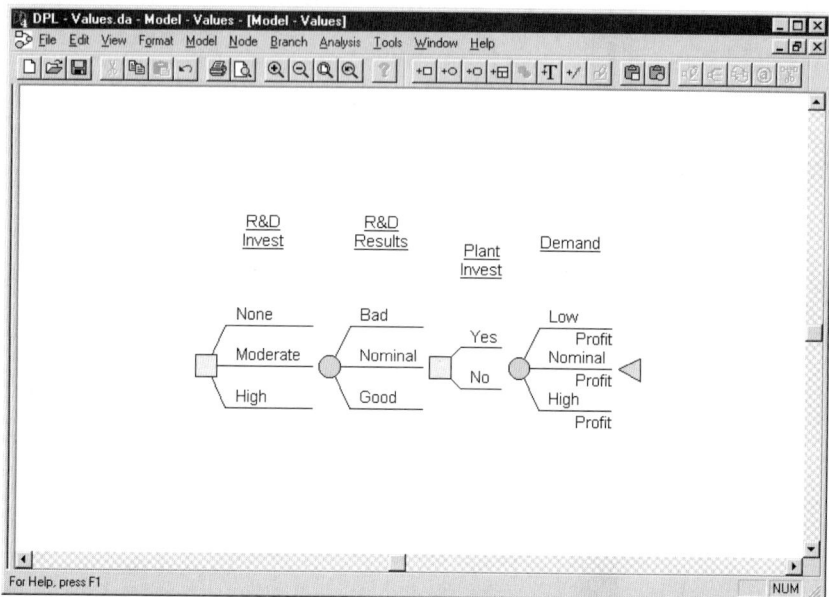

While promoting terms means not using the Profit node, you still want DPL to perform the same calculation at each endpoint, Sales Revenue less R&D Invest less Plant Invest. To do this, you'll tell DPL to "get" the value for Sales Revenue and "pay" the values for R&D Invest and Plant Invest as it goes through the tree.

Of the three variables, only Sales Revenue uses the value for Demand in its calculation. Therefore, it cannot be evaluated any earlier than Demand. As Demand is the last node in the tree, the get/pay expression for Sales Revenue must go on the branches of Demand.

▸ Double-click on the branches of Demand.

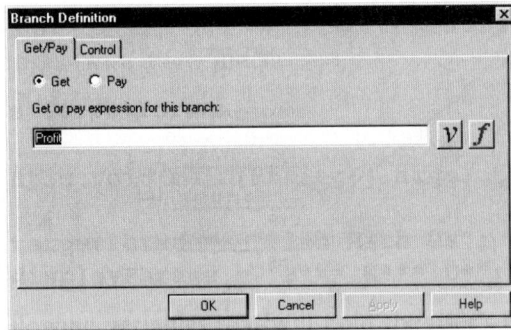

Profit is the current get/pay expression. It needs to be removed, and replaced with Sales Revenue.

▸ Click the variable icon.

▸ Click on Sales Revenue, then click OK.

DPL has replaced Profit with Sales Revenue. Notice that the radio button labeled "Get" is selected, indicating that the value of Sales Revenue will be positive.

▸ Click OK.

Accordingly, DPL has replaced Profit with Sales Revenue on the branches of Demand.

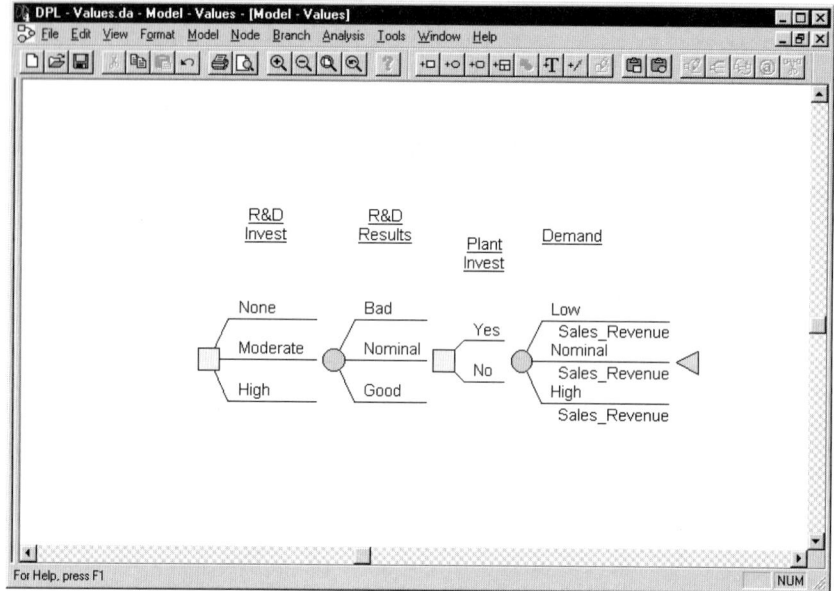

A get/pay expression does not have to be a single variable. If you had instead written the get/pay expression on the branches of Demand as "Sales Revenue - R&D Invest - Plant Invest", you would get the correct profit calculation. However, you would not gain any speed advantage because you did not promote terms.

As you move left in the decision tree, Plant Invest is the next node. The value for a node cannot be evaluated earlier than its instance in the decision tree. Therefore, the get/pay expression for Plant Invest should go on the branches of the Plant Invest node.

▶ Double-click on the branches of Plant Invest.
▶ Using the variable icon, enter Plant_Invest for the get/pay expression.

The money spent investing in the new plant is subtracted in the profit calculation, so Plant Invest is a "pay" expression.

▸ Click the radio button labeled "Pay".

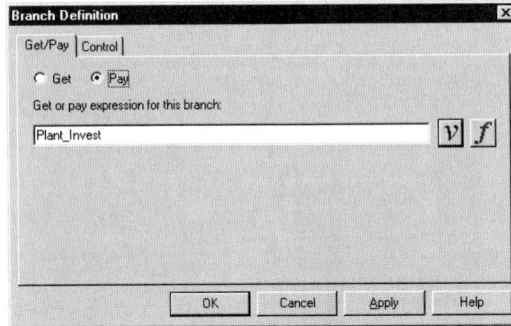

Branch Definition

Get/Pay | Control

○ Get ● Pay

Get or pay expression for this branch:

Plant_Invest

OK   Cancel   Apply   Help

▸ Click OK.

There are now two get/pay expressions in the decision tree.

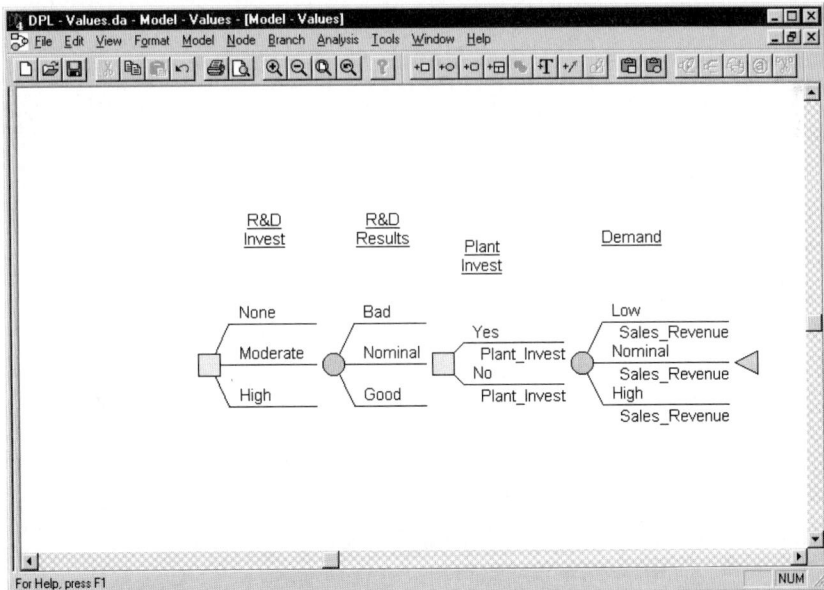

DPL - Values.da - Model - Values - [Model - Values]

File Edit View Format Model Node Branch Analysis Tools Window Help

R&D
Invest

R&D
Results

Plant
Invest

Demand

None

Moderate

High

Bad

Nominal

Good

Yes
Plant_Invest
No
Plant_Invest

Low
Sales_Revenue
Nominal
Sales_Revenue
High
Sales_Revenue

For Help, press F1   NUM

The final get/pay expression is R&D Invest. The value for R&D Invest is known when it is evaluated, so the get/pay expression can be placed on the branches of R&D Invest.

▸ Create a get/pay expression for R&D Invest on the branches of R&D Invest, repeating the process used for Plant Invest. Make sure to make it a "pay" expression!

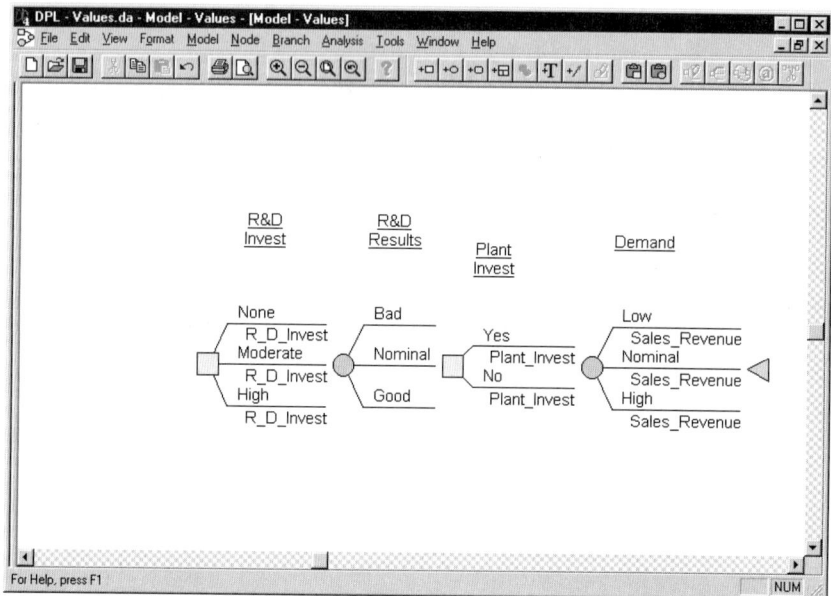

Get/Pay Expressions in Asymmetric Models

The model will now return the same calculations as it did when Profit was the only get/pay expression. However, these calculations are not quite accurate for the scenario in which no investment is made in R&D. If you run the model in its current form, it will tell you that if the "None" alternative of R&D Invest is chosen, the expected value is $100 million. That's pretty good, considering you didn't even develop a product!

To make the model accurate, you need to tell DPL that if you do not invest in R&D, the value for profit is 0. This can be done by making the "None" branch of R&D Invest an endpoint, and changing its get/pay expression to 0.

First, separate the "None" alternative of R&D Invest from the others.

▸ Double-click on the node (not the branches!) representing R&D Invest.

DPL displays the Instance tab of the Node Definition dialog box.

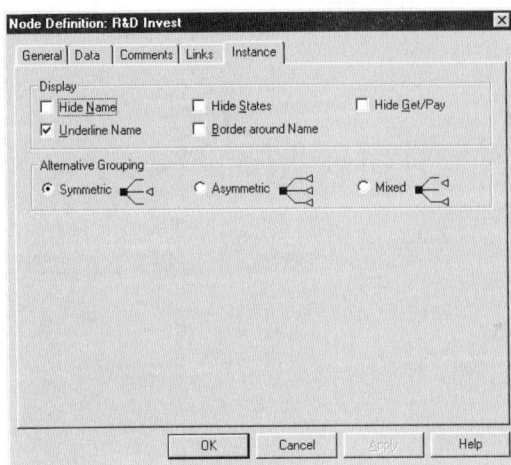

▸ In the "Alternative Grouping" section of the dialog box, click the radio button labeled "Mixed", then click Apply.

DPL displays the Select Groups dialog box.

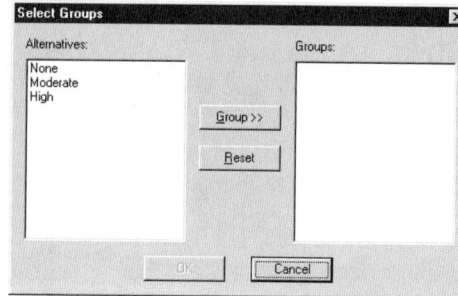

You want to create two groups of branches — one with just "None" and the other with "Moderate" and "High".

▸ Click on None, then click Group.
▸ Click on Moderate, then hold down the Ctrl key and click on High.
▸ Click Group.

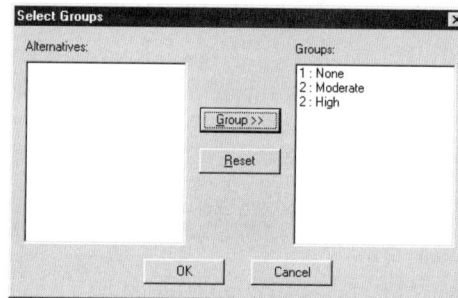

▸ On the "Groups" side of the dialog box, there is a 1 next to None, and a 2 next to both Moderate and High. This tells us that there are two groups. None is in the first group, and Moderate and High are in the second.
▸ Click OK.
▸ In the Instance tab of the Node Definition dialog box, click OK.

The R&D Invest node has the new branch configuration.

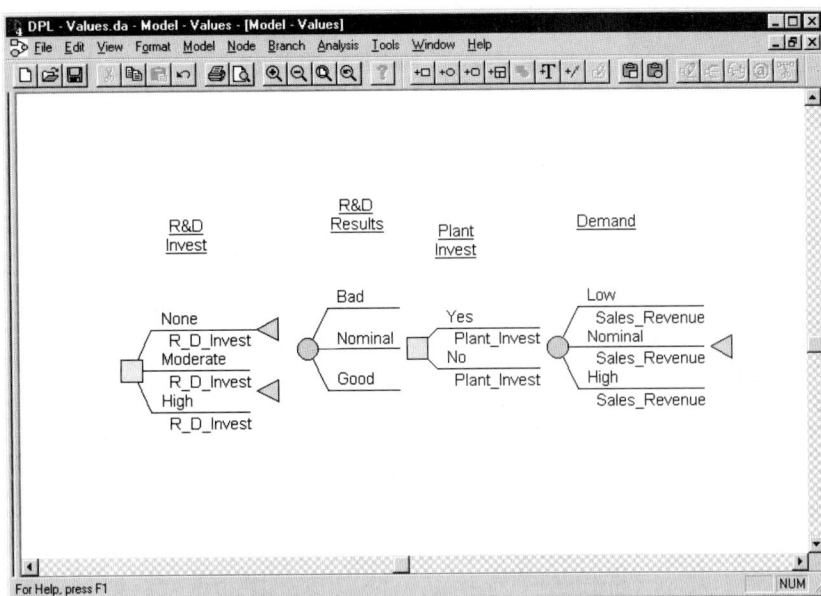

The None branch of R&D Invest is now separated from the other two branches, allowing it to have a different get/pay expression than they do. It's get/pay expression should be 0.

▸ Double-click on the None branch of R&D Invest.

▸ Type "0" to replace R&D Invest with 0 as the get/pay expression.

▸ Click OK.

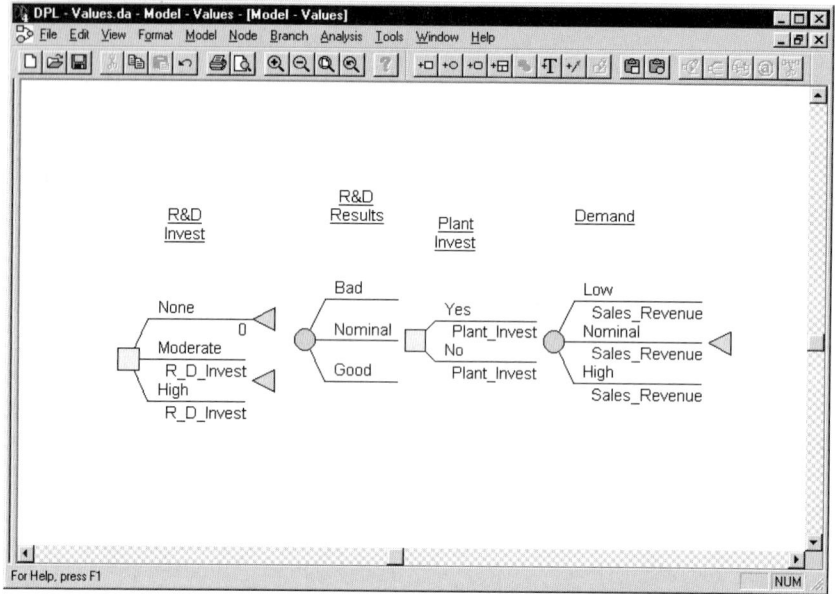

If either the Moderate or High alternatives of R&D Invest are selected, DPL needs to evaluate the entire decision tree. Therefore, the last step is to connect the Moderate and High branches to the rest of the decision tree.

▸ Click-and-drag the R&D Results node, placing it on the blue triangle between the Moderate and High branches of R&D Invest.

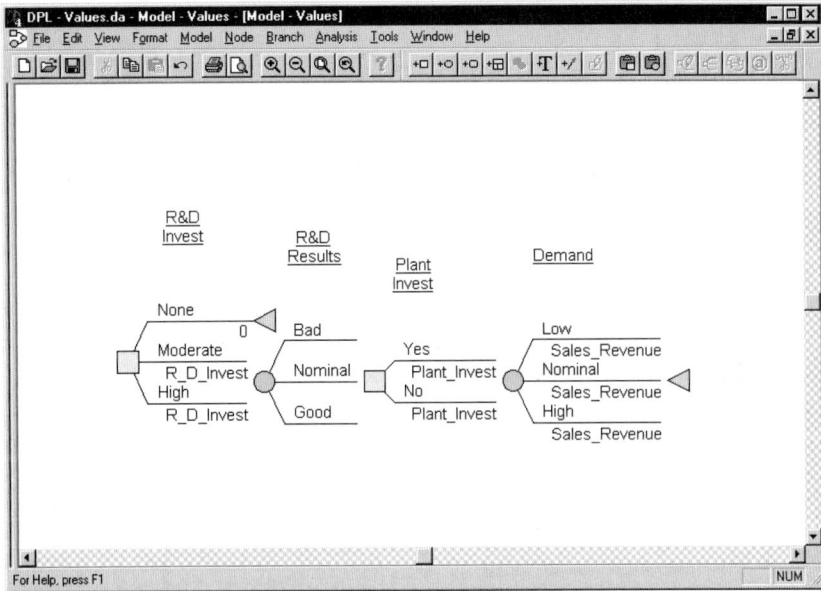

If you run the model, you will find that it still has an expected value of $24 million. However, the value corresponding to selecting the "None" alternative of R&D Invest is now 0.

In the previous chapter, we discussed the ways values, get/pay expressions and objective functions are used to calculate, assign, and process numbers for the endpoints of a decision tree. In the examples, we were using a single attribute, or measure of value, to compare scenarios and calculate expected value. The majority of DPL models are built this way.

There are times, though, when we find that a single attribute is limiting. The most common case is when we have a decision that must consider multiple criteria:

- Environmental cleanup decisions must consider cost, worker health effects, neighboring population health effects, and effects on the ecology of land and water;
- Locating an electric power company's operations control room must consider cost, the ability to keep the room secure, and convenience for employees; and
- Deciding whether to close a hospital or upgrade it to meet new earthquake codes must consider the cost to upgrade, whether surrounding facilities can provide the same services, whether facility employees can be transferred to quality jobs in other facilities, and whether the facility can be upgraded to an acceptable safety level.

There are other models where we'd like to make the decision based on one criterion, but keep our eyes on the impact on another variable:

- A capital investment decision might be made on the basis of expected net present value of profit over a 20-year time horizon — but we'd like to graph the distribution of the impact on stock price or employee head counts.
- A product line extension decision might be made on the basis of total profit— but we'd like to see the distribution on revenues for each individual product.

In these models, we need to be able to calculate more than one value for each endpoint of the decision tree, which allows us to graph their distributions, and then use an objective function to tell DPL which value, or combination of values, to use for calculating expected value and selecting the optimal decision policy.

Chapter 7

MULTIPLE ATTRIBUTE MODELS

Chapter 7: Multiple Attribute Models

How to Model Multiple Attributes

First, define the attributes. You can have up to 64 different attributes, but most models have between two and six.

For example, for an environmental cleanup model,

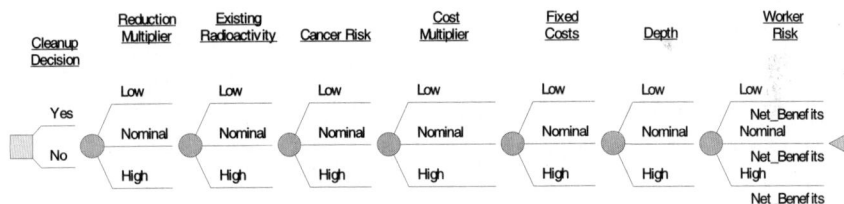

Cleanup Decision	Reduction Multiplier	Existing Radioactivity	Cancer Risk	Cost Multiplier	Fixed Costs	Depth	Worker Risk
Yes	Low	Low	Low	Low	Low	Low	Low
No	Nominal	Nominal	Nominal	Nominal	Nominal	Nominal	Net_Benefits Nominal
	High	High	High	High	High	High	Net_Benefits High
							Net_Benefits

you might define three attributes: Profit, Health, and Environ.

Second, define the objective function. The objective function uses the endpoint values of the attributes to calculate a single value to use when calculating expected value and selecting the optimal decision policy. It can be as simple as the sum of the attributes, or a complex weighting function. The objective function does not have to include all of the attributes.

Profit + 10 × Health + 5 × Environ

– or –

Profit (to see which decision policy makes the most money)

– or –

10 × Health + 5 × Environ (to see which decision policy is the most "socially responsible")

If you define the weights as value nodes in the influence diagram, you can do sensitivity analysis on them. This can be very useful when there is disagreement on the relative weights to be used.

Third, build an influence diagram. (Yes, even if you don't usually use them.) Define the decisions, chance events and values needed to handle each attribute. You can leave each attribute separate, or you can add a value node for "Total Value" or "Total Benefit" if you want to show that everything will be combined for calculating expected value.

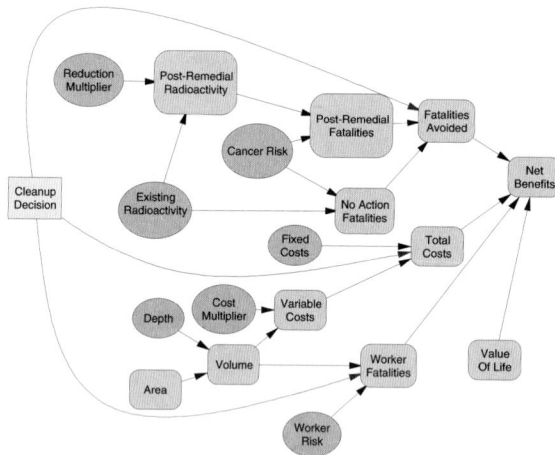

It is not necessary to have a value node in the influence diagram for each attribute, but we find that it helps considerably. Also, the name of the value node and the attribute do not have to match exactly.

Fourth, build a decision tree. The tree will be the same as for a single-attribute model, except for the get/pay expressions; whenever you define a get/pay expression, you must enter a number, variable name or formula for each attribute.

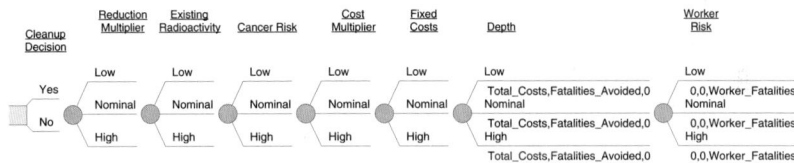

Cleanup Decision — Reduction Multiplier — Existing Radioactivity — Cancer Risk — Cost Multiplier — Fixed Costs — Depth — Worker Risk

Yes
No

Reduction Multiplier: Low / Nominal / High
Existing Radioactivity: Low / Nominal / High
Cancer Risk: Low / Nominal / High
Cost Multiplier: Low / Nominal / High
Fixed Costs: Low / Nominal / High

Depth:
Low — Total_Costs,Fatalities_Avoided,0
Nominal — Total_Costs,Fatalities_Avoided,0
High — Total_Costs,Fatalities_Avoided,0

Worker Risk:
Low — 0,0,Worker_Fatalities
Nominal — 0,0,Worker_Fatalities
High — 0,0,Worker_Fatalities

Using Variable Names in the Objective Function

The objective function has some special restrictions on the variables you can use.

You can use:

- The names of the attributes
- Constants
- The names of value nodes that do not depend, directly or indirectly, on the state of any event (decision or chance).
- A special variable that represents the probability of an endpoint (called pathprob).

Multiple Attributes and Run Time

Models with multiple attributes may take longer to evaluate then models with a single attribute for two reasons:

Bookkeeping. Keeping track of multiple attributes requires more calculations and a larger endpoint database than tracking a single attribute does. Each additional attribute adds a little to the run time.

Full enumeration. DPL's analytical engine makes use of some techniques to make evaluation faster. One of these, fastest exact, requires that the objective function be linear and predictable. While most multiple attribute objective functions satisfy this requirement, some do not. Because applying full enumeration incorrectly can get the wrong expected value, DPL's default behavior is to turn it off whenever multiple attributes are used. Since, in most cases, this is over-cautious, you should experiment with turning it back on to see if the answer is the same and run time is reduced.

Using a Constraint Function

DPL offers the ability to apply a constraint function to the evaluation of a decision tree. A constraint function allows you to dynamically change the objective function or delete branches from the tree during evaluation. This can be useful in a number of situations:

- If the value calculations for each endpoint take several seconds to run (such as if they require evaluating a large spreadsheet model), you can use a constraint function to trim off branches with zero probability, avoiding running the calculations for those endpoints.
- If you wish to constrain the optimization — say by maximizing a first attribute (profit) while remaining above or below a limit on a second attribute (such as capital budget or manufacturing capacity).

Defining a Constraint

First, define an if-then-else expression.

If   This defines the condition you wish to test. You can test for the path probability, the current value of a variable, or the current value of any attribute.

Then   This is the objective function to use if the condition is true.

Else   This is the objective function to use if the condition is false.

Either then or else can also include the halt( ) function, which tells DPL to stop evaluating the tree at this point, effectively deleting all of the branches that follow.

Second, specify whether the constraint function is to be applied at each node in the decision tree, at the endpoints, or both.

- **Apply to endpoints.** Apply the constraint at the endpoints if you simply wish to change the objective function dynamically, but don't wish to eliminate or prune any branches. For example, if you were allocating your department's capital budget, you would use the constraint function to identify those scenarios where the total capital exceeded the budget and apply a small penalty to model the fact that additional capital is available, but there is some effort and expense required to acquire it, so you would only do so if the benefits were significant. *Put the constraint function in the objective function field.*

- **Apply to nodes.** Apply the constraint at the nodes if you wish to prune branches and eliminate parts of the decision tree. For example, DPL evaluates the get/pay expressions and objective function for every path and endpoint in a decision tree, even if the probability for a path is zero. If these value calculations take a significant amount of time to run, as with a large spreadsheet, you might wish to use a constraint function to halt evaluation as soon as a path's probability reaches zero. The expected value will not change, but the run time may be shorter. *Put the constraint function in the constraint field.*

- **Apply to both nodes and endpoints.** Apply the constraint to both nodes and endpoints if you have a hard constraint that depends on the value of an attribute and you wish to halt evaluation of the decision tree wherever the constraint kicks in. *Put the constraint function in the objective function field and an asterisk (*) in the constraint field.*

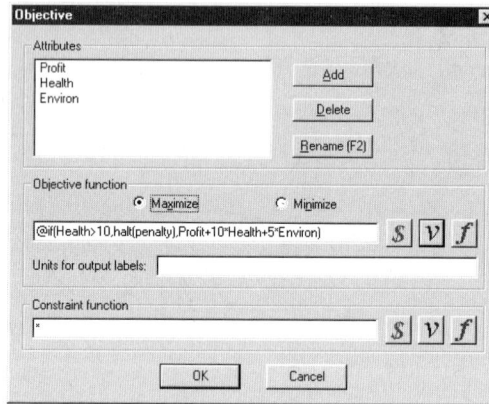

Constraints and Run Time

Models with dynamic constraint functions can rarely be evaluated using fastest exact, and will therefore run more slowly. DPL turns fastest exact off by default — you can experiment with turning it back on, if you wish.

In addition, models with constraints may have longer run times for sensitivity analysis and value of information and control evaluations as well, because DPL is likely to need to rebuild and recalculate parts of the decision tree.

A Few Words of Caution

Constraints are tricky things to model, because DPL takes them very seriously and very literally. This means that if you specify a constraint that eliminates, penalizes or disqualifies any path that exceeds a $500,000 budget, DPL will show you the optimal policy for $500,000 or less. It will not tell you that a much more profitable policy exists at $501,000, even though in the real world you could probably find that extra money fairly easily. For this reason, we strongly recommend that you run the model with the constraints eliminated or relaxed so you understand what effect the constraint has on the optimal decision policy and the riskiness of the outcome.

Another potential error is to confuse inputs with outputs. The inputs to a model are the full range of possible alternatives and outcomes; the outputs are the optimal decision policies. Constraints should only be put into the model to prevent DPL from choosing infeasible alternatives — it is DPL's job to choose the most attractive

Tutorial: Modeling Decisions with Multiple Objectives

A nuclear test site has soil contaminated with plutonium. You must decide whether the risks posed by radiation exposure justify spending the money necessary to reduce the amount of plutonium in the soil. There are three objectives to balance: The cost of cleaning up the site, the expected reduction in public safety as a result of the cleanup, and the risks posed to workers performing the cleanup. The model expresses cost in millions of dollars and both public and worker safety in number of fatalities.

‣ Open MUA1.da.

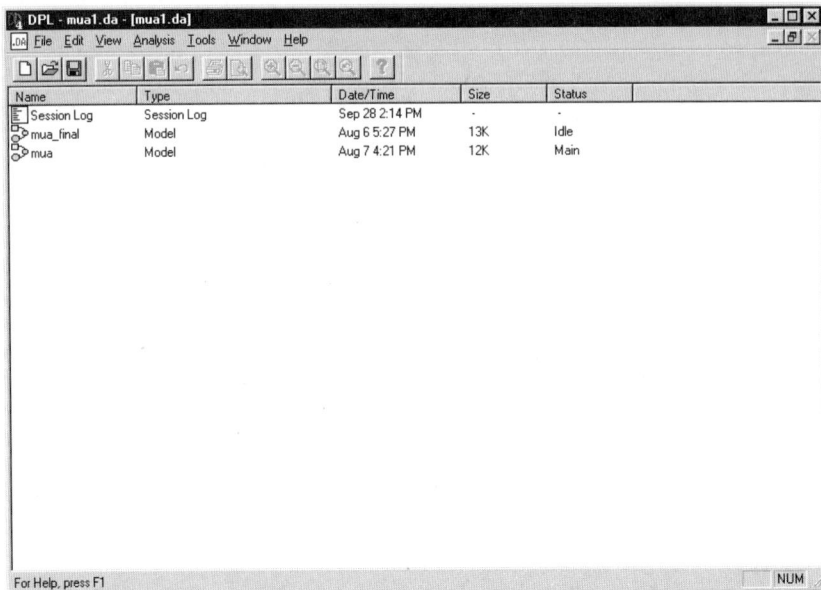

‣ Double-click on the model named "mua".

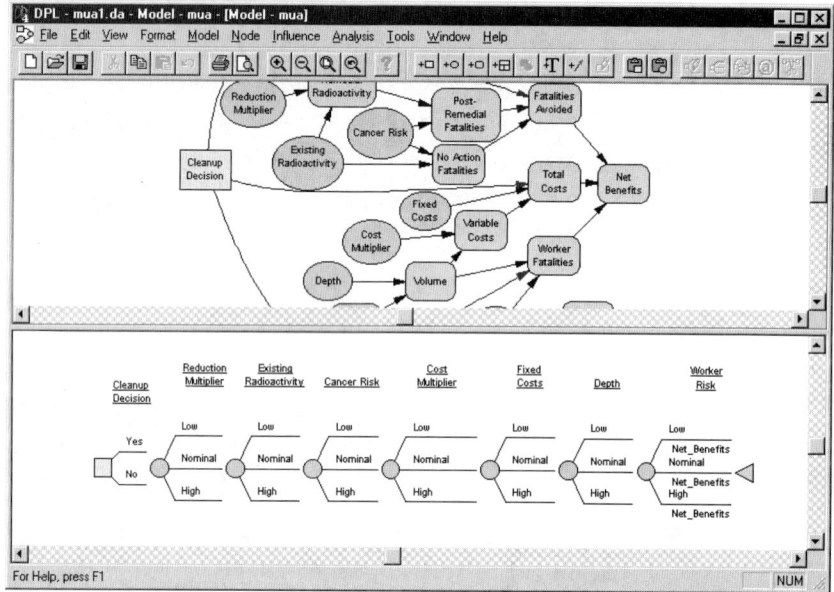

Currently, the only get/pay expression in the decision tree is the value node Net Benefits. Net Benefits is therefore the objective function; DPL is selecting the decision alternative which maximizes the expression contained in Net Benefits.

▸ In the Influence Diagram, double-click on the value node Net Benefits.

Node Definition: Net Benefits

General | Data | Comments | Links

Conditioning... | Print... | Page Setup... | Copy Picture... | Full Screen | Wheel | Reorder

(Fatalities_Avoided-Worker_Fatalities)*Value_of_Life-Total_Costs

Net_Benefits

es)*Value_of_Life-Total_Costs

OK | Cancel | Help

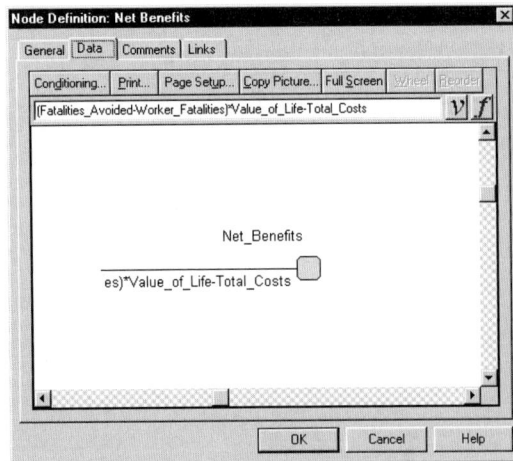

The expression for Net Benefits combines the three objectives: cost, fatalities avoided, and worker fatalities. The expression uses the value of life variable to enable the comparison of fatalities to dollars.

▶ Click OK.

This method of calculating the net benefits of cleaning up the site is perfectly acceptable. However, you could also utilize the multiple-attribute features of DPL, which will provide a number of advantages. By using multiple attributes, you can promote terms to speed up DPL analyses (see Promoting Terms). In addition, if you would like to run a distribution on one of the objectives, such as cost, it is much easier if you use multiple attributes.

First, tell DPL that your model has multiple attributes.

Creating Multiple Attributes and an Objective Function

▸ Select Model Objective.

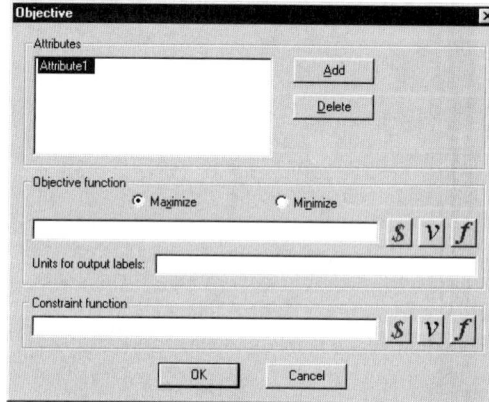

By default, models have only one attribute. You can add attributes in the Attributes section of the Objective dialog box.

▸ Click the Add button twice.
▸ Double-click on Attribute1 and type "Cost".
▸ Repeat to rename Attribute2 as "Public Safety" and Attribute3 as "Worker Safety".

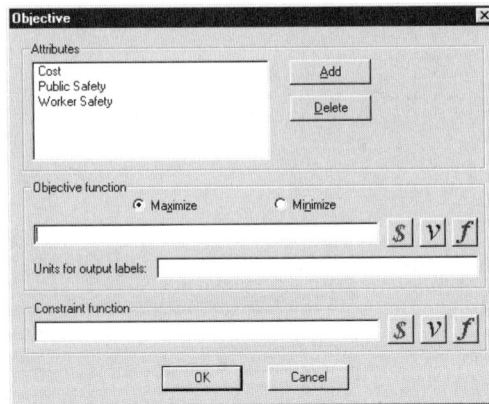

Next, create an objective function. The objective was previously expressed by the equation in Net Benefits. Because the objective has not changed, you can simply rewrite the equation as a function of the attributes. The equation will be:

▸ Type a "(" in the Objective Function entry bar.
▸ Click on the attribute icon.

DPL displays a list of the model's attributes.

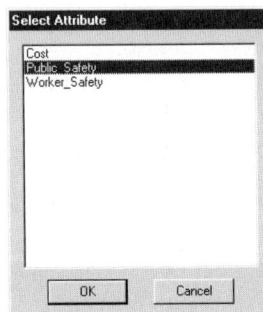

▸ Click "Public Safety", then click OK.

▸ Use the attribute icon and typing to complete the expression "(Public_Safety-Worker_Safety)*Value_of_Life-Cost".

▸ Use the variable icon to insert the variable Value_of_Life.

▸ Click OK.

Next, alter the get/pay expressions to reflect the multiple attributes.

Get/Pay Expressions with Multiple Attributes

Each get/pay expression must have three component expressions, each corresponding to an attribute. Separate the component expressions with commas.

Each of attributes is calculated completely in a single node, which simplifies the get/pay expressions. The attribute Cost is represented by the variable Total Costs, Public Safety by Fatalities Avoided, and Worker Safety by Worker Fatalities.

In order to speed computation, you want to evaluate each variable as early in the decision tree as possible (i.e. as far to the left as possible). However, DPL cannot evaluate a variable until all of its input variables are evaluated. For example, in order to evaluate the expression in the variable Fatalities Avoided, DPL has to have already evaluated Reduction Multiplier, Existing Radioactivity, Cancer Risk, and Depth. Because of this, the earliest you can place Fatalities Avoided in a get/pay expression is on the branches of Depth. This is also the earliest you can place Total Costs.

▶ In the decision tree, double-click on the branches of Depth.

DPL displays the Get/Pay tab of the Branch Definition dialog box.

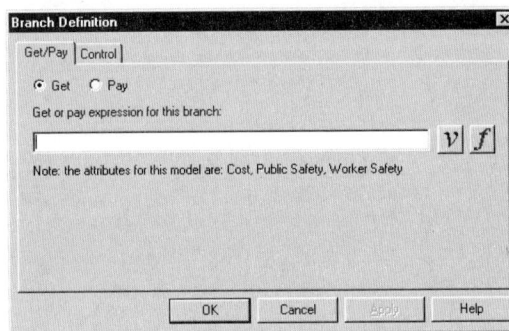

Under the entry bar, DPL shows the attributes in order. This order reports that the first component of the get/pay expression contributes to the attribute Cost, the second toward Public Safety, and the third toward Worker Safety. Because this get/pay expression will not contribute to Worker Safety, put 0 in the third component of the get/pay expression.

▶ Type the following expression into the entry bar, using the variable icon to insert the variable names: "Total_Costs,Fatalities_Avoided,0".

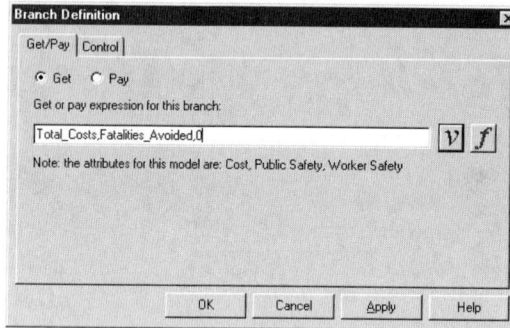

▸ Click OK.

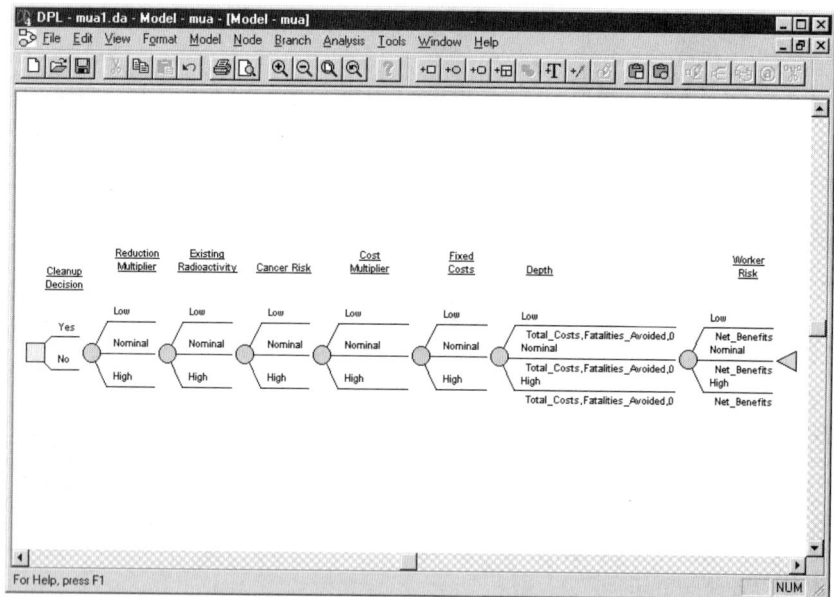

▸ Place Worker Fatalities in the get/pay expression on the branches of Worker Risk.

▸ Double-click on the branches of Worker Risk.

▸ Delete Net Benefits.

▸ Enter the expression "0,0,Worker Fatalities", using the variable icon to enter Worker Fatalities.

▸ Click OK.

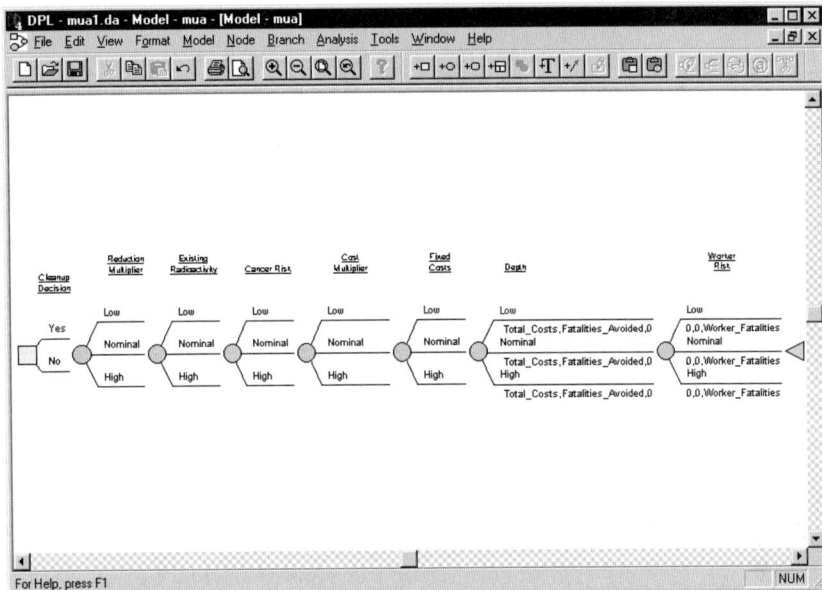

You are now ready to run a decision analysis. (For details on running a decision analysis, see the tutorial on analyzing models with multiple attributes in Chapter 10: Evaluation Methods.)

E very decision model has at least one scenario. Scenarios are built up from combinations of decision and chance events, but just defining these events isn't enough — we need to specify how they work together. There are two particular issues we need to address: timing, or the order in which events occur, and structure, which reflects whether all events occur in all scenarios. Simple timing and structure can be defined in either the influence diagram or the decision tree. More complex timing and structuring issues are most easily addressed in the decision tree.

Chapter 8

TIMING AND STRUCTURE

Chapter 8: Timing and Structure

Too Many Trees?

Before we discuss modeling with decision trees, we need to define a few terms. There are three tree-shaped graphic displays in DPL.

- **Data trees** are used for entering probability and value data for nodes. They are only accessible within the Node Definition dialog box. DPL uses a branched tree to display this information because it is an intuitive way to arrange data; an alternative format might be a table with multiple indexes.
- **Decision trees** are part of the decision model, along with influence diagrams. Decision trees are used primarily to define timing and structure. They are usually condensed for modeling ease, and do not show each path explicitly nor do they display specific probabilities or values.
- **Policy trees** are one result of a decision analysis run. They display every path of a decision tree explicitly, and include every probability, get/pay and endpoint value, and all expected values. They cannot be edited.

In this chapter we will be discussing ways to use decision trees to represent timing and structure.

DPL decision trees frequently are drawn using 'schematic,' or abbreviated, notation. A node, such as Sales, that has three outcomes, each of which is followed by the same chance node or subtree, such as Costs, can be represented in schematic notation as:

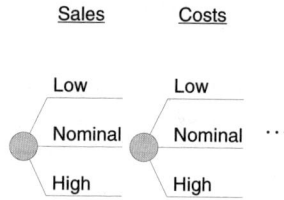

Rather than

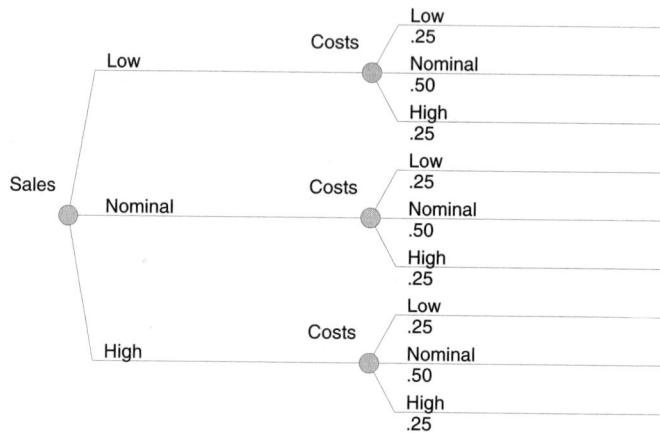

As you can see, schematic notation can significantly reduce the size of a decision tree.

The use of schematic notation has no impact on the data for either node. If the probabilities or values for Costs are defined to be conditional on the state of Sales, the appropriate numbers will be used when DPL evaluates the model. The policy tree will display the fully expanded tree.

Timing

Timing refers to the order in which decision and chance events occur. In general, the order in which a set of chance events occur doesn't matter mathematically — the expected value is the same.

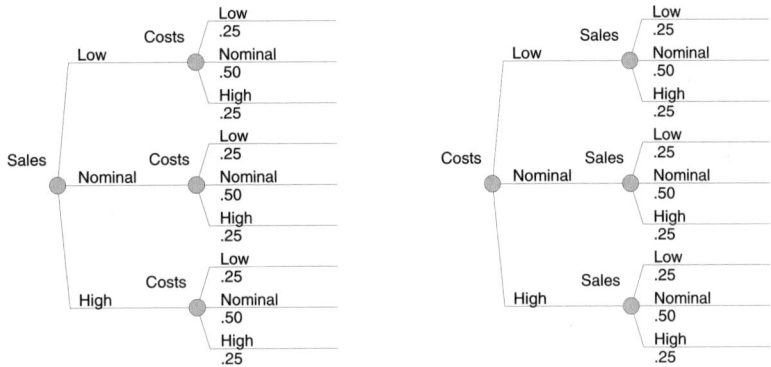

The same is true of a group of decision events.

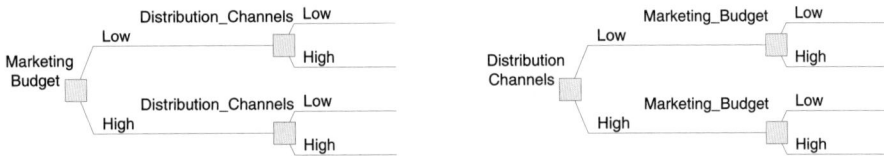

Order is very important when you mix decision and chance events. A decision before a chance event means you must make the decision before you know the outcome of the uncertainty. For example, you would have to make the decision whether or not to fund an R&D project before you find out how successful it will be.

```
                          R&D
                        Results
                Fund
                R&D

                            Low
                 Yes
                            Nominal  ◄
                  No
                            High
```

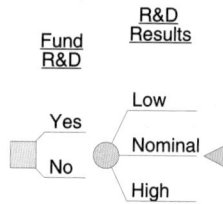

On the other hand, a decision that follows a chance event indicates that you get to wait and see how the uncertainty turns out before you have to make a decision. Continuing our example, while you have to make the R&D funding decision before finding out how well the program does, you may be able to postpone the decision about whether or not to put the new product into production until after the R&D is done.

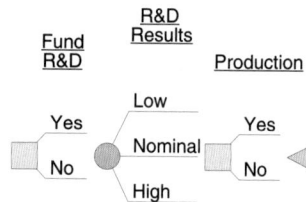

```
                           R&D
                  Fund   Results
                  R&D                    Production

                           Low
                 Yes                Yes
                         Nominal
                  No                No   ◄
                           High
```

If the decision occurs after the chance event, you can tailor your choice to each outcome of the uncertainty — if the R&D is a success, you can put the product into production, while if it is a failure, you do not go into production.

As you can see, the question of whether the production decision comes before or after the R&D chance event can make a big difference in the riskiness of this project. If the chance event comes first, then the downside of an R&D failure is only the R&D cost. If the production decision comes first, then an R&D failure loses **both** R&D Costs **and** Production Costs.

Timing and Conditional Chance Events

When one chance event conditions another, it is natural to place the conditioning node first in the tree and the conditioned node later.

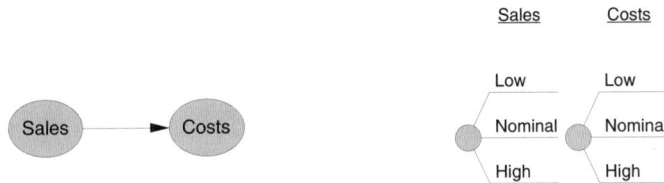

This is not necessary, however, because Bayes' rule allows us to "reverse the arrow" and calculate the probabilities the other way round.

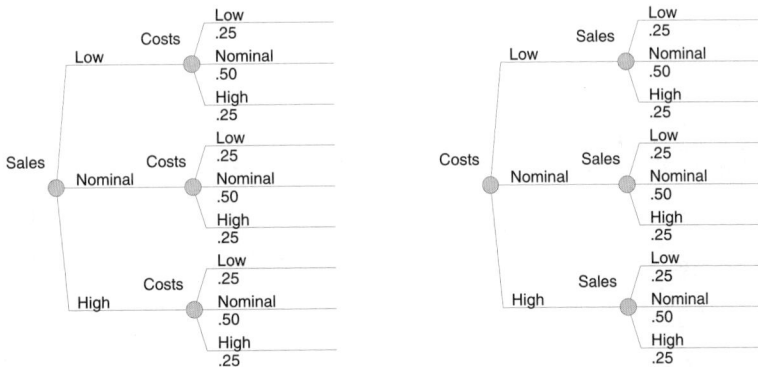

This is especially useful when we are modeling the value of imperfect information, such as introducing a new product into a pilot market before rolling it out worldwide. In this case, we would build the influence diagram this way, to indicate how reliable our test is.

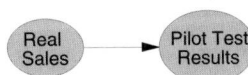

In the decision tree, we need to put the events in the other order, because we will get the pilot test results first.

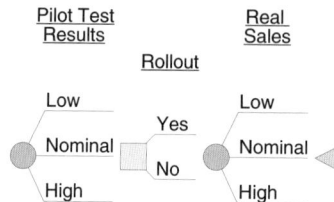

DPL will automatically apply Bayes' rule and calculate the correct probabilities.

Timing in a Decision Tree

In a decision tree, time moves from the root, at the left, toward the endpoints at the right, without looping. Therefore, an event that occurs earlier than another must appear to its **left**.

Timing in an Influence Diagram

An influence diagram doesn't naturally have a strong sense of time. Some people like to show timing by placing the nodes in order from left to right, following the decision tree convention. Others go top to bottom. The choice is yours. In this manual, when we wish to convey a sense of time, we will place nodes from left to right.

When you create an influence diagram, DPL builds a decision tree that is consistent with it. To build this "default" tree, DPL uses a few simple rules about timing:

- decisions come before uncertainties
- conditioning events come before the events they condition
- after that, nodes are placed in the order in which they were created.

If you wish to change the order of events, to move a chance event before a decision, for example, you can simply change the decision tree. It is **not** necessary to change the influence diagram.

This influence diagram
can be linked to either of
these trees (and others)

— or —

If you wish to reflect the timing in the influence diagram, you may add 'timing' arrows. The most common timing arrows lead from chance events into decisions. For example, we can modify the influence diagram for the pilot study decision. DPL will automatically update the decision tree.

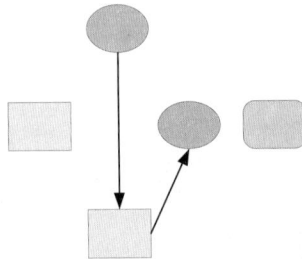

In this case, we also need a timing arrow from the second decision to the other chance event; otherwise, DPL will move it before the decision as well, following the rule that conditioning events usually precede the events they condition.

If you place timing arrows in an influence diagram and build a decision tree that is inconsistent with it, DPL will display the error message "All information arcs not satisfied" when you attempt to run an analysis.

Structure

In many decision models, we need to do more than simply specify the order of events. We need to indicate that some events just don't occur under or apply to some scenarios. We refer to this as the structure of the decision model.

Symmetry versus Asymmetry

A decision or chance event is symmetric if each of its alternatives or outcomes leads to the same sequence of events (in the same order). For example, decision A is symmetric because both alternatives lead to chance events B and C.

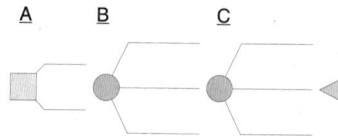

A decision or chance event is asymmetric if its branches lead to more than one sequence of events. For example, in our R&D example, the R&D decision is asymmetric because the Fund alternative goes on to the R&D Results chance event, the Production decision, and the Sales chance event, while the Don't Fund alternative stops without any of these subsequent events.

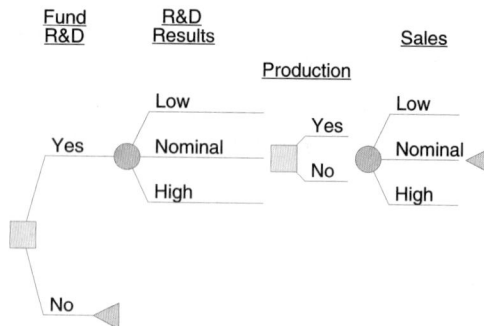

An asymmetric node may be completely asymmetric, meaning that the subtree following each branch is unique, or it may be mixed, meaning that some of its branches are symmetric, and one or more are unique. For example, chance event A shows mixed asymmetry because two of its branches lead to the same subtree, but the third branch leads to a different subtree.

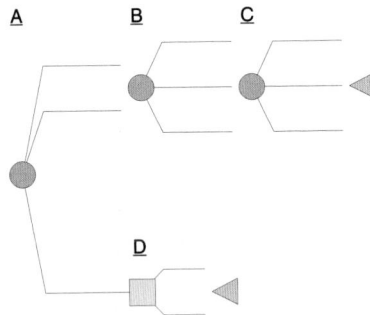

A decision tree or model is symmetric if it includes **only** symmetric nodes — every scenario (or path from the root of the tree to an endpoint) includes every event, in the same order.

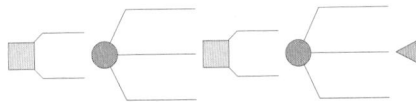

A decision tree or model is asymmetric if it includes one or more asymmetric nodes. An asymmetric decision tree may include symmetric nodes; in fact, most do.

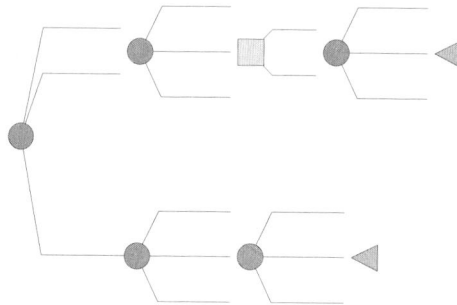

When to Use Asymmetry

Mechanically, symmetric decision trees are the easiest to work with. Influence diagrams are naturally symmetric, and building the decision tree is just a question of getting the events in the correct order. It is also much easier to link a spreadsheet to a symmetric decision tree. A symmetric decision tree will usually run faster than an asymmetric tree, unless the asymmetric tree has significantly fewer paths. Finally, a symmetric decision tree is usually more compact, making it easier to fit on the screen, a presentation slide, or a piece of paper. In general, we suggest that you start with a symmetric tree, only introducing asymmetry when you encounter a need for it.

You may need to use an asymmetric tree if:

- The symmetric tree is very large **and** using asymmetry will reduce the number of paths by half or more.
- The sequence of events that follows one branch of a decision or chance event is very different from those that follow another branch.
- The **order** of events is different for different branches of a decision or chance event. For example, to continue with our R&D model, suppose we have a decision about whether or not to conduct a pilot marketing test before deciding on a full rollout of our new product. Our decision tree would look like this:

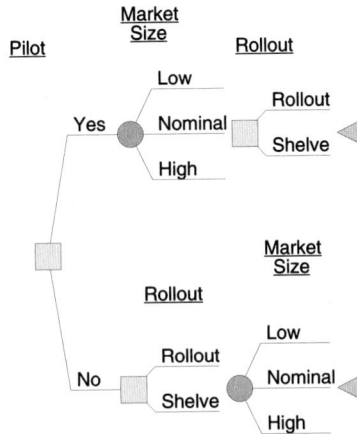

The influence diagram would look like this:

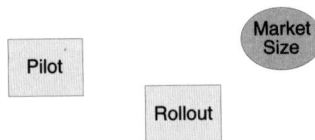

The biggest challenge when building asymmetric decision trees is handling the conditioning of values. If you build an influence diagram that calculates Profit based on all of the decision and chance events, but build a decision tree which includes paths that don't contain all of the decision and chance events, you may encounter the error message "Conditioning event <event name> not in a known state." Several methods for modeling this type of situation are discussed in Chapter 6 in the section called "Conditioning Event Not in a Known State."

Other Structuring Techniques

DPL provides a great deal of flexibility when structuring a decision tree. Some of these features allow you to build complex trees; others allow you to temporarily change the structure of a tree for debugging, sensitivity analysis, or other investigations.

Branch Control

Branch control allows you to force the decision tree to use one particular alternative of a decision or one particular outcome of a chance event. This overrides the probabilities you defined for the chance event or the optimal policy selection for the decision without changing the data in the influence diagram. The selected branch is highlighted in the decision tree.

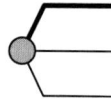

Branch control can be applied individually to each instance, or occurrence, of an event in the decision tree. This allows you to control an uncertainty to one branch in one part of a tree, control another instance of the uncertainty to a different branch, and leave a third instance uncontrolled.

Perform Links

Very few asymmetric decision trees are completely asymmetric. Most have a few asymmetric nodes and several symmetric ones. A fairly common structure looks like this:

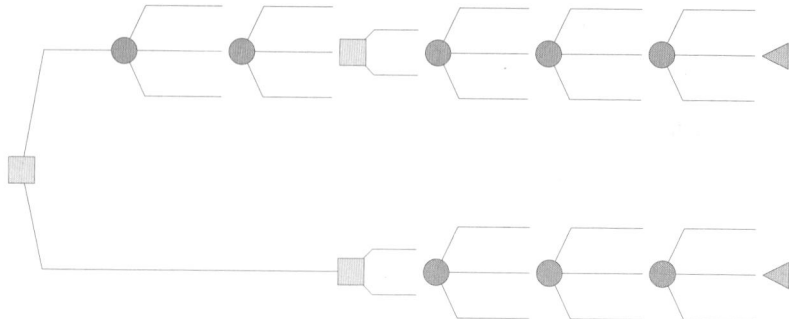

The first part of the tree is asymmetric, but starting with the second decision the second part is symmetric. Building the second set of nodes is tedious, and updating all sets, with a new event, for example, is a chore.

DPL offers a technique that allows you to build a subtree once, and then refer to it elsewhere in the decision tree. This is called a "perform" link.

A perform link references, or copies, the structure of a subtree from the label to the endpoints. A subtree may be referenced several times.

Perform links are helpful in three ways. First, they make modifying and maintaining a large decision tree easier. Second, they make a decision tree more compact, making it easier to view and print. And third, they can make a complex tree easier to understand and explain by showing the patterns within a model's structure.

Perform links can be useful in large symmetrical decision trees, as well. A symmetric decision tree that includes ten nodes, for example, can be hard to fit legibly on the screen or on a piece of paper because shrinking the tree to fit the width leaves the nodes very small and the text unreadable.

If you break the tree into two (or more) pieces, using perform links to hook them together, you can make more efficient use of the screen or page.

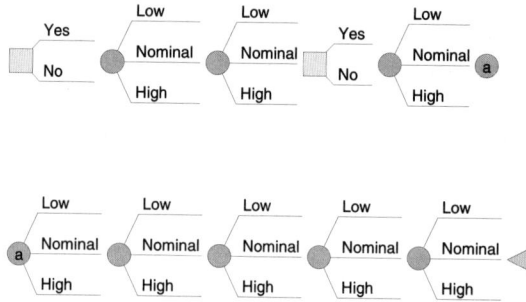

When you create a perform link in the model window, DPL automatically assigns a label. Because the labels are letters of the alphabet, you may have up to 52 separate links (lower and upper case letters). The labeling starts with the link closest to the upper left corner of the model, so if you create a second perform link, it may become 'a'; the first one may be renamed 'b.'

If you control a branch in a decision tree that includes perform links, you may get the error message "Perform label referenced before definition" when you attempt to analyze the model.

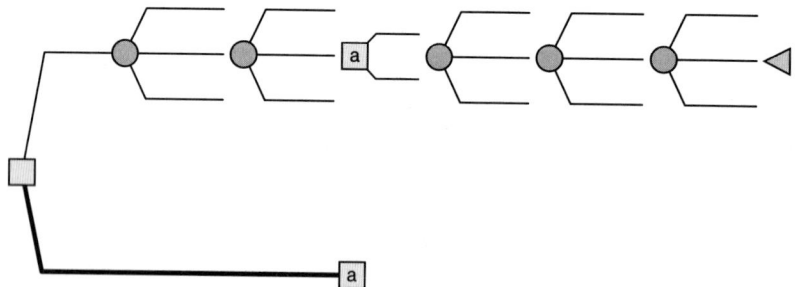

This problem occurs because DPL doesn't even build the "uncontrolled" subtree, so it doesn't have the performed subtree reference in mind when it does build the controlled branch. To fix this, simply detach the performed subtree, then add a perform link back to its original position.

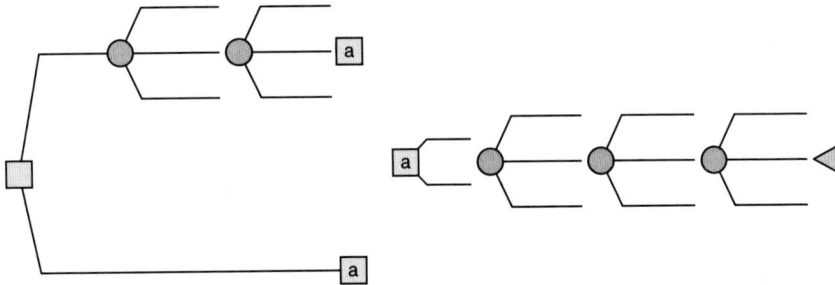

Leaving Nodes Out of the Decision Tree

It is not always necessary to include all nodes from the influence diagram in the decision tree. If an influence diagram chance node's only effect is to condition the probabilities of another chance event, it may be omitted from the tree. For example, consider a model with a chance event called Weather, which conditions the probabilities of the chance event Ice Cream Sales.

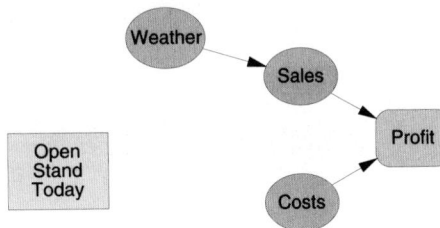

If we include Weather in the decision tree, our decision policy will have 28 (27 + 1) branches. The probabilities for Sales will be the conditional probabilities we entered in the influence diagram.

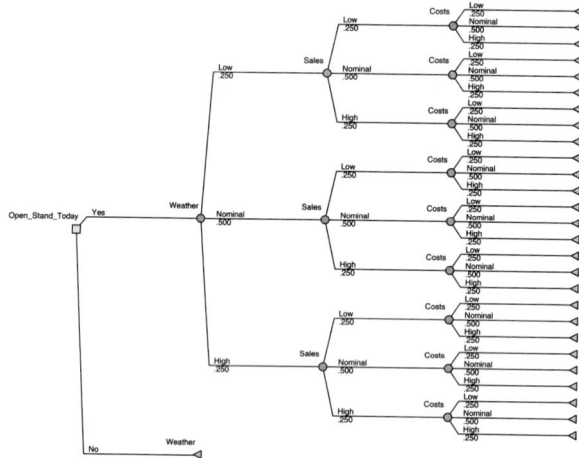

If we omit Weather, the decision policy will have 10 (9 + 1) branches. The probabilities for Sales will be the marginal probabilities DPL calculated from the probabilities for Weather and the conditional probabilities for Sales.

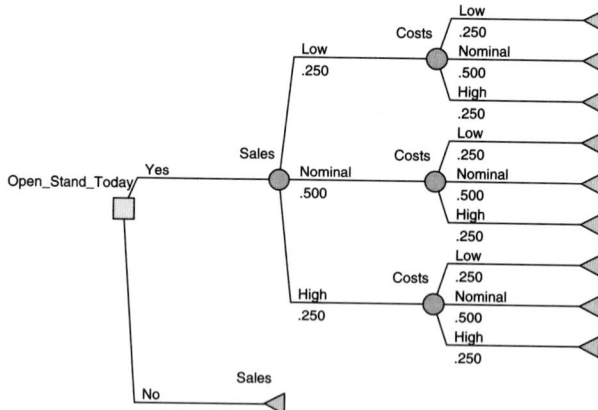

You cannot calculate the value of information or control for a chance event that has been omitted from the decision tree.

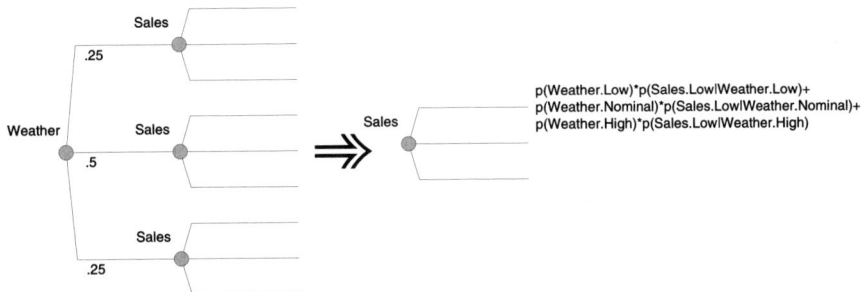

Setting the State of an Event

There are two ways to force a decision tree to use one alternative of a decision node or one outcome of a chance event. If you use Branch Control, the node appears as part of the decision tree structure, with one branch highlighted. Sometimes, though, you don't want or need the event to appear in the decision tree, but do need to have it forced to a particular state. An advanced feature, Branch Set, allows you to do this.

The most common use for Branch Set is to link an asymmetric decision tree to a spreadsheet. For example, consider a model in which one decision is a choice between signing a long-term fuel contract or buying fuel in the spot market. You might build a spreadsheet that included a flag for the choice -1 for contract, 0 for spot, for instance. In the calculation of fuel expenditures, you could use an if-then-else formula to use the contract price or the spot price, as needed.

If the spot price is a chance event linked to the spreadsheet, DPL will want to send its value to the spreadsheet every time it requests a calculation of the spreadsheet's value. But on those branches where the decision is to sign the contact, the spreadsheet doesn't need the spot price, so it would be nice to leave it out of the decision tree. If you build this asymmetric decision tree, you will get the message "Conditioning Event Spot Price not in a Known State." To satisfy DPL and continue to use an asymmetric decision tree, you can use Branch Set to set Spot Price to an arbitrary state.

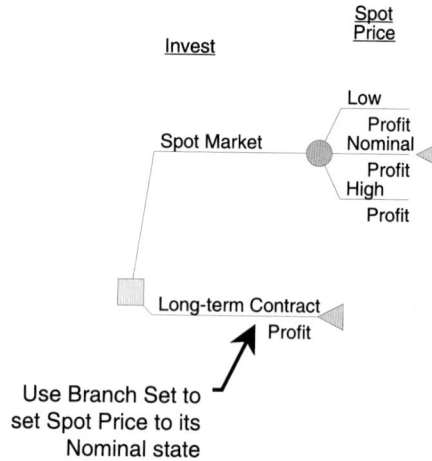

Invest

Spot
Price

Spot Market

Low
Profit
Nominal
Profit
High
Profit

Long-term Contract
Profit

Use Branch Set to
set Spot Price to its
Nominal state

<u>Dont Gamble</u>

Even with fast computers, it is not hard to build a decision tree that has so many paths that evaluation times are long enough to be annoying or impossible. The best prevention of this problem is aggressive and disciplined deterministic sensitivity analysis. But we all build models that are just too big, causing us to look for ways to reduce the size of the tree.

One technique is to use Branch Control to reduce one or more chance events from several outcomes to a single outcome. In a symmetric decision tree, for example, controlling a single three-state chance event to a single outcome reduces the number of paths by two-thirds. The main drawback of this technique is the need to choose one of the outcomes to represent the entire chance event; if the outcomes aren't symmetric, selecting a single outcome changes the expected value of the model.

Low [20]
.200 20
Costs Nominal [10]
[8] .400 10
Low High [0]
.250 .400 0

Low [50]
.200 50
Costs Nominal [40]
[38] .400 40
Sales Nominal High [30]
[38] .500 .400 30

Low [80]
.200 80
Costs Nominal [70]
High [68] .400 70
.250 High [60]
.400 60

—— **Expected value** ——

Low [20]
.200 20
Costs Nominal [10]
[10] .400 10
Low High [0]
.250 .400 0

Low [50]
.200 50
Costs Nominal [40]
Sales [40] .400 40
[40] Nominal High [30]
.500 .400 30

Low [80]
.200 80
Costs Nominal [70]
High [70] .400 70
.250 High [60]
.400 60

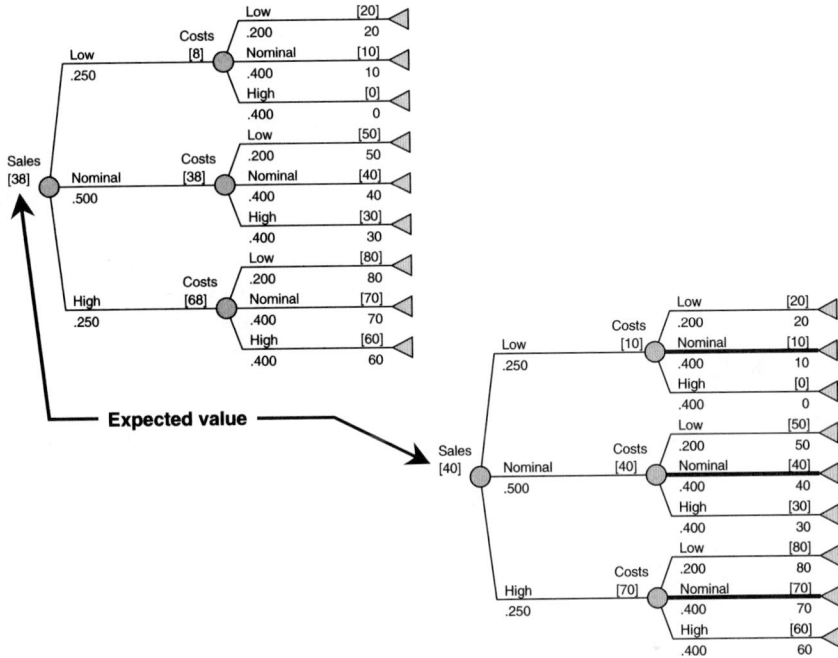

DPL offers another technique, called Dont Gamble, that allows you to reduce a chance event to a single outcome and preserve its expected value. An advanced feature, Dont Gamble, doesn't choose one of a chance event's outcomes, like Branch Control; instead, it replaces the event's outcomes with a single new outcome. The value associated with this outcome is the expected value of the event's original outcomes. A chance event that has had Dont Gamble applied has asterisks (*) in front of the outcome names in the model window, and an asterisk as the only name of the outcome in the policy tree.

Original decision tree

Decision tree with dont gamble

Original policy tree

Policy tree with dont gamble

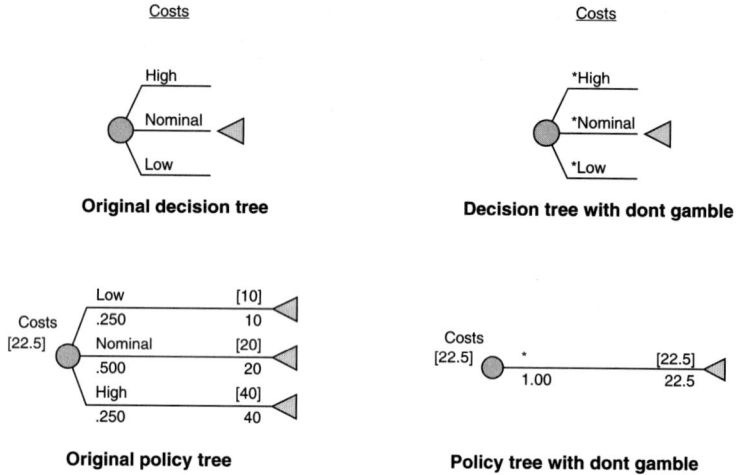

The advantage of Dont Gamble is that it usually preserves the expected value of the model. It can also be a slightly more subtle way to reduce the size of a decision tree than Branch Control. For example, consider an earthquake model.

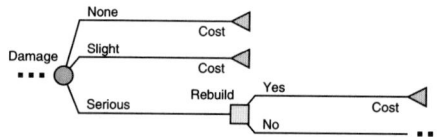

If we attempt to reduce the size of this model by using Branch Control, we have to choose a single outcome of this chance event. Losing the distinction between no earthquake damage and slight damage probably doesn't affect the credibility and usefulness of the model. But choosing to assume either that we never have a major earthquake, by controlling this event to either the none or slight states, or that we always have a major earthquake, by controlling to the major branch, is likely to distort our results.

Dont Gamble solves this problem by compromising between the need to reduce the size of the tree and the need to preserve the asymmetry. If we apply the Dont Gamble specification to this chance event, we will get a **two** outcome chance event.

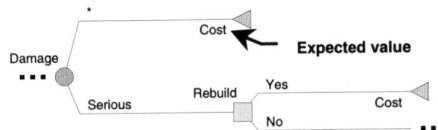

(A related feature, Always Gamble, is only used during an expected value tornado diagram analysis. Refer to the On-Line Help.)

Controlled Nodes

The good thing about asymmetric decision trees is that they enable us to build very large, realistic decision models. The bad thing about them is that they make tracking dependencies for value calculations harder. Chapter 6 includes a discussion of several ways to handle this problem in asymmetric models, and these techniques are sufficient in the majority of cases.

To help in the remaining cases, DPL offers an advanced feature called a controlled node. A controlled node appears in the influence diagram as a gray node with states. Its data may be conditioned by other nodes, and it may, in turn, condition other nodes. A controlled node's states do not have probabilities, so in the influence diagram, it looks much like a decision node.

A controlled node does not appear in the structure of a decision tree like decision and chance events do. Instead, it appears only in set statements on the decision tree's branches. This means that you, the modeler, explicitly control the state a controlled node is in at every point in the tree, rather than relying on probabilities, as with chance events, or the selection of an optimal policy, as with a decision. And, unlike decisions and chance events, a controlled node's state can change several times along a single decision tree path.

To demonstrate the use of controlled nodes, we will consider a decision by a pharmaceutical company whether to buy a patent for prescription medication or to develop a competitive product in-house. By purchasing the patent, we avoid the R&D expense, and we will be able to start selling the product sooner. Because of the entrance of generic products when the existing patent expires we think this market will only be profitable until it expires, even if we develop and patent our own product. The problem is that our lawyers tell us that the patent owner has a former partner who has threatened to sue for ownership of the patent. If he does, there will be settlement discussions, possibly a court case, and a judgment. Of course, even if the former partner wins, we can buy the patent from him.

The patent owner will pay all of his own legal fees, so our primary concern is how long this might delay our ability to put a product on the market. We have built the following influence diagram.

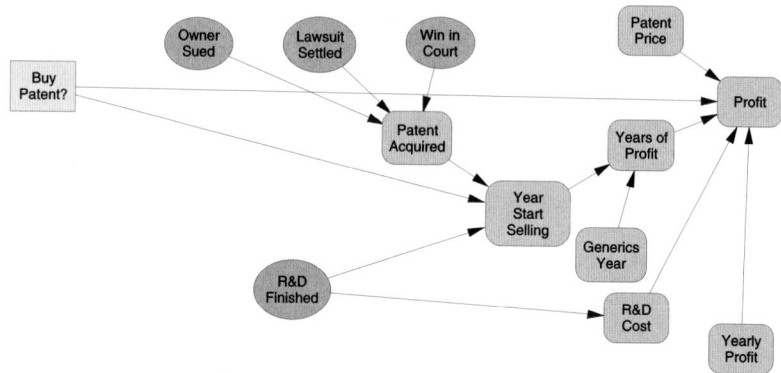

As you can see, our profit depends on the year we can begin selling the product, and that depends on the three lawsuit uncertainties. Without a controlled node, we might build a fairly symmetric decision tree that looks like this:

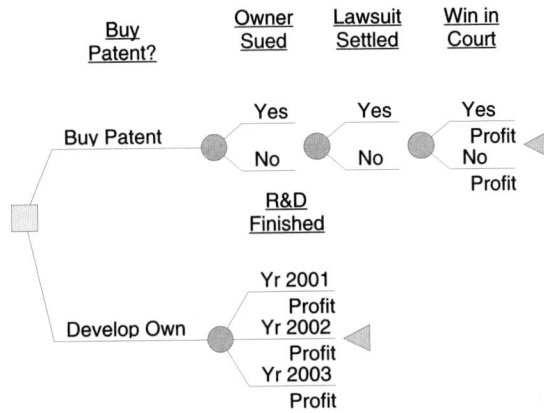

For this small example, this model is sufficient. However, there are two things we'd like to improve; in a larger, more complex model, these might be crucial. First, the node for the year the patent is acquired is conditioned by three chance events. They are only two-state events, but we will have to enter eight values, including values for impossible scenarios such as not getting sued, settling, and losing in court. Second, we have to make the buy patent subtree symmetric, including all three lawsuit nodes on every path.

We can replace the value node Year Patent Acquired with a controlled node.

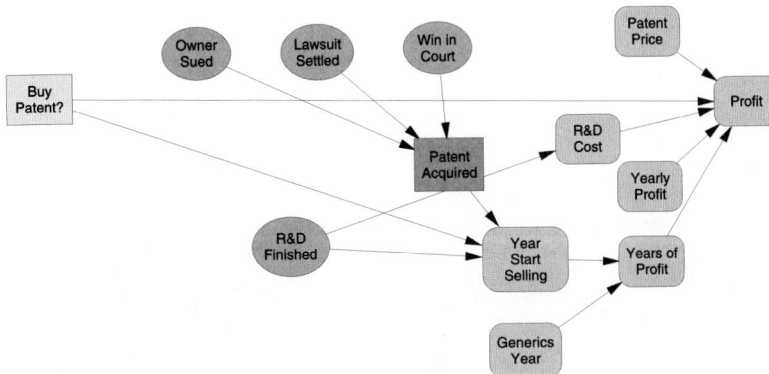

The arrows leading into Profit can be deleted, or not, depending on which you think makes the diagram clearer; the data for the controlled node is not conditioned on the chance events, making it quite easy to enter.

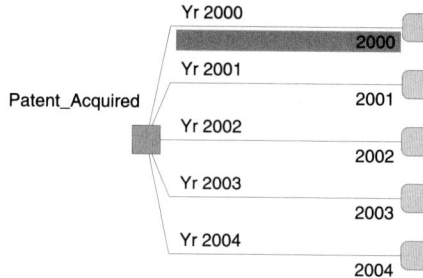

Yr 2000

    2000

Yr 2001

    2001

Patent_Acquired

Yr 2002

    2002

Yr 2003

    2003

Yr 2004

    2004

We can make the decision tree asymmetric now, reflecting the true structure of the events.

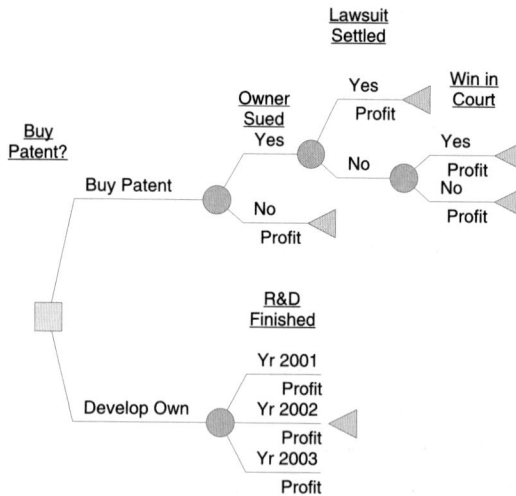

Lawsuit
Settled

                    Yes   Win in
Owner   Profit   Court
Sued   Yes
Yes   No   Profit
Buy   No
Patent?   Profit

Buy Patent   No
               Profit

R&D
Finished

               Yr 2001
               Profit
Develop Own   Yr 2002
               Profit
               Yr 2003
               Profit

Once the decision tree is structured, we start setting the controlled node. We start by setting it to its Yr. 2000 state, the most likely or default state, on the Buy Patent branch of the decision. This setting will apply to everything on this subtree, except where we change it.

So where do we change it? First, we know that if the owner is sued and the case is settled, we have a one-year delay, so we select the Yes branch of Lawsuit Settled and set the controlled node to Yr. 2001. Next, we know that if we go to court and win, we've had a three-year delay, so we set the controlled node to Yr. 2003. And finally, losing in court and having to buy the patent from the new owner adds another year of delay, so we set the controlled node to Yr. 2004. (You might find it easier to follow this logic in the text view of this model. Choose Tools Convert Model to Program, then scroll down to the section that follows the word "sequence:".)

(**Note:** This model also demonstrates a use of attributes in a single-objective model. If we define a second attribute called Year to Market, we can generate risk profiles that show how the two alternatives compare in terms of how soon we get our product on the market. The model in patwmua.da includes this modification.)

What advantages does this approach have? First, it frees the value model from having to track the asymmetries; if we were using a spreadsheet to calculate our profits, for example, we would only have to send one number for the year we acquire the patent rather than three or four variables. Second, it reduces the data entry in the influence diagram. And third, it gives us a way to specify the logic directly, rather than having to devise a complex set of rules or formulas to track asymmetries. Most DPL users never need controlled nodes, but every once in awhile you find you can't build a model without them!

A word of caution: controlled nodes should not (usually) be used to program a decision policy into a decision tree. There are two reasons. First, you don't want to over-constrain the model by telling DPL what decision to make — the decision policy should be the output of the analysis, not an input. Second, if you do choose to specify a decision policy (perhaps to generate a risk profile for comparison to the optimal policy recommended by the analysis) it is clearer to use decision nodes and Branch Control, which makes everything visible in the decision tree.

Tutorial: Timing and Structure

There are many tools in DPL for changing the structure of your model. Some of these tools are: timing arrows, reordering nodes in the decision tree, creating asymmetric decision trees, branch control, performing subtrees, setting states of nodes in the decision tree, and controlled nodes.

▶ Open "timing.da".

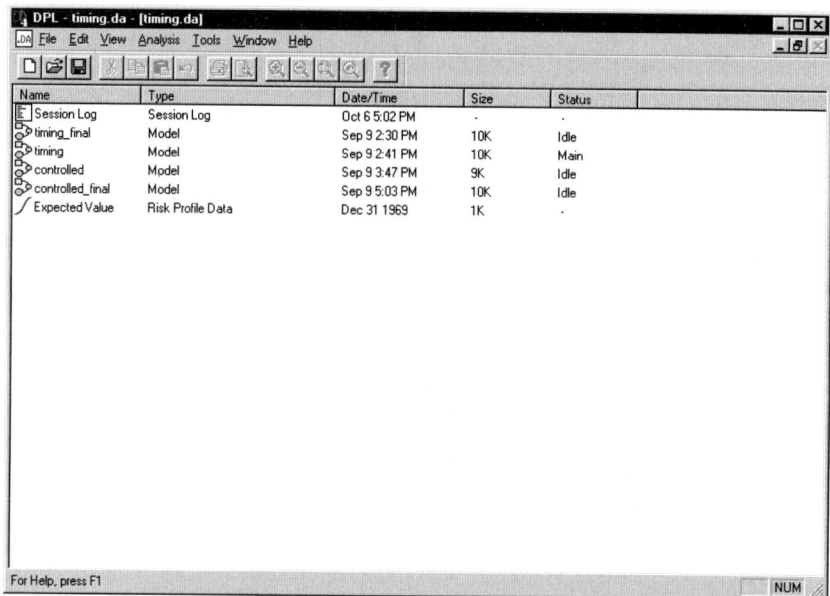

▶ Select Window Model-Timing.

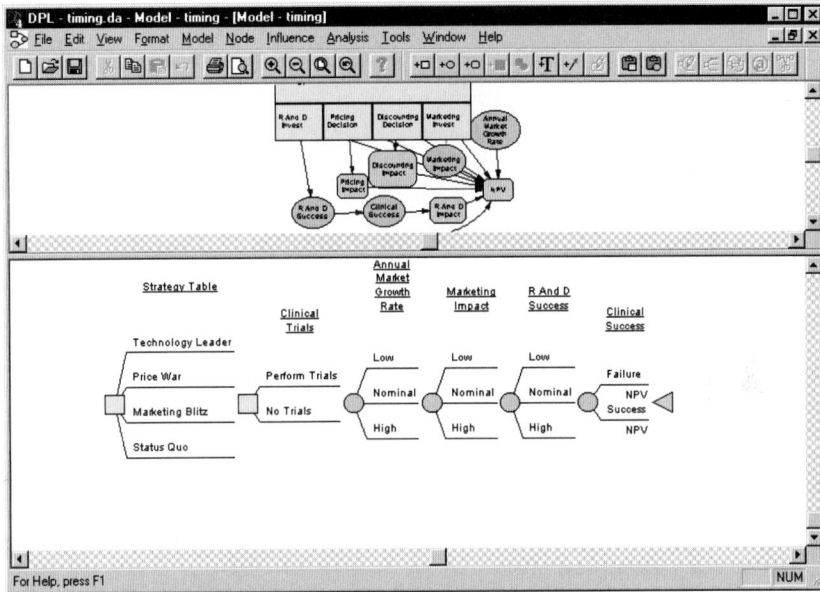

▸ Click on the Zoom Full icon.

▸ Click inside the decision tree pane.

▸ Click the Zoom Full icon again.

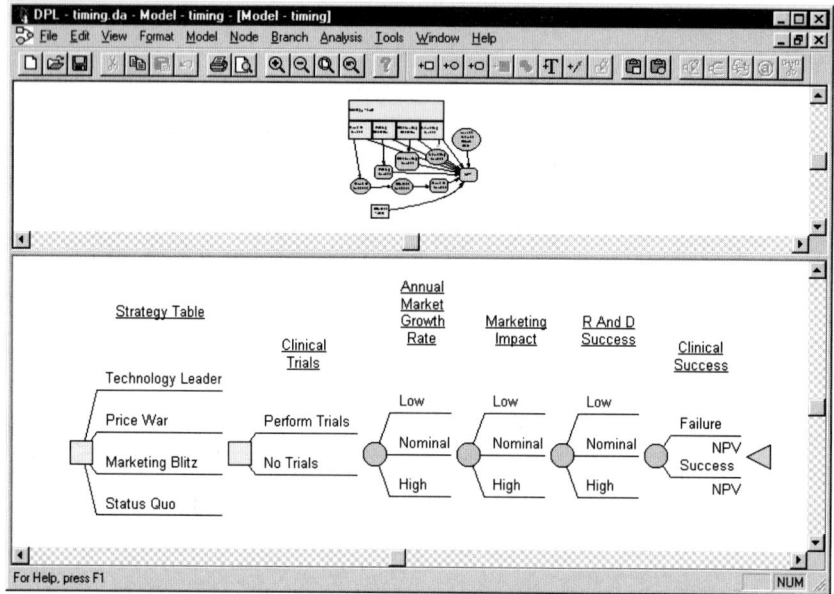

Timing Arrows

DPL uses arrows in the influence diagram to indicate both timing and conditioning. When you condition a node on another node, DPL automatically draws an arrow with a blue, burgundy, or green arrow between the nodes to indicate the type of conditioning. Sometimes you want the influence diagram to dictate in which order events will occur. For this you can use timing arrows.

The decision tree tells DPL the sequence in which events should be evaluated during the analysis. Currently, the decision Clinical Trials occurs after the Strategy Table and before the uncertainties. In reality, the Clinical Trials decision will be made after R And D Success and before Clinical Success. You could make this change by reordering the sequence of nodes in the decision tree (which you will do later on). However, you can also tell DPL this by using timing arrows.

▶ Click inside the influence diagram pane to make it active.

▶ Click on the "Create Influence" icon.

▶ Click on Clinical Trials, then on Clinical Success.

▶ Repeat the previous two steps to draw an arrow from R And D Success to Clinical Trials.

DPL has drawn the timing arrows, and accordingly has placed R And D Success before Clinical Trials in the decision tree.

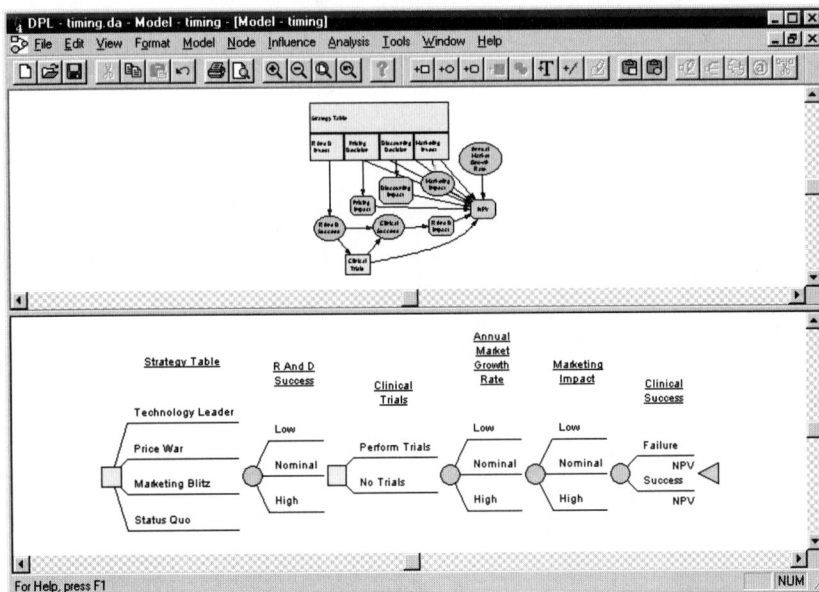

Reordering Nodes

If you decide not to perform clinical trials, you cannot have clinical success. You can reflect this by altering the decision tree. The first step is to reorder Clinical Success.

▸ Move the splitter bar to the top of the window to maximize the decision tree pane.

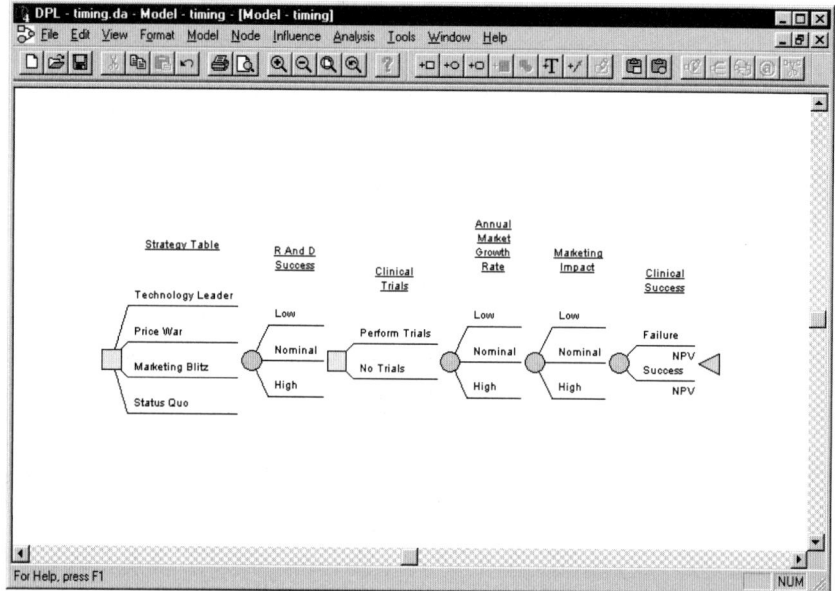

DPL has placed Clinical Success at the end of the tree. You can reorder the tree to place it after Clinical Trials.

▸ Click on the Clinical Success node.
▸ Select Node Reorder.
▸ Place the crosshairs over the Annual Market Growth Rate node and click the left mouse button.

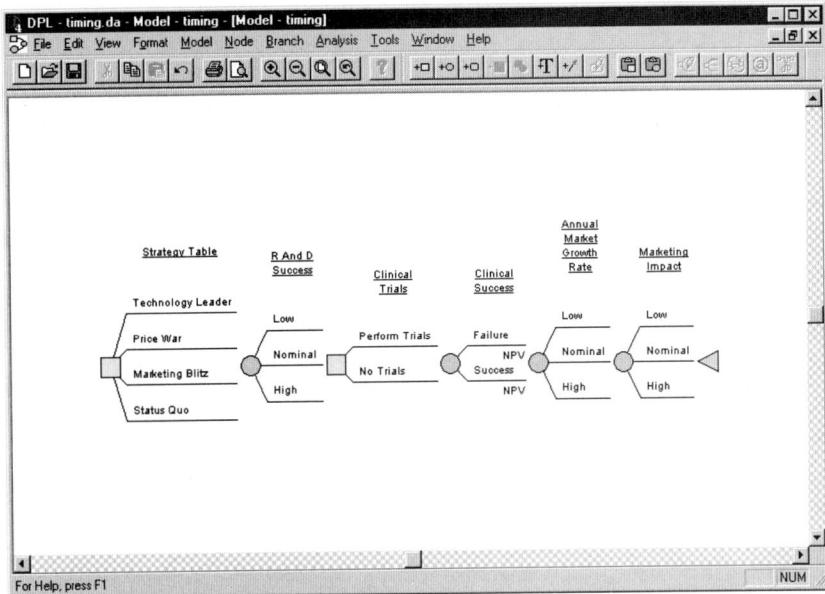

DPL has moved the get/pay expression NPV along with Clinical Success. It should be on the last node of the tree, Marketing Impact.

▶ Double-click on the branches of Clinical Success.

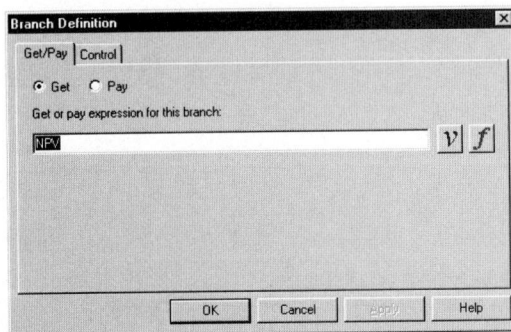

▸ Press the Delete key to remove NPV from the get/pay expression, then click OK.

▸ Double-click on the branches of Marketing Impact.

▸ Click on the Variable icon in the dialog box.

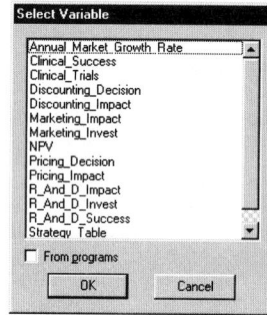

▸ Select NPV and click OK.

▸ In the Branch Definition dialog box, click OK.

The get/pay expression NPV is now on the branches of Marketing Impact.

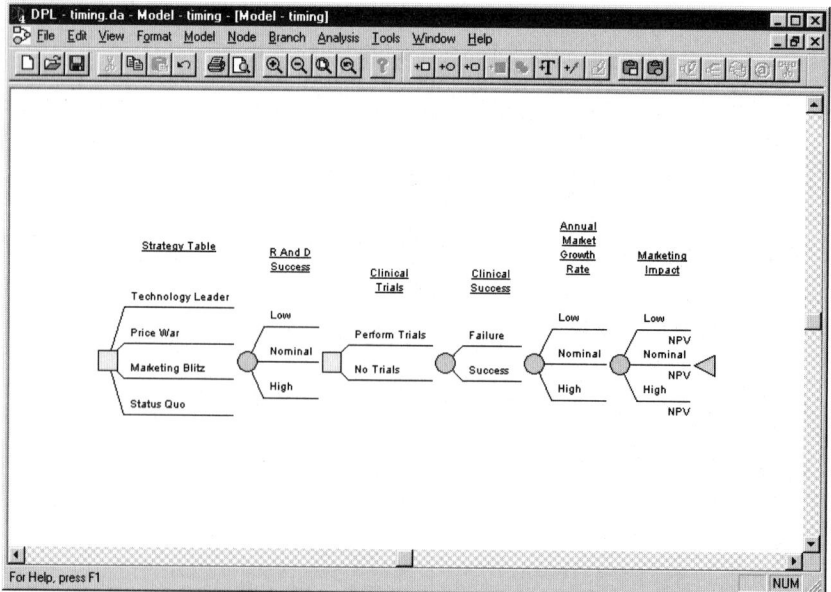

Asymmetry

If you decide to perform Clinical Trials, the probability distribution assigned to Clinical Success will tell you if they were successful or not. If you do not perform Clinical Trials, there is no way you can have Clinical Success. How Clinical Success is evaluated therefore depends on which alternative of Clinical Trials was chosen. You can model this by making Clinical Trials asymmetric.

▶ Double-click on the Clinical Trials node.

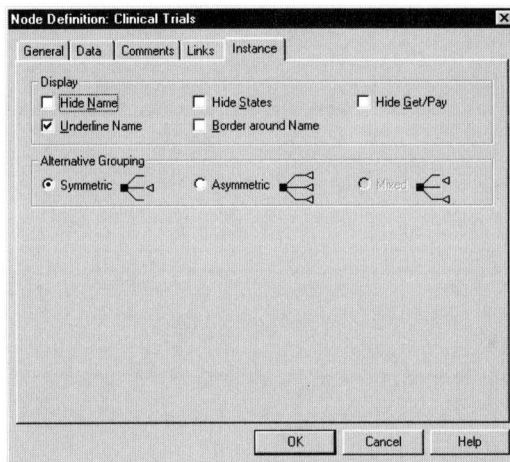

▶ Click on the radio button labeled "Asymmetric".
▶ Click OK.

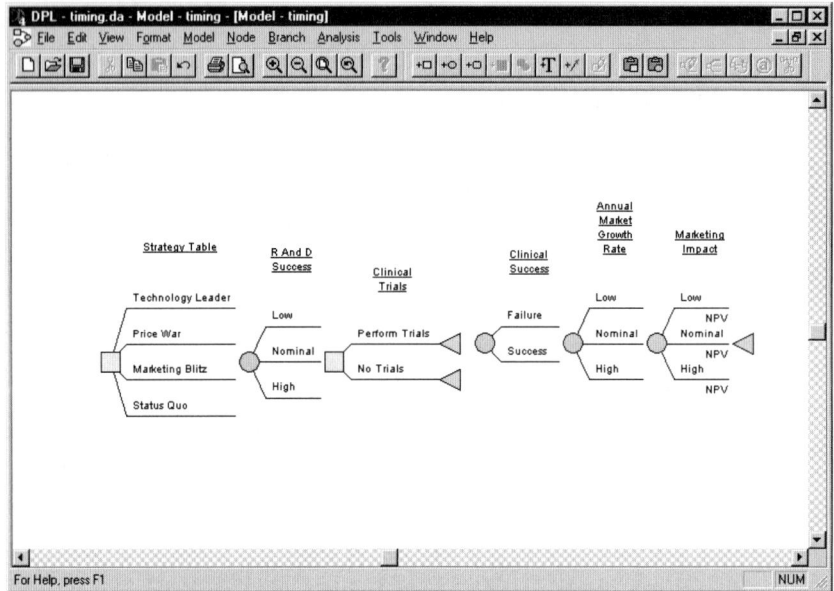

There is now a blue triangle on the end of each alternative of Clinical Trials. This tells you that each alternative can lead to different sequences of decisions and uncertainties (or subtrees).

Next you need to re-link the uncertainties to the Perform Trials branch of Clinical Trials.

▶ Click on the Clinical Success node and hold down the left mouse button.
▶ Move the crosshairs to the left until they are directly over the blue triangle at the end of the "Perform Trials" branch of Clinical Trials, then release the mouse button.

Now when DPL selects the Perform Trials branch of Clinical Trials, it will evaluate the remainder of the tree as it did before.

Adding Decision or Chance Instances to Decision Trees

Because DPL needs to evaluate Clinical Success in a special way if the No Trials alternative is selected, you need to add another instance of Clinical Success after the No Trials alternative.

▶ Select Node Add Chance Instance, or click on the Add Chance to Tree icon.

DPL displays the Select Event dialog box.

▶ In the Select Event dialog box, click on Clinical Success and click OK.
▶ Place the crosshairs over the blue triangle at the end of the "No Trials" branch of Clinical Trials and click the left mouse button.

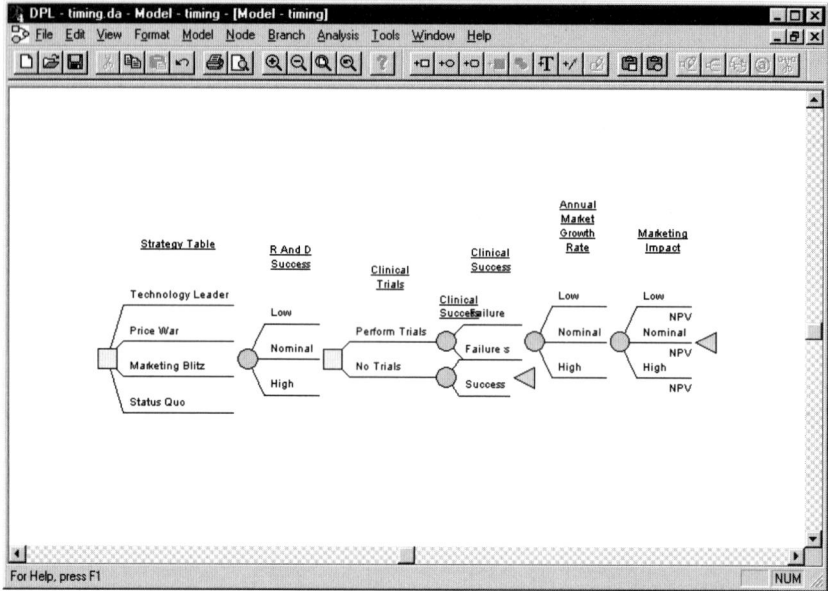

The two branches of Clinical Trials are too close together to view the nodes clearly. We will move the No Trials branch down to make the nodes more visible.

▶ Click on the Clinical Success node at the end of the No Trials branch of Clinical Trials and hold down the left mouse button.

▶ Drag the node down and release the left mouse button to place it. If necessary, move the text "Clinical Success" as well.

▶ Select View Redraw to refresh the diagram.

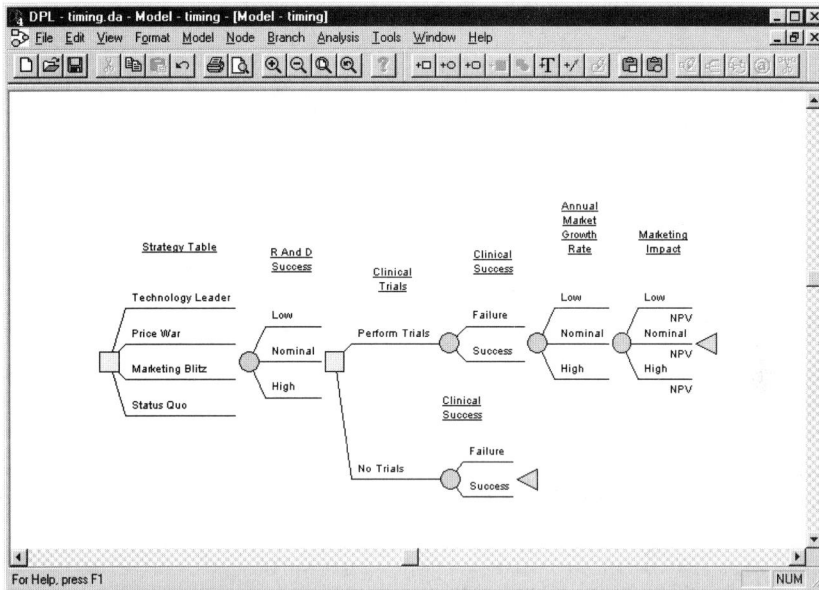

Branch Control

If clinical trials are not performed, there is no way to achieve clinical success. In terms of the model, this means that if the No Trials alternative of Clinical Trials is chosen, the outcome of Clinical Success has to be "Failure". You can tell DPL this by using Branch Control.

▶ Click on the branches of the instance of Clinical Success at the end of the No Trials branch of Clinical Trials.

▶ Select Branch Control.

DPL displays the Control tab of the Branch Definition dialog box.

▶ In the Control State section of the dialog box, select "Failure" from the drop-down menu.

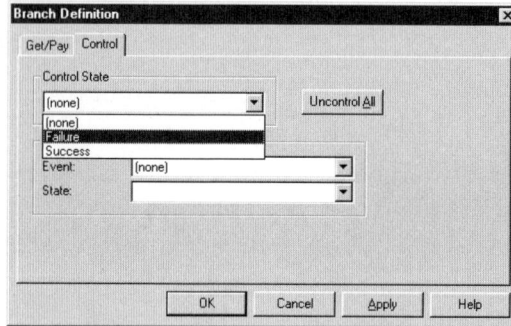

▶ Click OK.

The branch "Failure" is now highlighted, indicating that DPL will assign this outcome a probability of 1 during analysis.

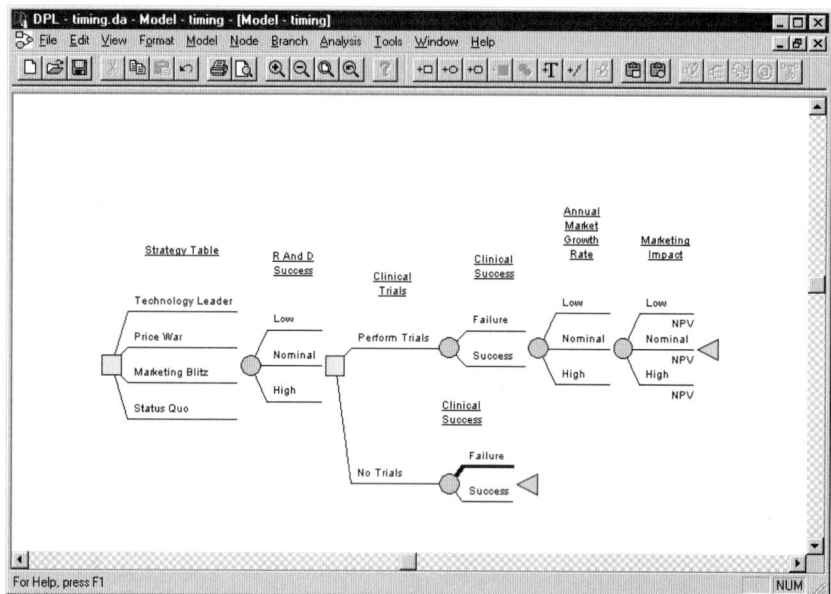

Perform Subtree

The blue triangle after the controlled instance of Clinical Success tells DPL there are no more nodes to evaluate. In reality, you want DPL to then evaluate Annual Market Growth Rate and Marketing Impact, just like it will with the other instance of Clinical Success.

One way to solve this problem is to simply add new instances of Annual Market Growth Rate and Marketing Impact and place them after the controlled instance of Clinical Success. Alternatively, you can use the Perform Subtree function, which is generally faster and simplifies the decision tree.

▶ Click on the Annual Market Growth Rate node.
▶ Select Node Perform Subtree.
▶ Place the crosshairs over the blue triangle at the end of Clinical Success and click the left mouse button.

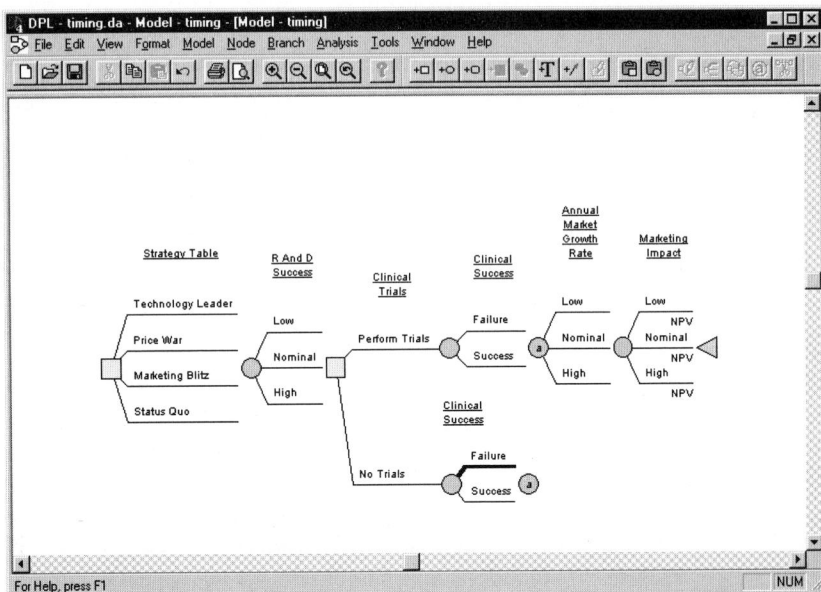

The "a" at the end of the controlled instance of Clinical Success tells DPL to go to the other "a" and evaluate the tree from there until the end.

Set Event

Looking at the decision tree, it is clear that if you do not perform Clinical Trials you will end up with a Failure outcome for Clinical Success. This is an intuitive result, and therefore there is really no need to have the second instance of Clinical Success in the tree for illustrative purposes. However, if you delete the second instance of Clinical Success, DPL will not know the state of Clinical Success when it evaluates the paths which include the "No Trials" alternative of the Clinical Trials decision, and will therefore give you an "Influencing Event is Not in a Known State" error when you try to run an analysis.

The solution is to tell DPL the states of Clinical Success without cluttering up the decision tree. This is done using the Set Event feature.

First, you'll need to delete the second instance of Clinical Success and re-create the Perform Subtree link.

- ▶ Select the instance of Clinical Success at the end of the "No Trials" branch of Clinical Trials.
- ▶ Select Node Delete.

DPL has deleted the instance of Clinical Success.

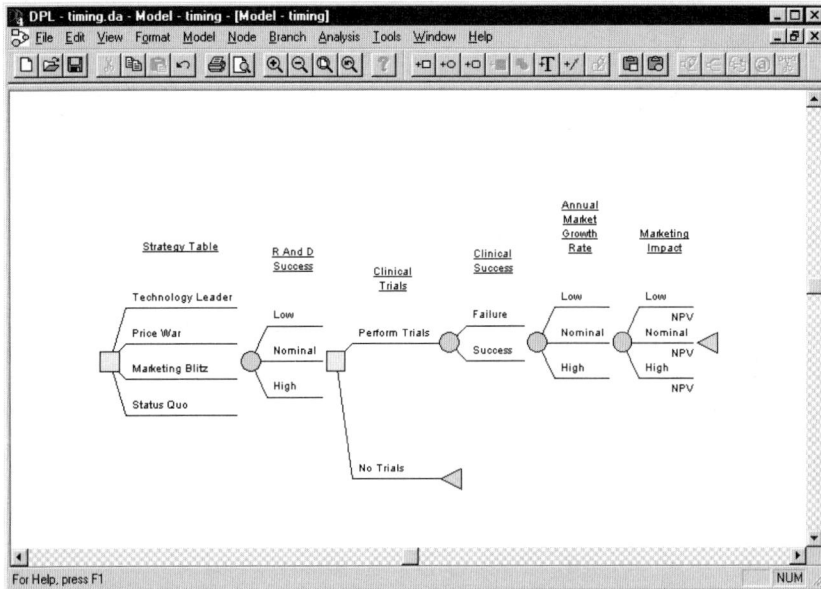

In deleting the Clinical Success instance, DPL also deleted the Perform Subtree link. The end of the "No Trials" branch needs to be re-linked to the subtree.

▶ Click on the Annual Market Growth Rate node.
▶ Select Node Perform Subtree.
▶ Place the crosshairs over the blue triangle at the end of the "No Trials" branch of Clinical Trials and click the left mouse button.

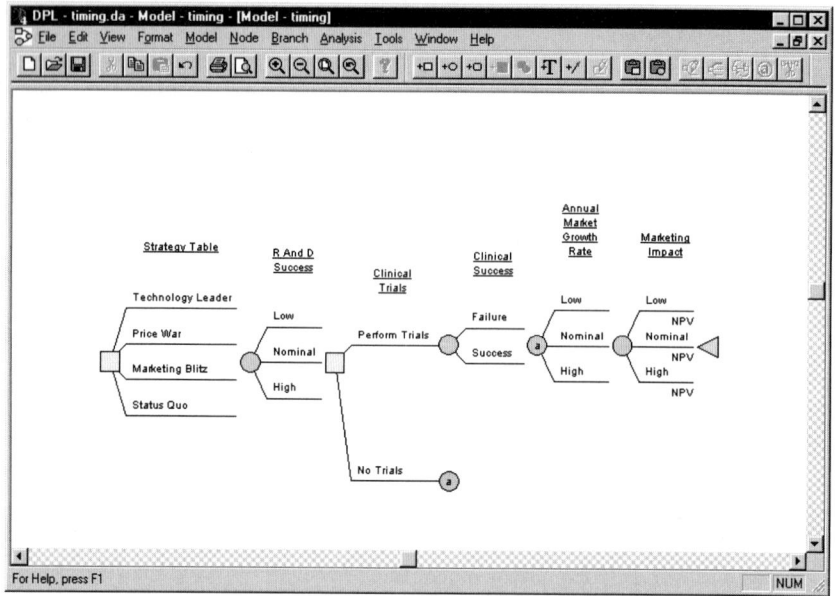

Now DPL will only evaluate Clinical Success if the "Perform Trials" branch is selected. If you run the model as it stands, DPL will not know the state of Clinical Success if "No Trials" is selected. The Set Event feature will correct this.

The Set Event feature is an advanced feature which can be enabled from the Options dialog box.

▶ Select Tools Options Advanced.

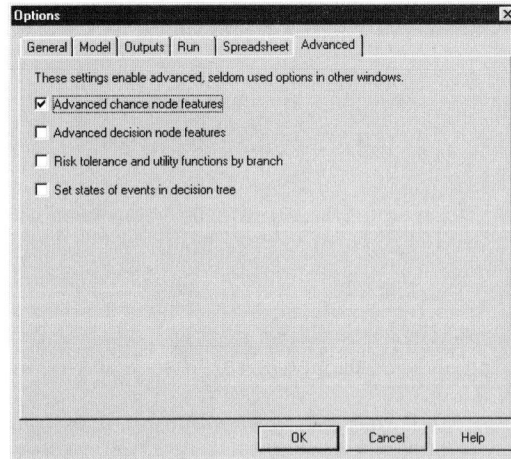

▶ Check the box labeled "Set states of events in decision tree".

▶ Click OK.

▶ Click on the "No Trials" branch of Clinical Trials.

▶ Select Branch Control.

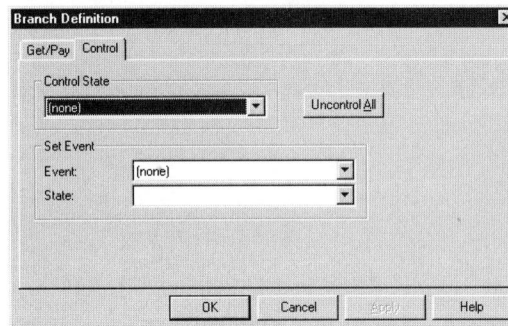

DPL displays the Control tab of the Branch Definition dialog box. The Set Event section of the dialog is now displayed because it was enabled in the Options dialog box.

▸ In the Set Event section, use the drop-down menu to select Clinical Success.

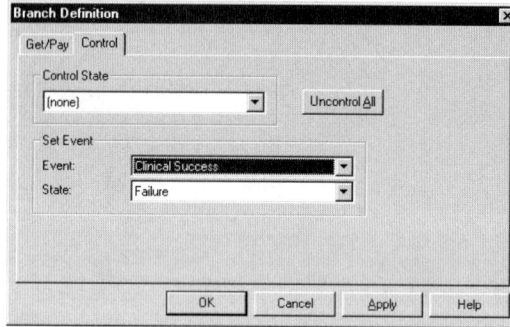

By default, DPL selects the first state of the node, "Failure", as the set state for Clinical Success. If you wanted to set Clinical Success to "Success", you could select "Success" from the "State" drop-down menu instead.

▸ Click OK.

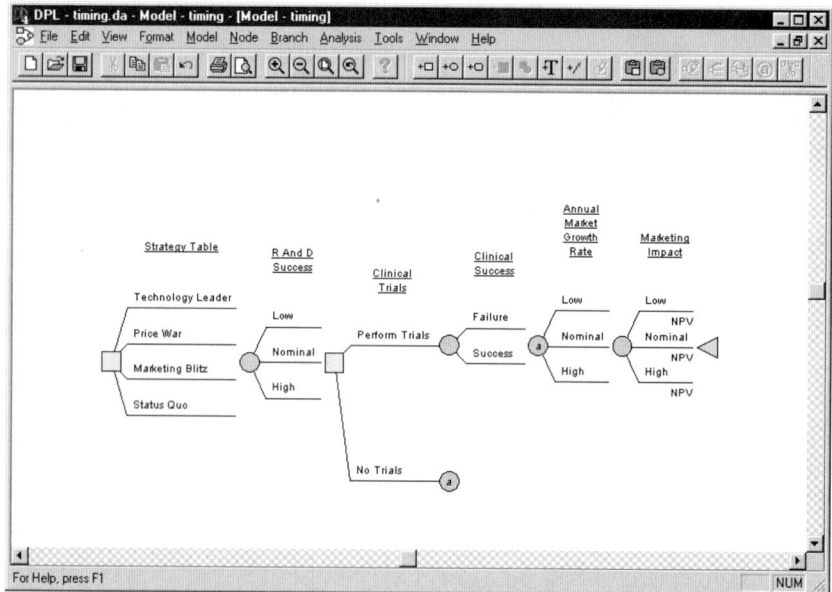

Functionally, the decision tree is identical to the one you started with. The Set Event serves the same purpose as branch control, the difference being that there is no graphical indication in the decision tree that you have used the Set Event feature. While it makes the decision tree appear less cluttered, it is easy to forget where in the tree you have set events. This is potentially dangerous, and is the reason why Set Event is an advanced feature!

Controlled Nodes

Controlled nodes can save you from entering repetitive data, and can allow you to change a value numerous times along a single path in the decision tree. Controlled nodes are generally most useful when they can help simplify extremely complex models. In order to keep this example simple, the model is somewhat simple, and using a controlled node does not provide a great deal of benefit. For a better understanding of when to use a controlled node, see the example given in the chapter.

▶ Select Window timing.da.

▶ Right-click on the model titled "controlled".

▶ Select "Make Main" from the context menu.

▶ Double-click on the model titled "controlled".

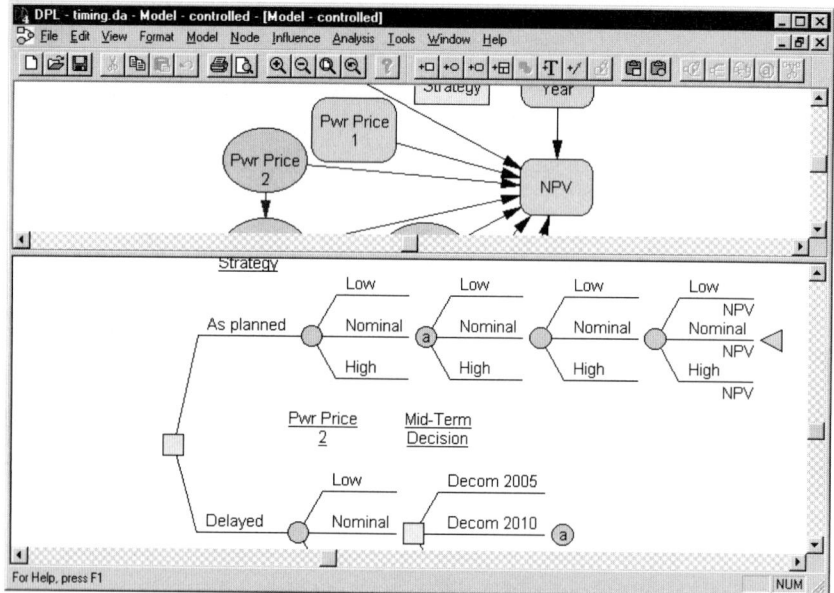

▶ Move the splitter bar down to the bottom of the window to maximize the influence diagram pane.

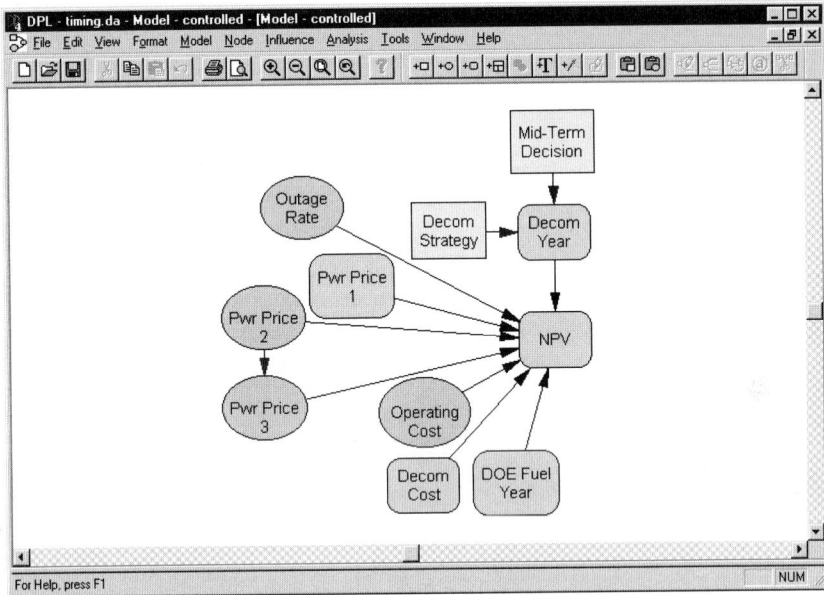

This model represents a decision of when to decommission a nuclear power plant. The node Decom Year represents the year of decommissioning.

▶ Double-click on the value node Decom Year.
▶ Click Full Screen.

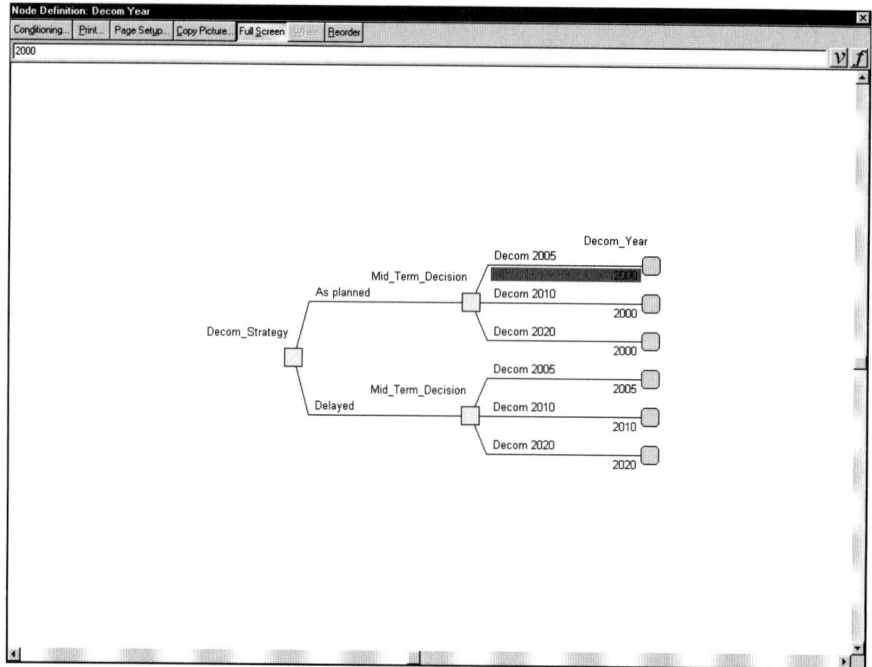

Decom Year depends on the two decisions, Decom Strategy and Mid-Term Decision. If the decommissioning strategy is "As Planned", the decommissioning year is 2000, regardless of the alternative selected for Mid-Term Decision. If the decommissioning strategy is "Delay", the alternative selected for Mid-Term Decision dictates the year of decommissioning.

With only six different values, the data entry for this node is hardly unbearable, and that in and of itself would not call for changing Decom Year into a controlled node. However, for the purposes of this example, and to be able to change the value of Decom Year along a path in the decision tree, Decom Year is a good candidate for a controlled node.

▶ Click Full Screen.

▶ Click OK.

▶ With Decom Year still selected, click the Change Node Type icon.

DPL displays the Node Type dialog box.

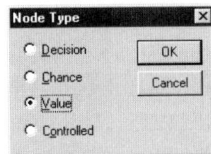

▶ Click the radio button labeled "Controlled", then click OK.

DPL displays the General tab of the Node Definition dialog box.

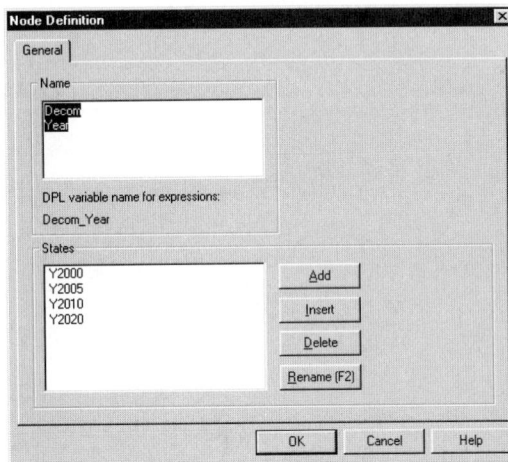

The node and state names do not need to change.

▶ Click OK.

DPL displays controlled nodes as gray rectangles in the influence diagram.

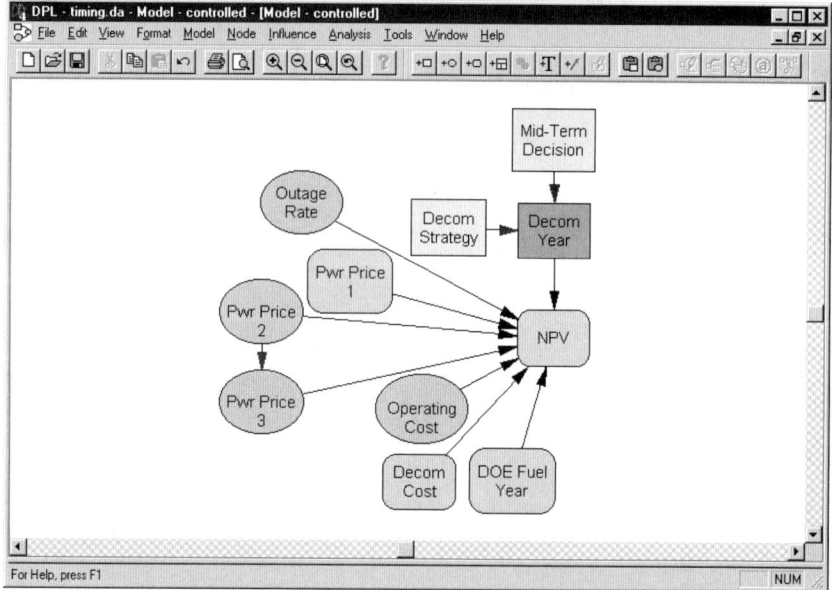

▸ Double-click on Decom Year.

DPL displays the Node data dialog box.

▸ Click Full Screen.

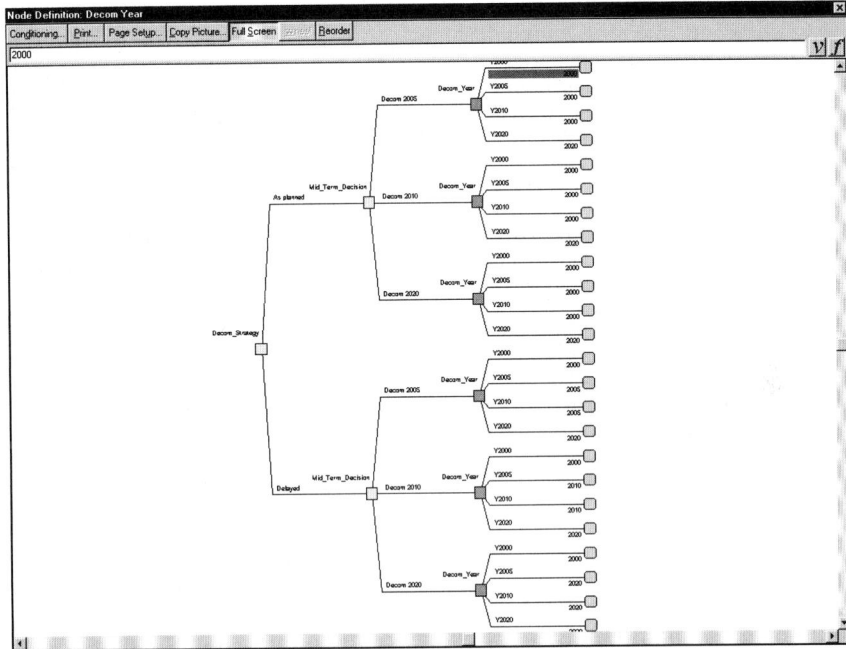

There is no need for Decom Year to be conditioned any longer.

▸ Click Conditioning.

DPL displays the Conditioning dialog box.

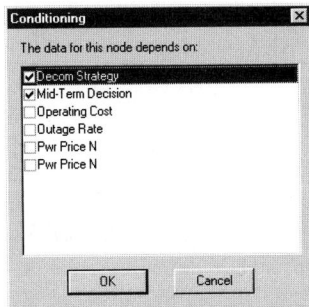

▶ Uncheck the checked boxes labeled "Decom Strategy" and "Mid-Term Decision".

▶ Click OK.

The node data is now greatly simplified.

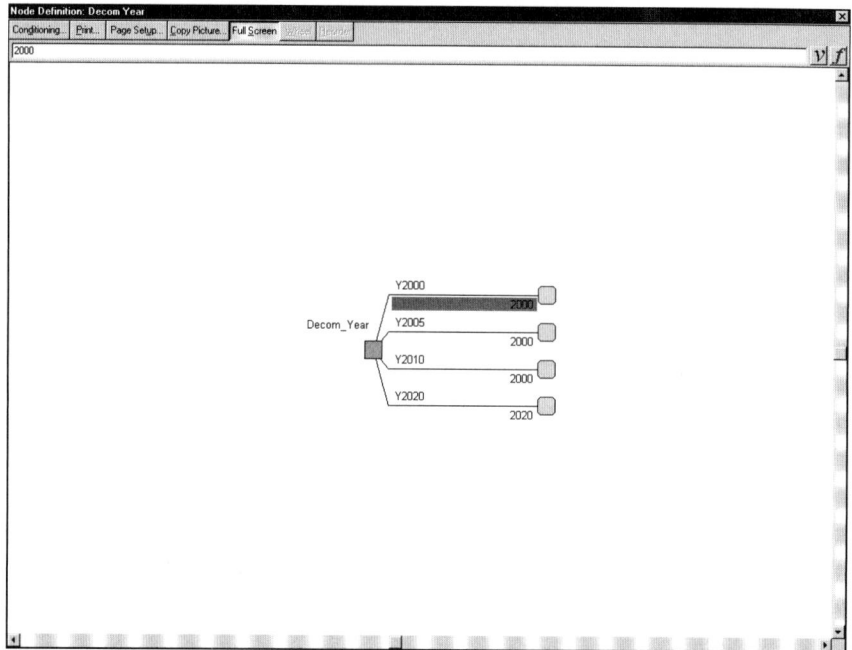

Next, the values representing the decommissioning year should be changes to reflect the different states of Decom Year.

▶ Click underneath the branch labeled "Y2005".

▶ Type "2005".

▶ Press Enter to move to the next line.

▶ Enter values of 2010 and 2020 to represent the states "Y2010" and "Y2020", respectively.

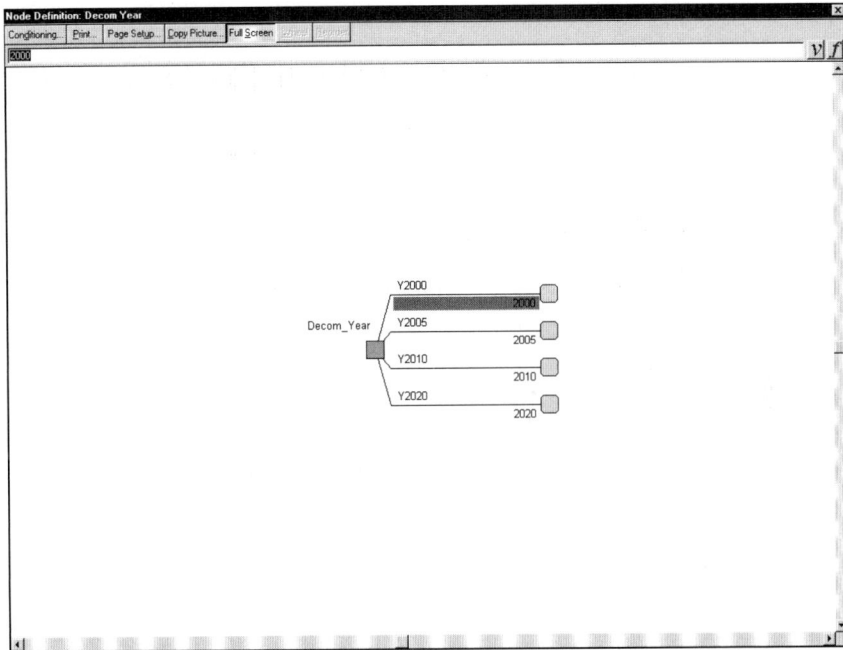

▸ Click Full Screen.

▸ Click OK.

Now the controlled node Decom Year is ready to be referenced in the decision tree.

▸ Select View Decision Tree.

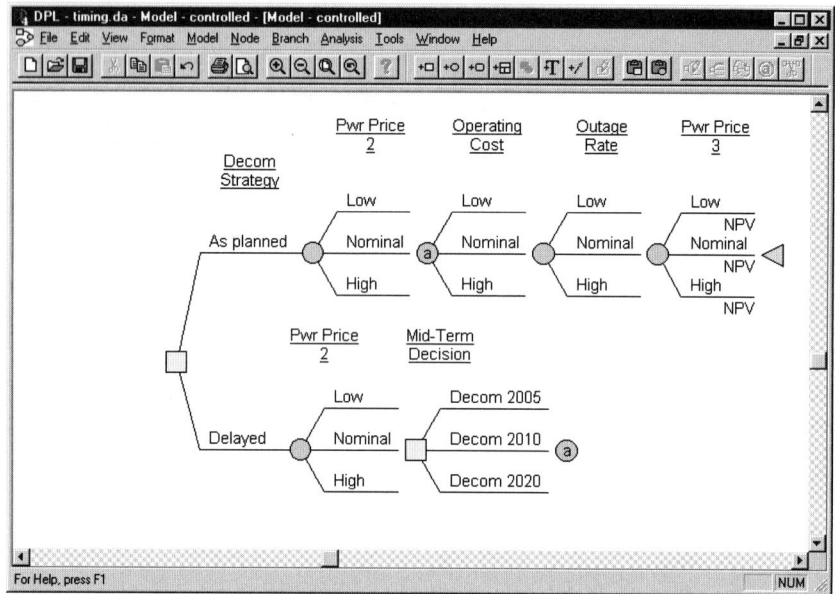

Controlled nodes do not appear in the decision tree. Instead, they are set on branches of other nodes. The process for setting the state of a controlled node is identical to that used earlier in the "Set Event" section. (**Note:** If you did not follow the steps in the Set Event section of this tutorial on page 308, you must select Tools Options Advanced and check the box labeled "Set states of events in decision tree" before continuing.)

When Decom Year was a value node, it was conditioned by the two decision nodes. Accordingly, the values for the controlled node Decom Year will be set on the branches of the decision nodes.

If the "As Planned" alternative of Decom Strategy is selected, the plant will be decommissioned in the year 2000. Therefore, Decom Year should be set to the state "Y2000" (which has a value of 2000) on the "As Planned" branch of Decom Strategy.

▸ Select the "As Planned" branch of Decom Strategy.

▸ Select Branch Control.

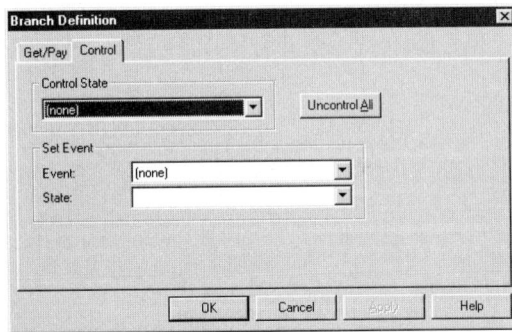

▸ In the Set Event section of the dialog box, use the drop-down menu to select Decom Year.

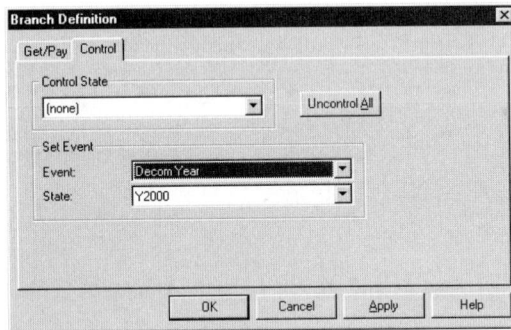

By default, DPL selects "Y2000" as the set state for Decom Year.

▸ Click OK.

As discussed in the section "Set Event", DPL does not indicate graphically that nodes are being set in the decision tree.

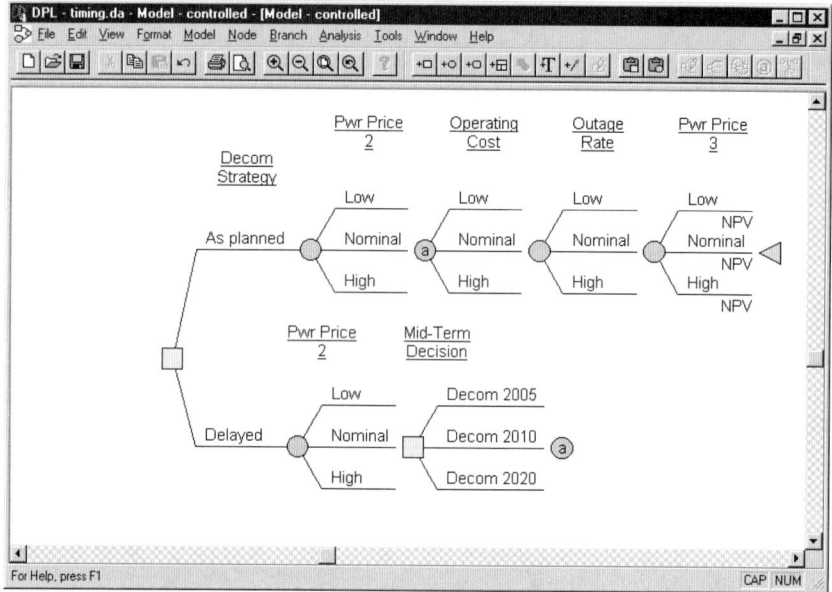

Similarly, Decom Year should be set to 2005 on the "Delayed" alternative of Decom Strategy.

▸ Select the "Delayed" branch of Decom Strategy.
▸ Select Branch Control.

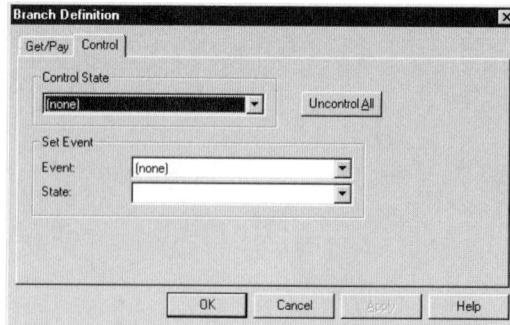

▸ In the Set Event section of the dialog box, use the drop-down menu to select Decom Year.

▸ Use the drop-down menu corresponding to "State" to select the state "Y2005", which has a value of 2005.

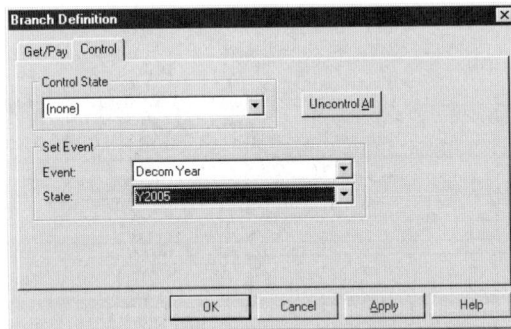

▸ Click OK.

If the "Delayed" alternative of Decom Strategy is selected, the Mid-Term decision will be made, which may change the decommissioning year. The next step is to reflect this in the model.

▸ Click on the branches of Mid-Term Decision.

DPL highlights all three branches of Mid-Term Decision.

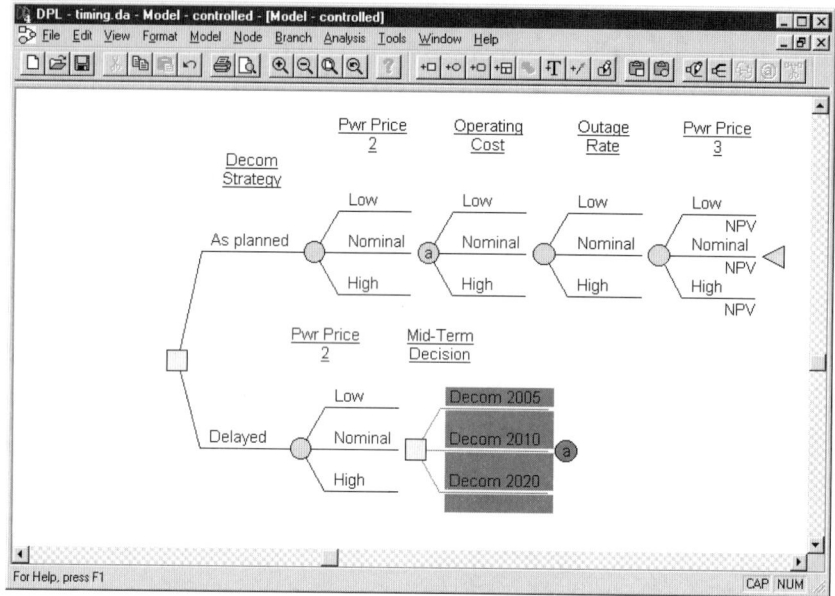

Mid-Term Decision is currently a symmetric node, which means that all its branches have to share the same Set Event setting. Because the different alternatives of Mid-Term Decision should set Decom Year to different states, Mid-Term Decision needs to be asymmetric.

▸ Double-click on the yellow square representing the Mid-Term Decision instance.

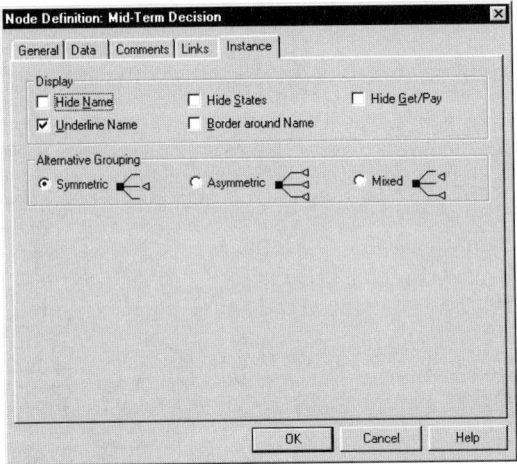

▸ In the Alternative Grouping section, select Asymmetric.

▸ Click OK.

DPL displays Mid-Term Decision as asymmetric.

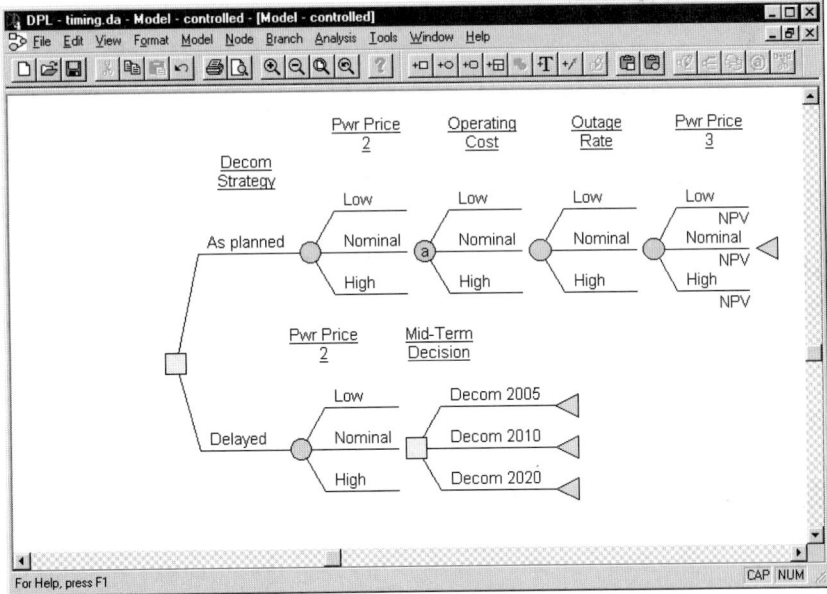

Because Decom Year has already been set to 2005 on the "Delayed" branch of Decom Strategy, there is no need to re-set it on the "Decom 2005" branch of Mid-Term Decision. It only needs to be set on the "Decom 2010" and "Decom 2020" branches.

- ▸ On the "Decom 2010" branch of Mid-Term Decision, set Decom Year to the state "Y2010".
- ▸ Repeat the previous step to set Decom Year to the state "Y2020" on the "Decom 2020" branch of Mid-Term Decision.

In making Mid-Term Decision asymmetric, the Perform Subtree link to Operating Cost was lost. The final step is to re-create this link.

- ▸ Click on the green circle representing Operating Cost.
- ▸ Select Node Perform Subtree.
- ▸ Place the crosshairs over the blue triangle at the end of the "Decom 2005" branch of Mid-Term Decision, then click the left mouse button.
- ▸ Repeat the previous steps to place Perform Subtree links on the end of the "Decom 2010" and "Decom 2020" branches of Mid-Term Decision.

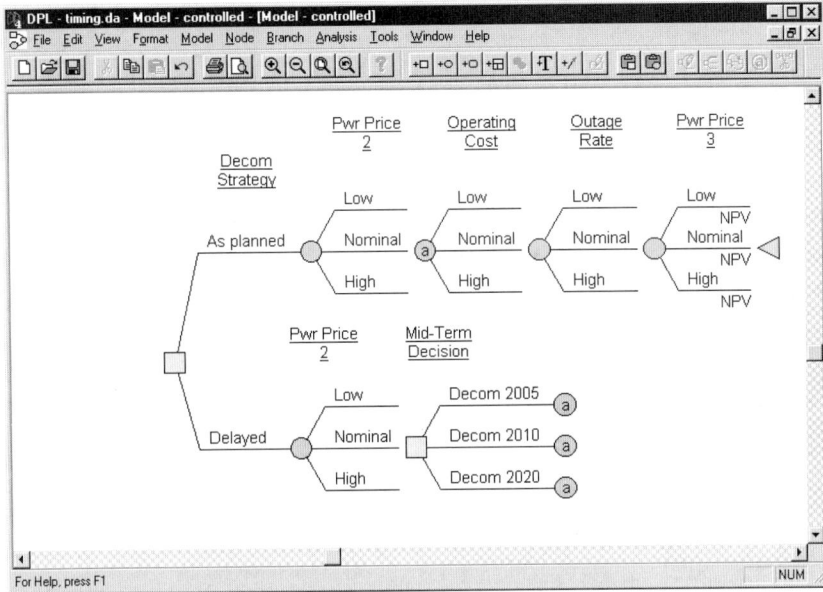

The decision tree is now complete, and will return the same answer as it would have if we had run when Decom Year was a value node with a symmetric decision tree.

As discussed in the chapter on Value Nodes, Get/Pay Expressions and Objective Functions, it is necessary to assign a value, or score, to each scenario in a decision model. These values are used to generate risk profiles and calculate expected values and optimal policies. For some models, it is straightforward to assign values to the scenarios. For other models, only a few simple calculations are necessary, and these are easily built directly into the DPL model.

For many models, however, calculating the scenario values is complex. This is especially true when the decision involves projects and events that will unfold over a period of many years and requires calculations of taxes, depreciation, and cash flow. Modeling these decisions using only DPL's value nodes would require huge influence diagrams with hundreds of value nodes. It is much simpler to use a spreadsheet, which is a familiar and convenient format, to perform these calculations.

Chapter 9

SPREADSHEETS

Chapter 9: Spreadsheets

Spreadsheet Links

How Does a DPL/Spreadsheet Link Work?

When you build a model and link it to a spreadsheet, you are linking specific DPL variables to specific spreadsheet cells. Each link is a one-way channel for sending data.

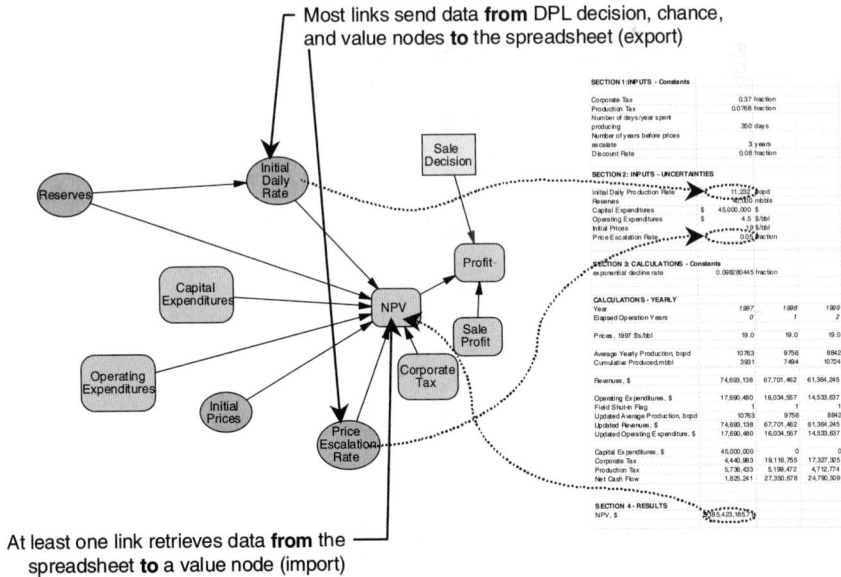

Most links send data **from** DPL decision, chance, and value nodes **to** the spreadsheet (export)

At least one link retrieves data **from** the spreadsheet **to** a value node (import)

During a decision or sensitivity analysis run, DPL essentially sends data to the spreadsheet and requests a calculation once for every unique endpoint, or scenario. For each endpoint, the export links send the spreadsheet the data appropriate for that scenario, using all available conditioning and formulas in the influence diagram.

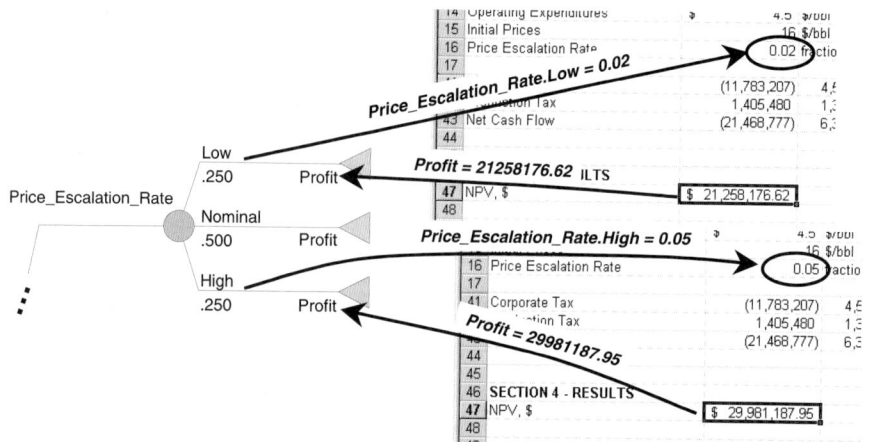

Preparing a Spreadsheet for Linking

DPL can link to almost any spreadsheet you can build, but there are a few guidelines that make linking and analysis easier:

- Identify and name the cell(s) that contains the output(s) of the spreadsheet. This (or these) is the final value the spreadsheet calculates, such as total profit, net present value of 10-year cash flow, etc. There must be at least one output cell, which usually contains a formula.
- Identify and name the cells that contain the inputs to the spreadsheet. These are the cells that define the scenario for which the spreadsheet is to calculate the output value. They are usually cells for input such as price per unit, capital cost, sales, market share, or number of years to complete construction. They must contain constants, and there must be at least one such cell.

- Identify and name cells that contain important intermediate calculations. These cells will be linked to dummy (intermediate) value nodes in the influence diagram; these nodes have no effect on the calculations, but may make the influence diagram easier to understand.
- Make sure the spreadsheet can calculate the output value for every scenario in the model, including scenarios that involve selling out, having demand exceed production capacity, or annual profits going negative for several years in a row.
- Design the spreadsheet to define and calculate one scenario at a time. For example, if Market Share is a three-state uncertainty in the DPL model, the spreadsheet should not calculate the output value for all three outcomes at once, but for one; DPL will tell the spreadsheet which one, get the output value for it, and then set up the next scenario. Think of the spreadsheet as a black box that the DPL model is using.
- Don't put scenario data, such as the three values for Market Share, in the spreadsheet. Put in one value, name the cell Market Share, and link it to the chance node in DPL, which should have the three values. You can't do sensitivity analysis on values stored in the spreadsheet.
- Group the input and output cells together to facilitate tracking, debugging, and documenting.
- Save the spreadsheet before building a model from it, DPL reads from the file on your hard drive, not what's displayed in Excel.

Linking a DPL Model to a Spreadsheet

There are two main ways to link a DPL model to a spreadsheet. When you are first building a model, the fastest way to link several variables is to have DPL build a deterministic influence diagram from the spreadsheet, then edit the model to add decision and chance events. After this, if you find you need to add or change a link, you can create individual links directly.

Building a Deterministic Influence Diagram from a Spreadsheet

First, start with a new, blank model window.

Second, identify the spreadsheet you wish to base your DPL model on.

Third, indicate which cells in the spreadsheet should have influence diagram nodes to link to. There are a few rules:

- You must have at least one export link (to a cell with a constant) and one import link (to a cell with a formula).
- DPL can only create nodes for cells with names. To link to an unnamed cell, you will have to create a link manually, as described in the next section.
- You can't "double dip." For any one scenario, DPL will send one set of data to the spreadsheet and receive one set of calculated values back, completing a loop. It cannot send some data, receive a calculated value, then send some more data and receive additional calculations back **for the same scenario.** If you need two loops, you'll need two spreadsheets.

Fourth, clean up the influence diagram. When it creates value nodes linked to the spreadsheet, DPL will place them in roughly the same relative positions as the cells in the spreadsheet, which usually makes an odd-looking diagram! Format Arrange Nodes will give you a good start.

Fifth, test the model by running a decision analysis. Because this is a deterministic model and has no decision tree, DPL will present a dialog box asking which variable you wish to calculate. Indicate the appropriate output variable. The "expected value" should match the calculated value of the original spreadsheet.

Linking One Spreadsheet Cell to a Node in the Influence Diagram

After you have built a model from a spreadsheet and have progressed with your modeling and analysis, you may need to create another link or two.

- **If the cell has a name**, choose Tools Add Linked Nodes. DPL will offer either to add a new value node for every named, unlinked cell in the spreadsheet or to present a selection box of currently unlinked, named cells.
- **If the cell does not have a name**, you can paste the link manually. First, select the cell in the spreadsheet and choose Edit Copy (or Ctrl-C). Then, go to the Links tab of the Node Definition dialog box of the appropriate influence diagram node and click the Paste Link button. Note: without a cell name, DPL will link to the absolute reference (e.g., E19), which will **not** change if you move things around in the spreadsheet.

Managing Spreadsheet Links

Once you've linked a spreadsheet to a DPL model, you need to be able to view, verify, and change them.

At the Model Level. To see a complete list of all spreadsheets linked to a model, select Model Links. Here you can see complete path information, as well as a count of the number of linked nodes. It is from here that you can convert a linked spreadsheet to DPL code (discussed later in this chapter.)

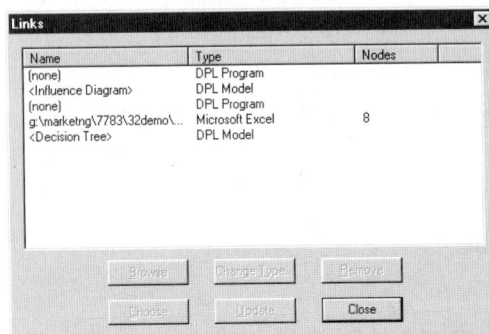

At the Node Level. To view the details on an individual linked node, go the Links tab of the Node Definition dialog box. Here you can see the spreadsheet and cell the node is linked to, and change it if you wish.

Editing a Linked Model

With only a few exceptions, a model linked to a spreadsheet can be edited and analyzed like any other model.

Adding New Nodes. You can add as many new nodes as you need. These nodes can condition the nodes that have export links to the spreadsheet. These nodes may not condition the nodes that have import links to the spreadsheet. The data for a new node may include a node with an import link in a formula.

Changing Linked Value Nodes to Decisions or Chance Events. You can change any node with an export link to any other type of node — decision, chance or controlled. You may **not** change a node with an import link; import links must be value nodes.

Adding or Deleting Influence Arrows. You can add or delete any influence arrow into or out of any node with an export link as you wish, except for the arrows leading into a node with an import link. You can add and delete influence arrows out of a node with an import link as you choose.

Arrows leading into nodes with import links are special. Such an arrow indicates that DPL needs to send data from the conditioning node before asking the spreadsheet to recalculate and bringing the result back to the import node. If there are **no** arrows into an import node, DPL assumes that **all** linked export nodes condition the import node. If there are any arrows into an import node, DPL assumes that **only** those nodes with arrows condition the import node.

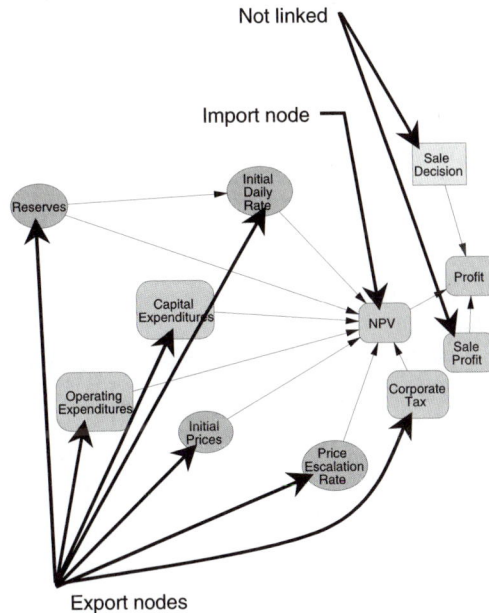

Editing Node Data. A node with an export link contains data just like any other node. This data can include constants and formulas, and may refer to other nodes in the model. A node with an import link **must not contain data** — DPL is going to fetch the data from the spreadsheet.

Running an Analysis

- DPL cannot reset the spreadsheet to its original state after an analysis; the spreadsheet will be left with the numbers from the last path calculated. If you wish to retain the original information, do not save changes when you leave the spreadsheet.
- If the spreadsheet is open in Excel when you run an analysis, DPL will use the version that is currently displayed. If the spreadsheet is not open, DPL will ask Excel to open the file on disk.

- If you wish to perform sensitivity analysis on a variable, it must be defined in your influence diagram; numbers in spreadsheet cells that are not linked to DPL are not available.
- You may notice that models linked to spreadsheets take longer to run than models of comparable size built entirely in DPL. This is because DPL is unable to optimize the calculations that the spreadsheet is doing.

Solving Problems

Because using spreadsheet links means having three software packages (Windows, DPL, and the spreadsheet) work together, the process works best if all three can "concentrate" on just this analysis. Try to have plenty of available memory and minimize the number of other applications, windows, and spreadsheets that are open at the same time.

You may see two common error messages:

- **DDE Time Out.** After DPL requests a value from the spreadsheet, it waits for a response. By default, it will wait 20 seconds before notifying you that there appears to be a problem. If you know that your spreadsheet application will require more than 20 seconds to recalculate the spreadsheet, you can tell DPL to wait longer in the spreadsheet tab of the Tools Options dialog box.
  If you see this message and you know that the spreadsheet has had enough time to recalculate, move to the spreadsheet application and check for error messages there.
- **DDE Server could not supply data.** This message means that DPL asked for a value and the spreadsheet application replied with an error code. The message box should include a "topic" and an "item" to help you look for the problem. Common problems include having incorrect path information or cell names.

Programming Your Own Links

The DPL-to-spreadsheet link uses a Microsoft Windows communication protocol called Dynamic Data Exchange (DDE). The links created by Tools Create Model from Spreadsheet, Add Node from Spreadsheet, and Paste Spreadsheet Link are handled automatically by DPL, and are all most DPL models will require. However, you can define your own links using the initiate, poke, request, execute, and terminate functions. This will allow you to run macros in a spreadsheet or link a DPL model to a Windows application other than Microsoft Excel or Lotus 1-2-3. Refer to On-Line Help or call DPL technical support for more information.

Tutorial: Spreadsheets

An analysis often begins with a spreadsheet model, and then DPL is used to evaluate the spreadsheet model probabilistically. There are a number of ways to create links between DPL and a spreadsheet. You can create a new model from a spreadsheet, you can link individual nodes in an existing DPL model to a spreadsheet, or you can automatically add additional linked nodes to an existing DPL model.

DPL can link to spreadsheet models created in either Microsoft Excel or Lotus 1-2-3. (The examples are based on Excel.)

The Spreadsheet

▸ Start Microsoft Excel (version 5.0 or higher).

Excel.exe

▸ Open PHARMA.XLS.

This is a greatly simplified six-year cashflow model, which calculates the profit generated by a pharmaceutical company from a drug it produces.

The first worksheet, titled "Inputs", contains the decisions, inputs, simple calculations, and fixed data. The inputs section contains the variables to be modeled in DPL. In order to link a cell to DPL, the cell must be named. Accordingly, all of these cells have names. A list of which cells are named can be found in the Define Name dialog.

▸ Select Insert Name Define.

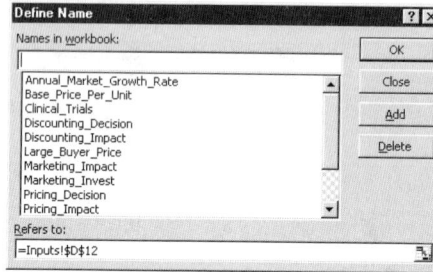

The second worksheet, titled "Calcs", contains the cashflow model.

▶ Click on the Calcs tab.

The Revenues section of the cashflow model calculates the revenues generated by the drug based on the market size, market share, and sales.

▶ Scroll down to view the Costs section of the worksheet.

The Costs section analyzes both the capital and operating costs associated with the drug. The profit for each year is also calculated here. Below the Costs section is the Profit calculation, which is just over $1 billion for the base case. The cell containing the Profit calculation is also named, and will be the value on which DPL makes decisions.

From a DPL standpoint, a spreadsheet has three kinds of variables: Import, Export, and Intermediate (or "Dummy") variables. Understanding the distinction is extremely helpful to understanding how DPL communicates with spreadsheets. The majority of nodes in your model will be Export variables. During a run, the values in these nodes are sent to the spreadsheet, and overwrite whatever data is currently contained in the corresponding cells. Accordingly, Export variables in DPL are input variables in the Excel model. In the spreadsheet, they are simply numbers which are used in the calculations. In this spreadsheet, the variables in the "Inputs" and "Decisions" section of the "Inputs" tab will be Export variables in DPL. Intermediate, or Dummy, variables, are nodes in DPL which represent calculations that are performed in the spreadsheet model. These nodes are simply placeholders in DPL that make it easier to understand the flow of data in the model. Because the calculations are actually performed in Excel, Intermediate nodes do not have an impact on the calculations during a DPL analysis. The final variable type is an Import variable. An import variable is a value that DPL will calculate in the spreadsheet and then bring back into DPL (a "final" calculation). Most models only have one Import variable, although you can have as many as you like. In this model, Profit is the only variable which DPL will designate as an Import variable.

Creating New DPL Models from Spreadsheets

The first method of linking DPL models to spreadsheets is simply using a spreadsheet to create a new model in DPL. (**Note:** This feature only works with Excel.)

▶ Start DPL.

DPL opens to a view of the Project Manager, Session Log, and Model Window.

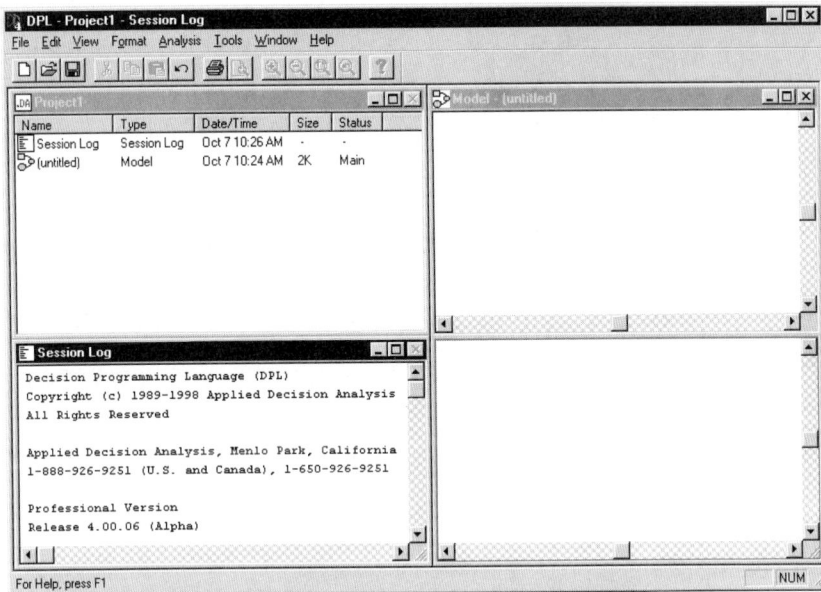

▶ Click the Maximize icon in the upper-right hand corner of the Model Window.

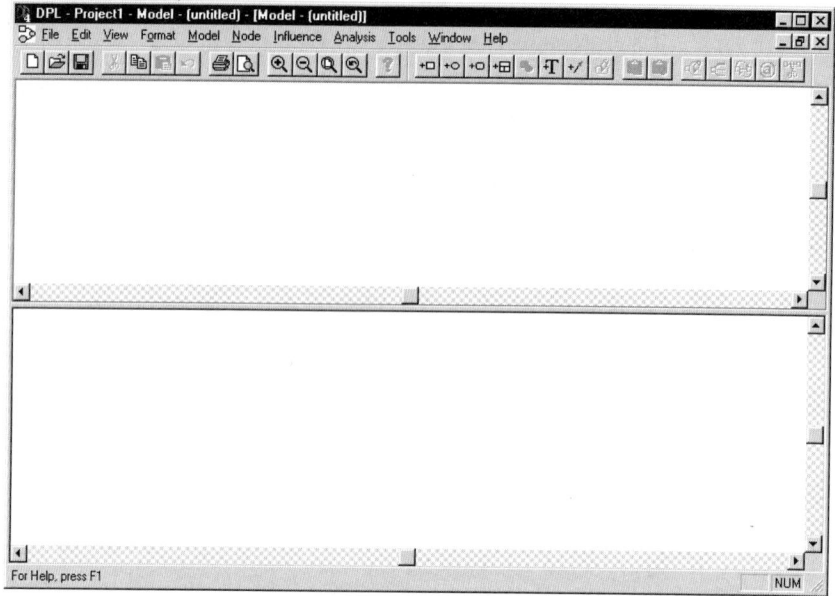

Models are created from Excel in the Influence Diagram, so it's helpful to first maximize the size of the Influence Diagram pane.

▸ Click-and-drag the splitter bar down to the bottom edge of the window.

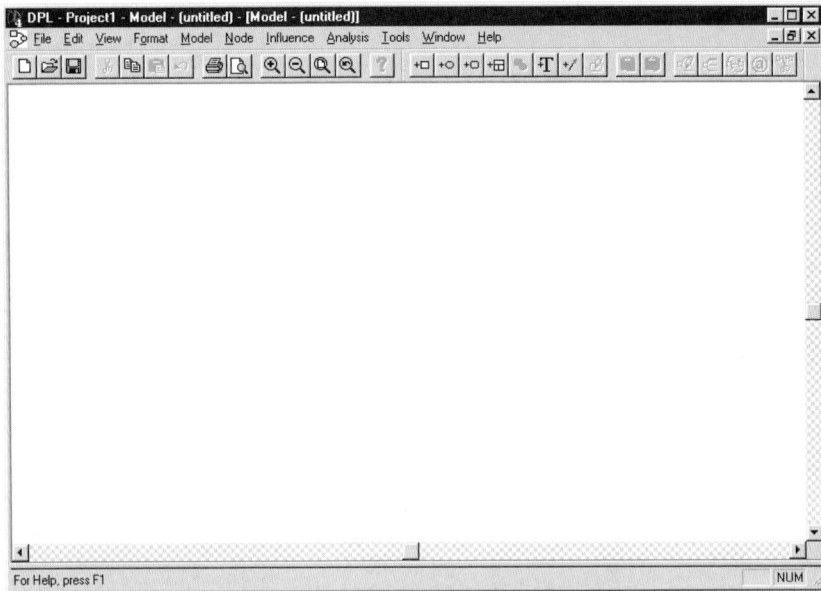

▶ Select Tools Create Model from Excel.

DPL displays the Create Model from Spreadsheet dialog box.

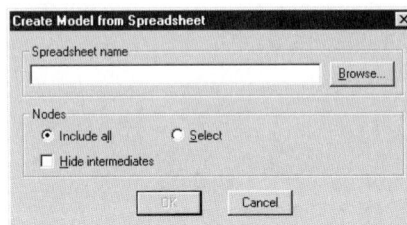

▶ Click Browse.

DPL displays the Input File Name dialog box.

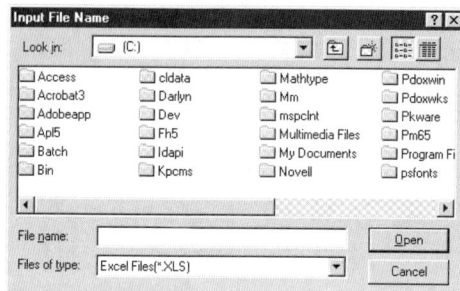

The spreadsheet has the path "c:\DPL\Tutorials\Pharma.xls".

▶ Select the spreadsheet, then click Open.

The Nodes section of the dialog box allows you to do two things. The first is to choose either "Include all" or "Select". Choosing "Include all" tells DPL to create a node for each named cell in the spreadsheet. If you choose "Select", DPL will display a list of all named cells in the spreadsheet, and you can choose which ones you would like to model in DPL. For a small model such as this, "Include all" is the better option. (If there are nodes which later you decide you do not want to model in DPL, it is very easy to delete them.)

The second option is a check box labeled "Hide intermediates". If this box is checked, DPL will not create nodes for Intermediate (or Dummy) variables representing calculations in the spreadsheet. While Intermediate nodes can make understanding the flow of data in the model easier, they can also confuse things by cluttering up the Influence Diagram.

▶ Check the box labeled "Hide intermediates".

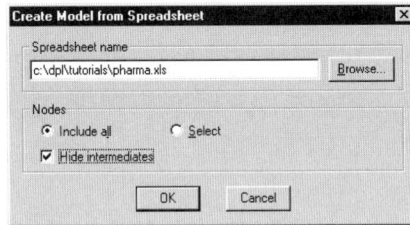

▶ Click OK.

DPL creates a node for each named cell and arranges the nodes to reflect the relative position of the corresponding cells in the spreadsheet. DPL also draws arrows between the nodes to reflect their relationships among the cells in the spreadsheet.

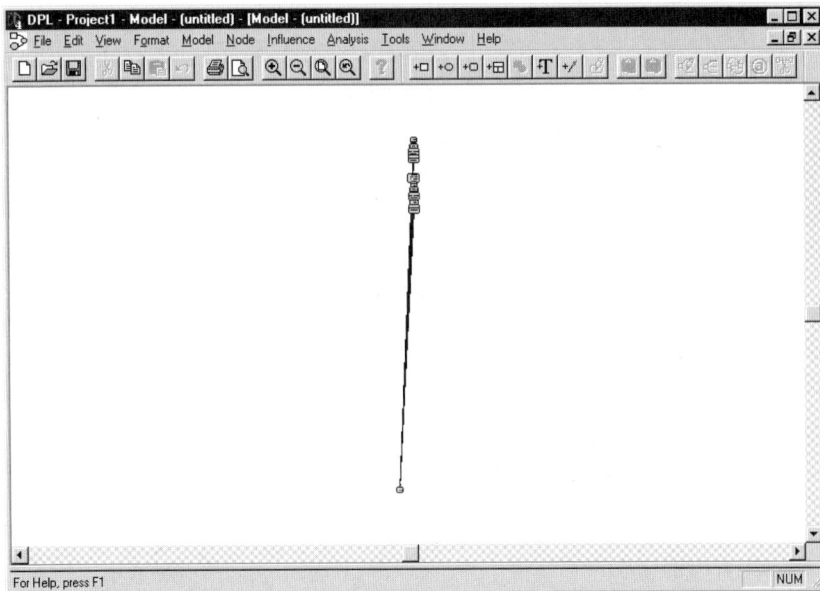

It is helpful to re-arrange the diagram.

▶ Select Format Arrange Diagram Left-to-Right.

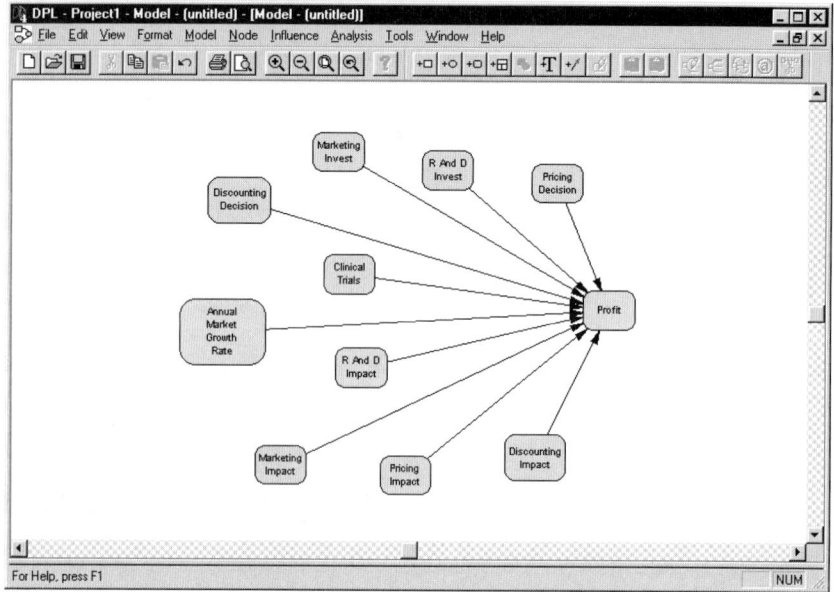

Each node has been given the same name as the spreadsheet cell to which it is linked.

Which Nodes are Linked?

There are two levels at which you can check which nodes are linked to a spreadsheet: at the model level and at the node level.

At the model level, DPL keeps track of spreadsheets linked to the model and how many nodes are linked to each spreadsheet.

▶ Select Model Links.

DPL displays the Links dialog box. The "Name" column tells you the filename and path of the spreadsheet to which the model is linked, the "Type" column tells you it's an Excel spreadsheet, and the "Nodes" column tells you how many nodes are linked to the spreadsheet.

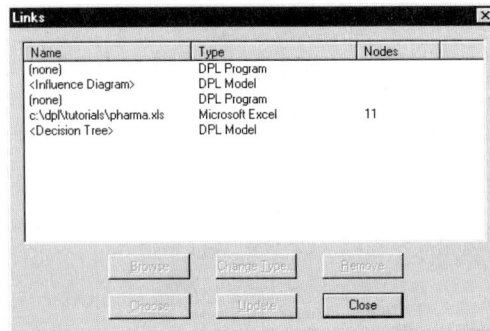

▶ Click Close.

You can also find out if an individual node is linked.

▶ Double-click on Pricing Decision.
▶ In the Node Definition dialog box, click on the Links tab.

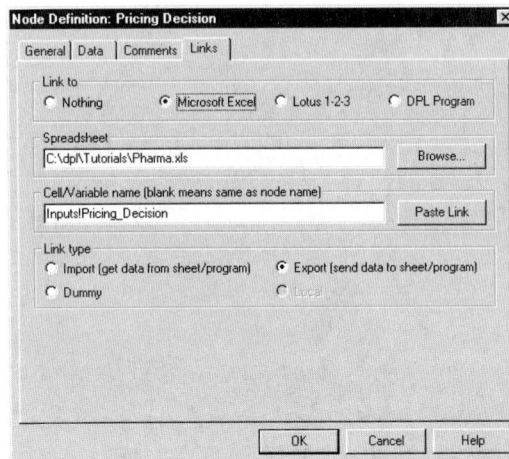

The Links tab gives you detailed information about the node's link:

- The "Link to" section tells you it's linked to Excel. If you wanted to unlink the node, you would simply click the "Nothing" radio button. When you clicked OK, DPL would remove the link.
- The "Spreadsheet" section tells you the filename and path of the spreadsheet. If you moved the spreadsheet or wanted to link the node to a different spreadsheet, you could type in the file name of the new spreadsheet location, or click Browse to find it.
- The "Cell/Variable name" section tells you the cell the node is linked to in the spreadsheet is located on the worksheet named "Inputs", and that the cell name is "Pricing_Decision".
- The "Link type" section tells you that Pricing Decision is an Export variable (and therefore the data in Pricing Decision will be used as an input to the spreadsheet). You can change the type of link to Import by clicking on the "Import" radio button, and then clicking OK. (**Note:** An Import node gets its data from the spreadsheet, so it cannot contain data in DPL. If you change the link type from Export to Import, DPL will warn you that doing so will erase the node's data.)

▸ Click OK.

While the Create Model from Excel function is a good way to quickly create a new model from an existing spreadsheet, it can only be used from a blank Model Window. If the Model Window is not blank, DPL will prompt you that it will erase all data from the window when it performs the function. There are other features which can add linked nodes to an existing model.

(**Note:** A copy of the imported version of this model can be found in the file "spreadsheets.da". The model is titled "Create Model".)

Adding Linked Nodes to an Existing Model

If you want to add linked nodes to an existing model and/or want to add linked nodes from Lotus 1-2-3 or a DPL Program, you can use the Add Linked Nodes feature.

Add Linked Nodes is also helpful if you delete linked nodes, then later decide you want them back in the model.

The Influence Diagram currently has 11 nodes.

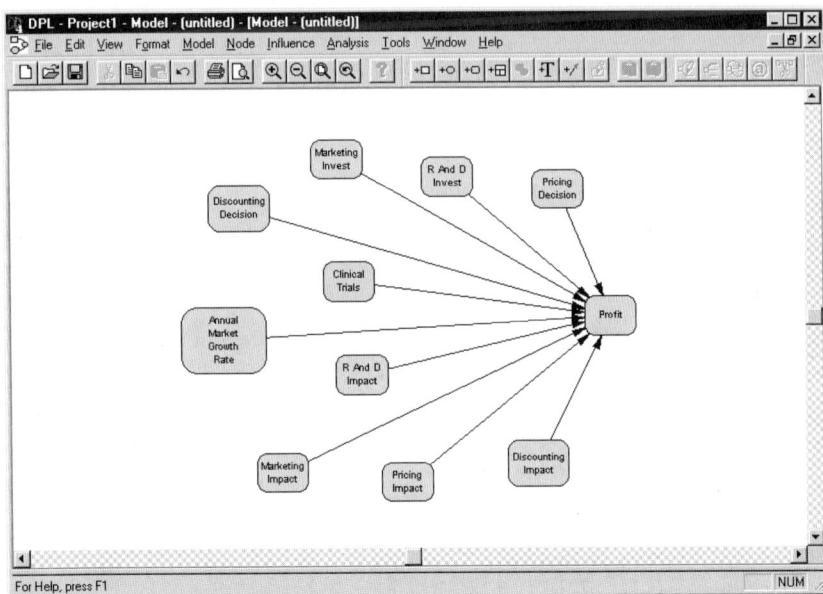

▸ Select the node Annual Market Growth Rate.
▸ Select Node Delete (or press the Delete key).
▸ Repeat the previous steps to delete the node Clinical Trials.

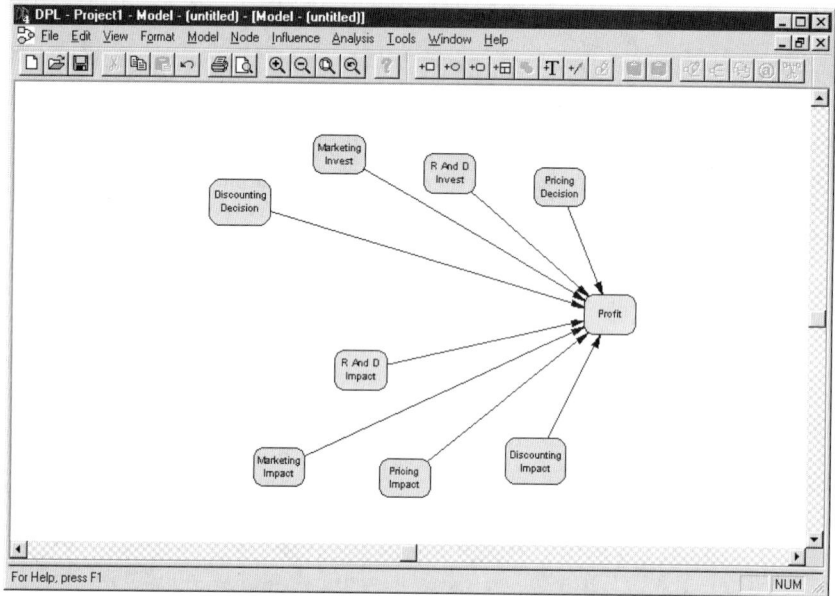

Let's say you want to add those nodes back into the model and have them linked to the spreadsheet. Of course, the easiest way to add these nodes back into the model is to use the "Undo" function. If you've made major modifications to the model since deleting the nodes, this would be an unsatisfactory solution! There are two ways to do this: the first is by using Add Linked Nodes, and the second is by using Paste Link, which will be discussed in the next section.

▸ Select Tools Add Linked Nodes, then select From Excel.

DPL displays the Add Nodes from Spreadsheet dialog box.

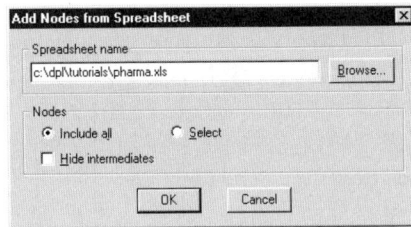

The default spreadsheet name is the spreadsheet to which the current nodes are already linked. The Nodes section has the same options as the Create Model from Spreadsheet dialog box, which was discussed earlier in "Creating New DPL Models from Spreadsheets." Choosing "Include all" will tell DPL to create nodes for all named cells in the spreadsheet which are not already linked to a node in DPL. (If a named cell is already linked to a node, DPL will not create a duplicate node.) Choosing "Select" will tell DPL to show you a list of the available named cells.

▸ Click the radio button labeled "Select", then click OK.

DPL displays the Select Nodes dialog box.

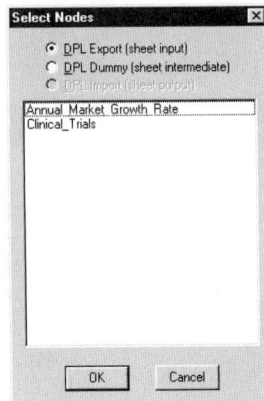

There are three radio buttons at the top: "DPL Export," "DPL Dummy," and "DPL Import." "DPL Export" is currently selected, indicating the variables shown would be created as Export nodes. DPL will only create nodes for the variables you select.

▸ Click on Annual_Market_Growth_Rate.

▸ Holding down the "Ctrl" key, click on Clinical_Trials.

DPL highlights the two variable names to indicate that they have been selected.

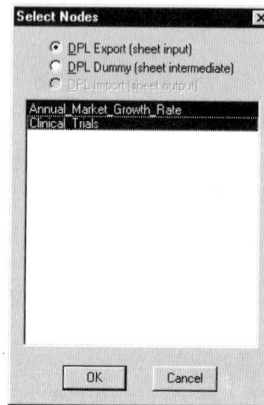

▸ Click on DPL Dummy.

There are also two variables which would be created as Dummy Nodes. If you wanted to create nodes for these variables, you would select them as well.

The radio button for "DPL Export" is grayed out. This tells you that DPL did not find any named cells which were potential DPL Export variables.

▶ Click OK.

DPL creates the new nodes in the Influence Diagram, and draws arrows from the nodes to Profit to indicate the relationship.

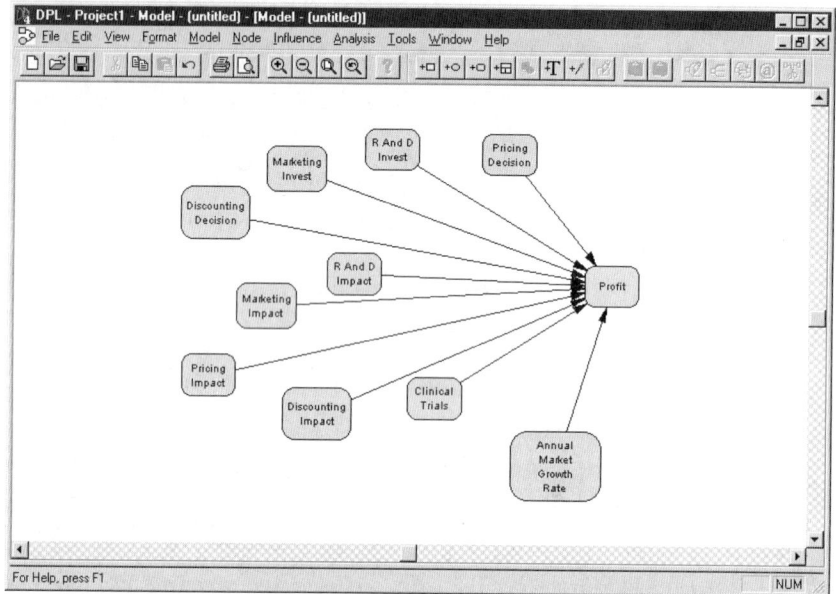

Final Note: If you decided that there were additional cells in the spreadsheet which you wanted to model in DPL, you would first name those cells in the spreadsheet (remember to save it afterwards!), then use the Add Linked Nodes feature.

Removing Links

As discussed earlier, a node's link can be removed in the Links tab of the Node Definition dialog box.

▸ Double-click on the node Clinical Trials.
▸ Click on the Links tab.

The Links tab displays the spreadsheet and cell to which the node is linked.

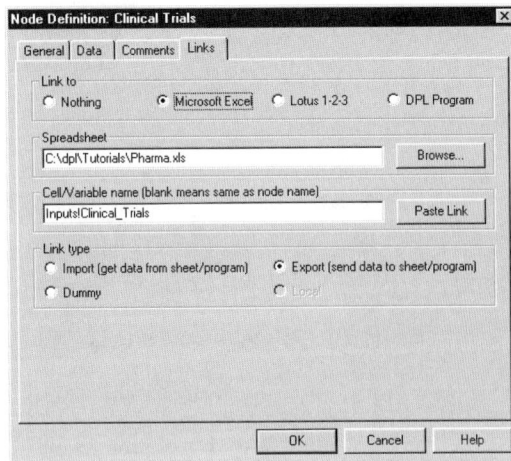

▶ In the "Link to" section, click the radio button labeled "Nothing".

DPL grays out the spreadsheet link, indicating that if you click OK at this point, the link will be broken. When you click OK, the link information will be deleted, and the node's data will not be sent to the spreadsheet during an analysis.

▶ Click OK.

The link can be re-instated using the Paste Link function, as described in the following section.

Paste Link

While the Create Model from Spreadsheet and Add Linked Nodes features are very useful for creating new linked nodes, they are not useful if you have existing nodes which you wish to link to the spreadsheet. For this, the Paste Link feature is most useful.

The link for Clinical Trials was removed in the previous section. You can use Paste Link to re-instate it. To use Paste Link, you "copy" the cell in the spreadsheet and "paste" the link into DPL.

▶ Double-click on the node Clinical Trials.

▶ Click on the Links tab.

The Links tab shows that the node is not linked.

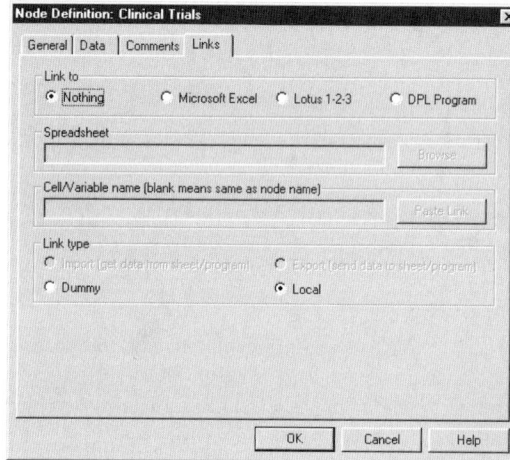

▶ Use the taskbar or Alt-Tab to switch to the Excel spreadsheet "Pharma.xls".

The cell named "Clinical_Trials" contains the value for the Clinical Trials Costs in millions of dollars. The current value is 200.

▶ On the sheet named "Inputs", select the cell named Clinical Trials (cell D9).

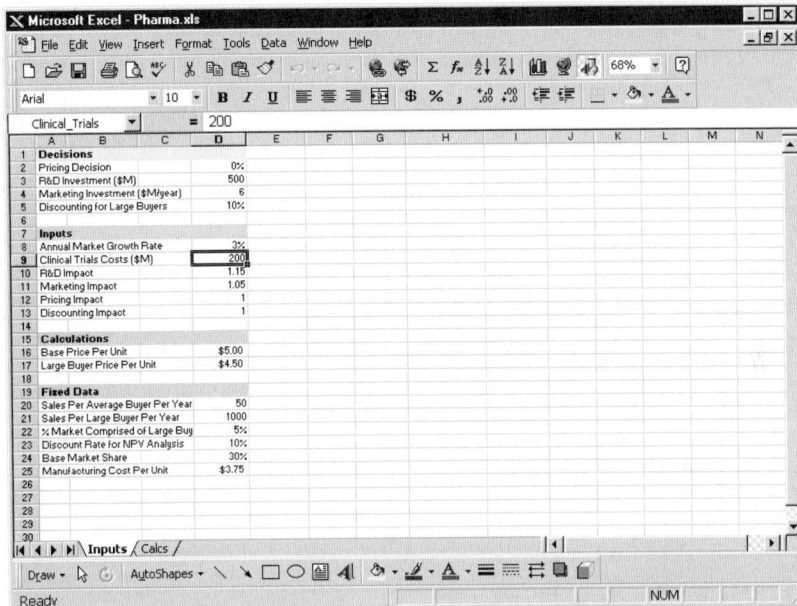

- From the Edit menu, select Copy.
- Use the taskbar or Alt-Tab to switch back to DPL.
- In the "Link to" section of the Links tab, click the radio button labeled "Microsoft Excel".
- Click Paste Link.

DPL copies the spreadsheet and cell information to the appropriate places in the dialog box.

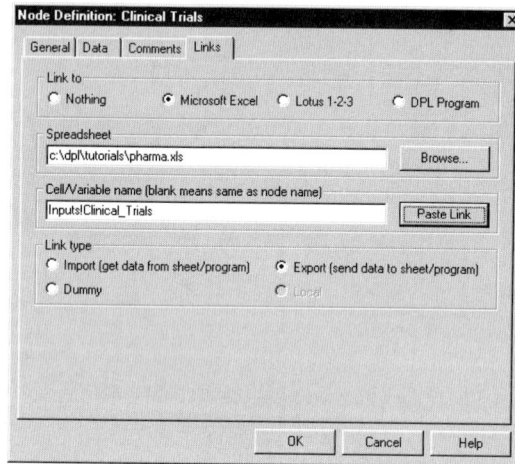

While the Paste Link feature copies the link information to the node, it does not copy the value. This is because it you were linking the cell to a chance or decision node, copying the same value to each state wouldn't make sense. Because Clinical Trials already contains the correct data, the node's data does not need to be updated.

▶ Click OK.

In addition to enabling you to build a wide variety of models, DPL provides a complete toolkit for evaluating them. This chapter explains these methods in detail, including calculating the expected value of a model, simulating a model, evaluating a model with multiple objectives, and using a utility function to model risk preference. Later chapters describe each of DPL's decision and sensitivity analysis outputs.

Chapter 10

Chapter 10: Evaluation Methods

How DPL Analyzes a Model

When you tell DPL to evaluate a model, whether it is to run a decision analysis or sensitivity analysis of any type, DPL follows the steps outlined below. Actually, depending on the type of run, DPL may rearrange or combine some of these steps. But this list should help you understand what's happening.

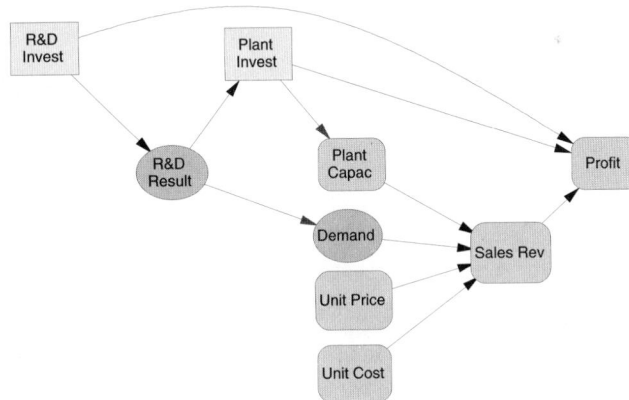

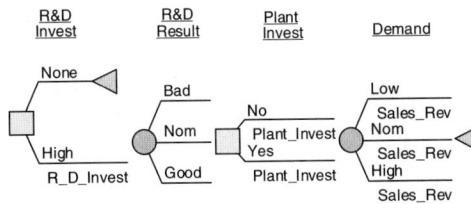

First, DPL builds a fully specified decision tree from the influence diagram and the schematic decision tree in the Model Window (or the program); this tree explicitly defines all possible scenarios.

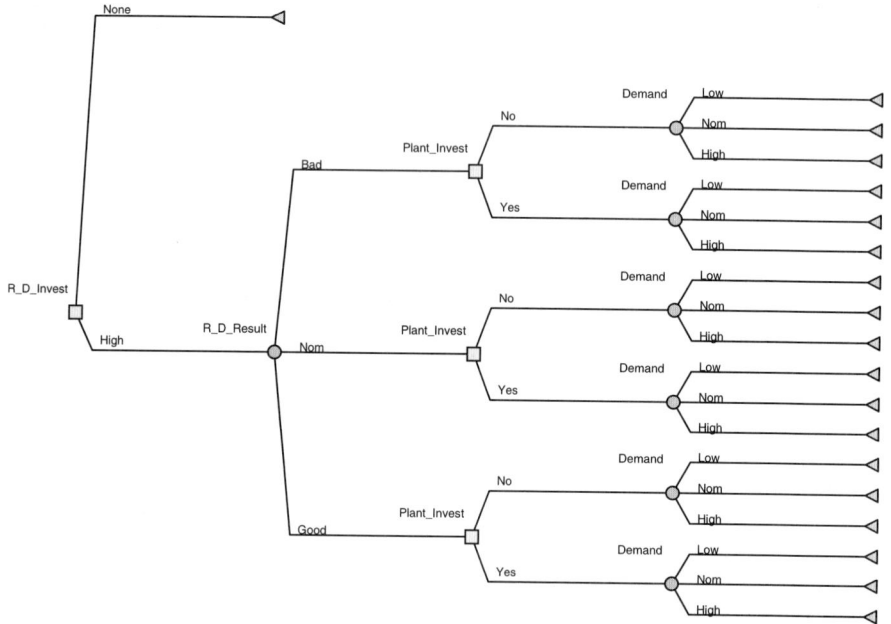

Second, DPL makes any structural modifications needed for simulation, sensitivity analysis, or value of information or control. These modifications are discussed in detail in the simulation section of this chapter and in subsequent chapters.

Third, DPL fills in the tree with probabilities and values, replacing all variable names with the appropriate numbers.

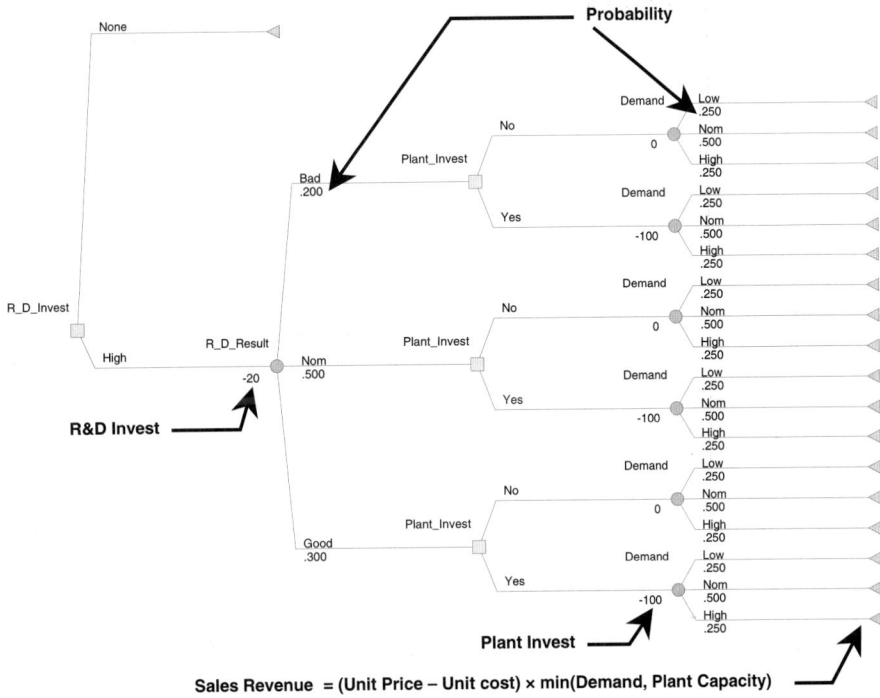

Probability

None

Demand — Low .250 / Nom .500 / High .250

No — 0

Plant_Invest

Bad .200

Yes — -100

Demand — Low .250 / Nom .500 / High .250

Demand — Low .250 / Nom .500 / High .250

No — 0

Plant_Invest

R_D_Invest

R_D_Result

High — -20

Nom .500

Yes — -100

Demand — Low .250 / Nom .500 / High .250

R&D Invest

Demand — Low .250 / Nom .500 / High .250

No — 0

Plant_Invest

Good .300

Yes — -100

Demand — Low .250 / Nom .500 / High .250

Plant Invest

Sales Revenue = (Unit Price – Unit cost) × min(Demand, Plant Capacity)

Fourth, DPL rolls forward through the tree, evaluating all formulas and running any external models to get the attribute values for each endpoint.

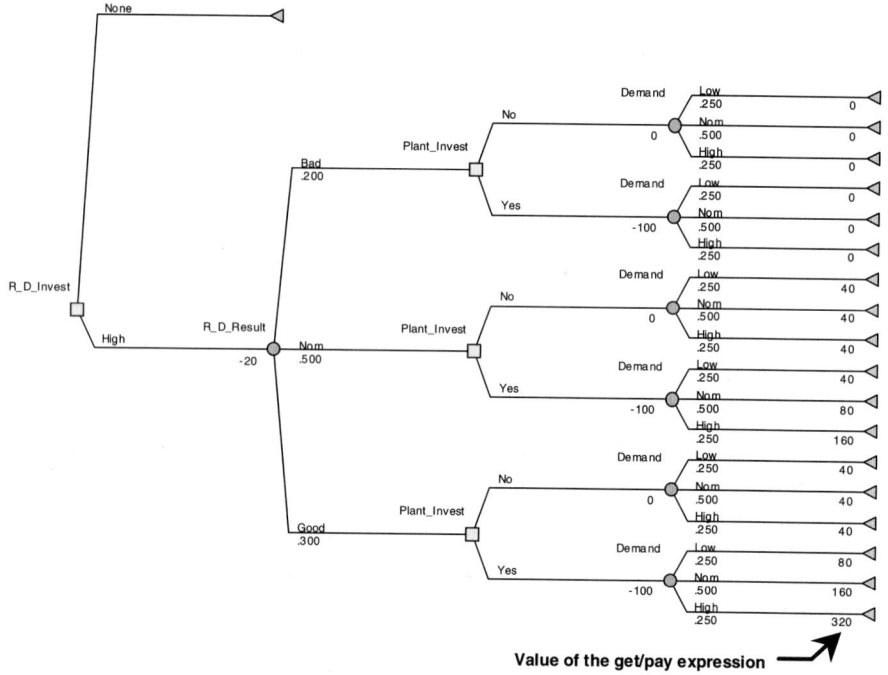

Value of the get/pay expression ➚

Fifth, DPL calculates the objective function value for each endpoint.

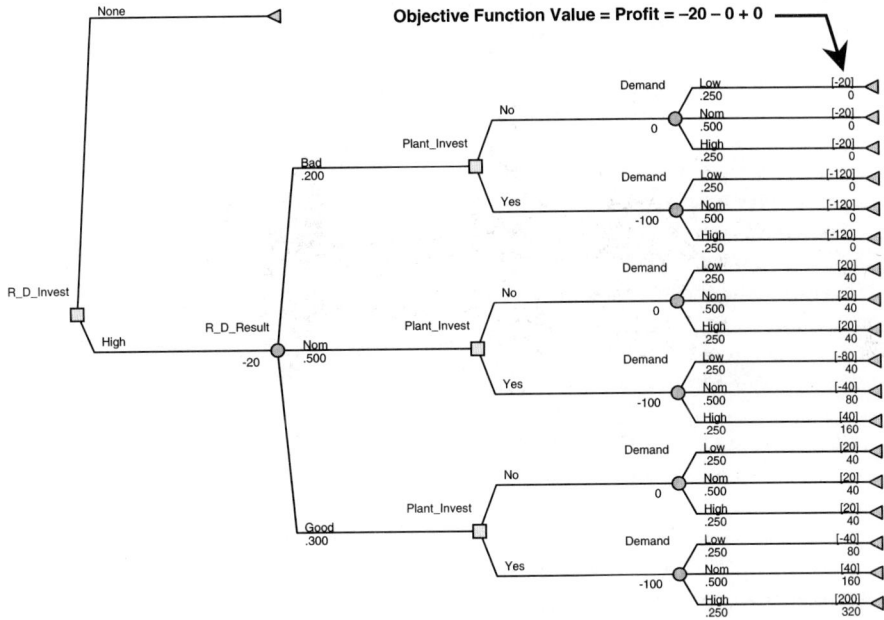

Sixth, DPL calculates the certain equivalent of the objective function value, if a utility function has been specified. See the example later in this chapter.

Seventh, DPL rolls back through the decision tree, calculating expected values and, if appropriate, certain equivalents.

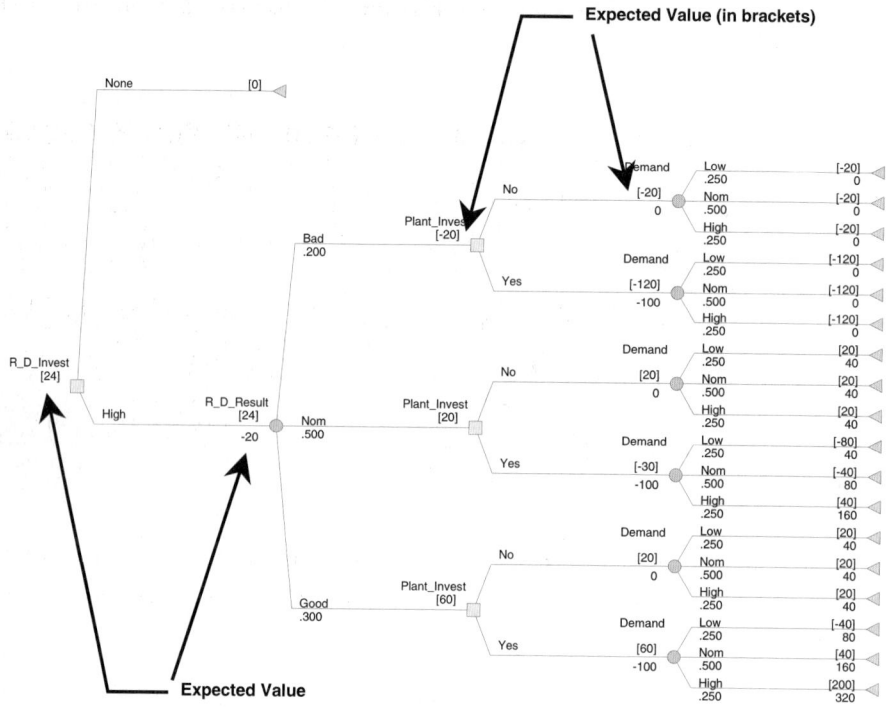

Expected Value (in brackets)

Expected Value

Eighth, DPL reports the results in the requested format — policy tree, risk profile, expected value of perfect information and control bar chart, tornado diagram, etc.

How DPL Calculates Expected Value

To calculate expected value, DPL starts at the endpoints of the fully expanded decision tree and moves to the left, node by node.

For a chance event, the expected value is calculated by multiplying the probability of each outcome by the value of that outcome, and summing all the outcomes.

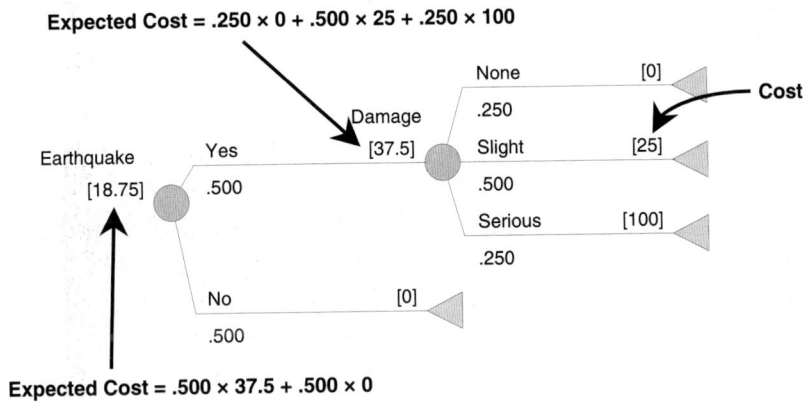

Expected Cost = .250 × 0 + .500 × 25 + .250 × 100

Earthquake
[18.75]

Yes
.500

Damage
[37.5]

None
.250
[0]

Slight
.500
[25] — Cost

Serious
.250
[100]

No
.500
[0]

Expected Cost = .500 × 37.5 + .500 × 0

For a decision node, the expected value is calculated by choosing the alternative with the best value, where best is the highest value, if the objective function is to be maximized, or the lowest value, if the objective function is to be minimized.

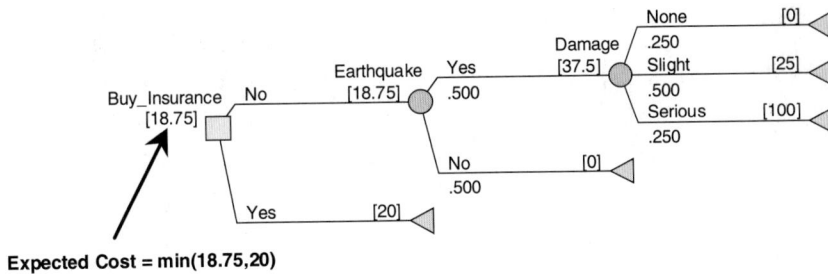

Buy_Insurance
[18.75]

No

Earthquake
[18.75]

Yes
.500

Damage
[37.5]

None
.250
[0]

Slight
.500
[25]

Serious
.250
[100]

No
.500
[0]

Yes
[20]

Expected Cost = min(18.75,20)

Tutorial: Calculating the Expected Value of a Model

This example shows you how to have DPL calculate the expected value of your model and write it to the Session Log. See the examples in the chapters on Decision Policies and Risk Profiles to learn how to have DPL display the expected value in the Decision Policy and Risk Profile Chart windows.

▸ Open expected_value.da.

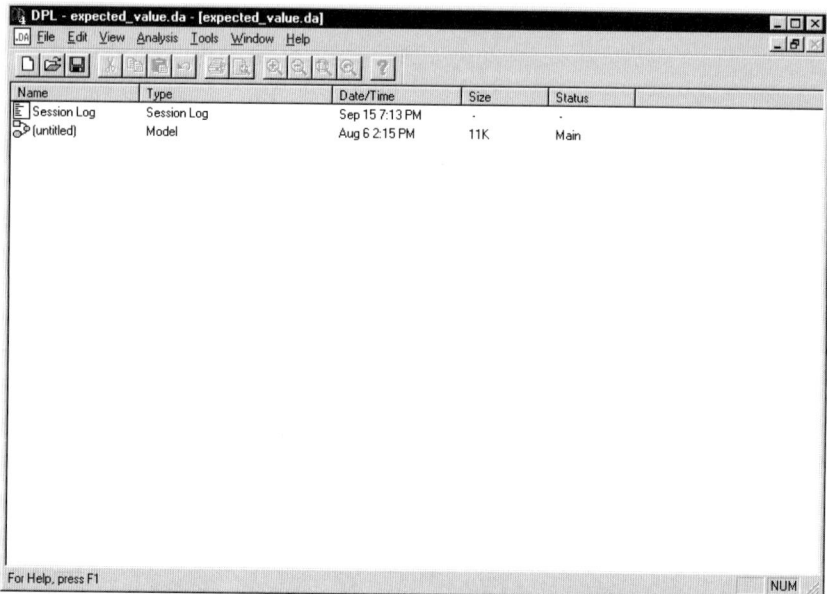

Name	Type	Date/Time	Size	Status
Session Log	Session Log	Sep 15 7:13 PM	.	.
(untitled)	Model	Aug 6 2:15 PM	11K	Main

▸ Select Analysis Decision Analysis.
▸ If both Excel and the spreadsheet model are not open, DPL will open them.

DPL displays the Decision Analysis Options dialog box.

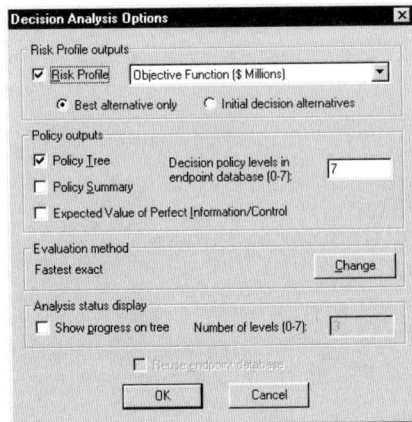

▸ If any of the outputs are checked, click on them to uncheck. The output boxes are labeled "Risk Profile," "Policy Tree," "Policy Summary," and "Expected Value of Perfect Information/Control".

▸ Click OK.

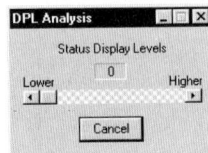

DPL displays the Status Display Level dialog box while it runs the analysis. You could use the slide bar to have DPL show which scenario it is currently evaluating. However, this slows down the analysis, so leave it at 0 for this simple model.

When DPL finishes the analysis, it creates an Endpoint Database, which contains the results, in the Project Manager.

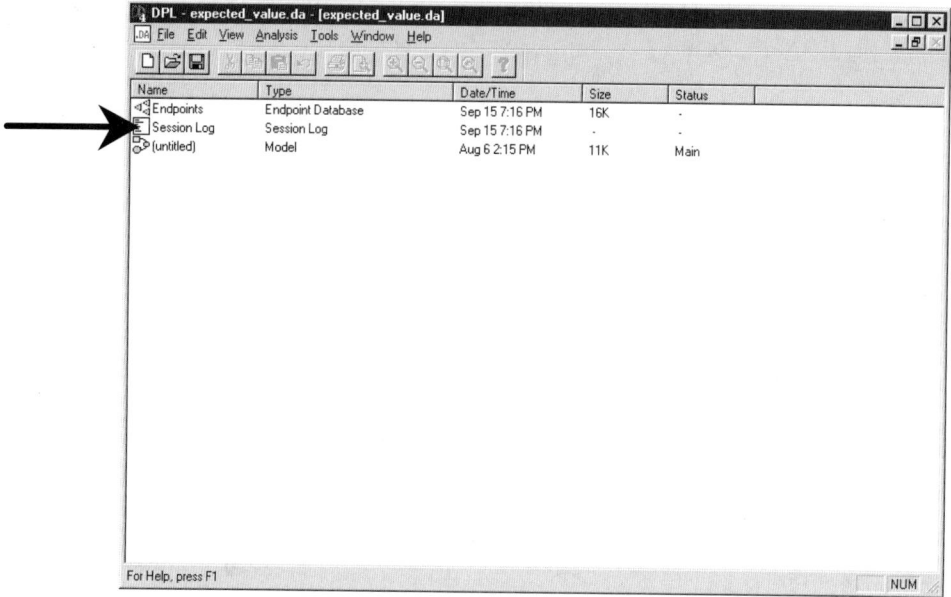

▸ In the Project Manager, double-click on the Session Log icon to display the Session Log.

```
DPL - expected_value.da - Session Log - [Session Log]              _ □ ×
File  Edit  View  Format  Analysis  Tools  Window  Help             _ ₧ ×

Decision Programming Language (DPL)
Copyright (c) 1989-1998 Applied Decision Analysis -- All Rights Reserved
Applied Decision Analysis, Menlo Park, Calif.  (888) 926-9251


Professional Version
Release 4.00.04 (Alpha)


19:17:55  Compiling: (untitled)
Number of paths = 324
19:17:55  Complete

19:17:56  Analyzing...
19:17:58  Complete

Expected value = 1442.2




For Help, press F1                                             NUM
```

The Session Log reports the number of paths in the model, the start and end time of
the analysis, and the expected value. In this case, the expected value is 1,442.2 (the
units are $ Millions).

How Can DPL Run So Fast?

The DPL engine includes several techniques for improving the performance of the evaluation, reducing run times. Most of these are simply computer science tricks that make the calculations more efficient but have no effect on the numerical results. Others make some assumptions about a model's structure to speed calculation.

Evaluation Methods

DPL offers three main evaluation methods. These methods can be applied to any model (with a couple of exceptions noted below), regardless of whether you are using one attribute or several and whether you are calculating expected value only, or expected value and certain equivalent.

Full Enumeration. This is the plainest, safest, and slowest calculation method. DPL explicitly builds and evaluates each and every path through the tree. This is the default evaluation option for models that use constraints or DDE functions.

Fastest Exact. This is the fastest way to run a full decision analysis. DPL checks the decision tree for sections that are structurally identical, which eliminates some duplicate calculations and speeds things up. This is the default for most models.

You can usually use this method if your model has only one attribute, no constraint function, and no DDE functions. You may not be able to use fastest exact if your model has a constraint function. If the function simply prunes irrelevant paths (such as those with zero probability) you can still use fastest exact. If the constraint eliminates otherwise relevant paths, and won't do it in exactly the same way everywhere (e.g., if the constraint depends on a value that depends on decisions or chance events earlier in the tree), you can not use fastest exact.

DPL's defaults are set to be very conservative, so, for example, if your model has any constraint function, the evaluation method will be set to full enumeration. You can experiment with your models, comparing the results with full enumeration and fastest exact, to see if you can safely use the faster option.

Simulation. This is usually the fastest way to evaluate a model, but you sacrifice some accuracy. DPL builds the full decision tree, but only evaluates a randomly selected subset of the paths. Simulation is discussed in detail later in this chapter.

You can choose the evaluation method each time you run a decision or sensitivity analysis. When you save a model, it remembers the evaluation method used last.

Other Evaluation Options

DPL has some other techniques that speed the evaluation of some of the common elements of decision models.

Expression Optimization speeds the calculation of formulas. Use this when you have a formula-intensive model, especially a converted spreadsheet.

Series Optimization speeds the evaluation of DPL variables defined as series. Use this if your model includes a converted spreadsheet.

Lottery Optimization speeds the calculation of chance events. Use this if your model has many chance events and conditional probability distributions.

Expression and series optimization are on by default. Lottery optimization usually requires a lot of memory, so it is off by default. You can change these settings using the Tools Options dialog box.

How DPL Runs a Simulation

Although DPL's engine has many tricks to improve performance, there are some models that are simply too big to run a complete decision analysis in a reasonable amount of time. (Depending on your circumstances, a "reasonable" amount of time may mean "less than an hour," "no more than overnight" or "within my lifetime.") In these cases, you have several choices — you can find a faster computer, you can reduce the size of the model, or you can run a simulation.

When DPL runs a simulation on a model, it starts by building the same decision tree it would build for a regular decision analysis. But instead of evaluating every path, it evaluates a subset of the paths, which reduces run time. It also reduces the accuracy of the results, although in practice it's remarkable how often the reduction in accuracy is small. You can control the run time/accuracy trade-off by specifying the number of paths that are evaluated — more paths improve the accuracy but increase evaluation time.

Samples

When you tell DPL to run a simulation on a model using 1,000 paths, DPL must somehow choose which 1,000 paths to evaluate. What method does it use?

One very simple approach might be to divide the number of simulation paths into the total number of paths and evaluate every nth path. For example, to simulate a 1,000,000 path tree using 1,000 paths, DPL could simply start at the top of the tree and evaluate every 1,000th endpoint. This might be a good approach if every endpoint had an equal probability of occurring, but such models are very rare.

Instead, DPL uses a combination of three techniques that use the probabilities in the model to select a representative subset of paths — to ensure that a path with a higher probability of occurring is more likely to be included in the simulated subset than a path with a lower probability. These techniques involve starting at the root of the decision tree with a pool of samples equal to the desired number of simulation paths, and moving through the tree node by node, allocating the samples according to the node probabilities.

- **At the beginning of the tree.** At the beginning of the decision tree, there are lots of samples and a few branches, so DPL uses the probabilities on the branches to directly allocate the samples, a technique called "distributed sampling." For example, if the simulation is started with 1,000 samples and the first chance event in the tree has branches with probabilities of {0.25, 0.5, 0.25}, DPL will allocate 250 samples each to the first and last branches and 500 to the second branch, as shown below.

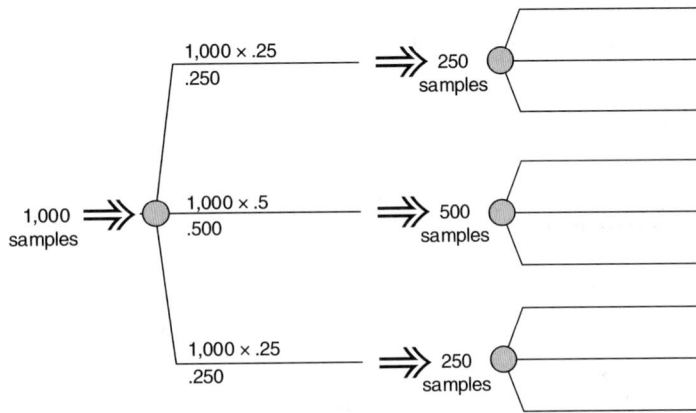

- **In the middle of the tree.** At some point in the middle of the decision tree, the number of samples allocated to a path gets too small to allocate evenly to the branches of the next chance event. When the number of samples is less than or equal to three times the number of branches of the next chance event, DPL allocates them to the branches using a technique called modified Monte Carlo (a form of stratified sampling). The number of samples assigned to each branch is determined by multiplying the number of samples by the branch probability and truncating the result to an integer. Any remaining samples are allocated randomly, as explained next.

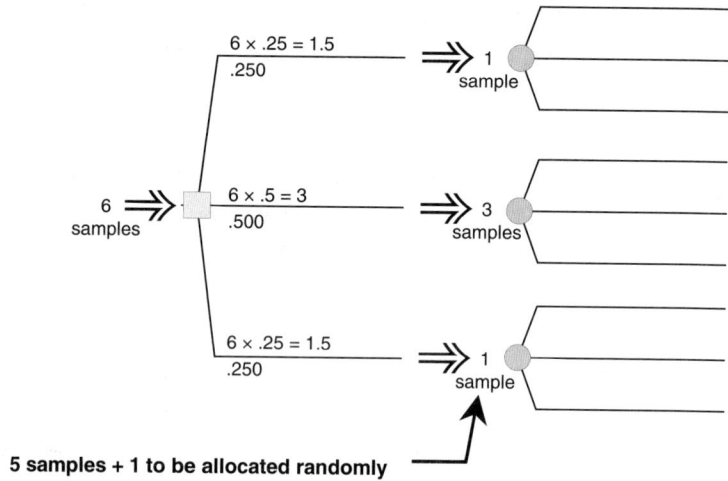

```
                        6 × .25 = 1.5           ⟹ 1
                          .250                     sample

        6 ⟹               6 × .5 = 3             ⟹ 3
     samples                .500                     samples

                         6 × .25 = 1.5          ⟹ 1
                           .250                     sample
```

5 samples + 1 to be allocated randomly

- **Towards the endpoints.** In many places in the decision tree, the number of samples allocated to a branch is only 1, but there are several chance events left before the endpoints. In this case, DPL chooses a single branch at each chance event, using a random number between 0 and 1 and the probabilities on the branches of the chance event.

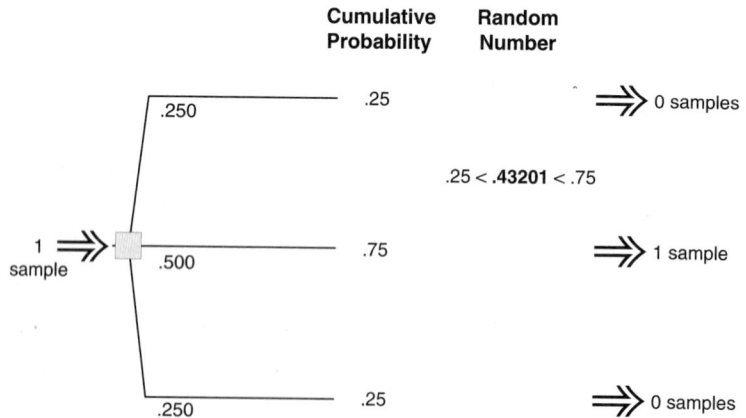

	Cumulative Probability	Random Number

```
                                        .25              ⟹ 0 samples
              .250
                                              .25 < .43201 < .75

    1 ⟹                                   .75              ⟹ 1 sample
  sample     .500

              .250
                                        .25              ⟹ 0 samples
```

Because the probability of any random number between 0 and 1.0 is equally likely, a branch with a probability of .5 is twice as likely to be allocated the sample as a branch with probability .25. This simple technique is called Monte Carlo simulation.

There is a fourth simulation technique available in DPL, called 'dont gamble,' which can be used toward the endpoints in place of the simple Monte Carlo sampling. If you recall, simple Monte Carlo sampling handles the task of simulating a chance node with one sample by randomly selecting one branch. Dont gamble sampling replaces the event with a one-branch event with the same expected value. This technique may improve the accuracy of the simulation's expected value, but it may also underestimate the variance of the outcome distribution. It is best to use this on models with mostly simple, unconditioned probability distributions, as a complex expected value calculation at each chance node may slow the evaluation significantly.

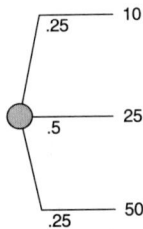

| Original Event | Monte Carlo | Dont Gamble |

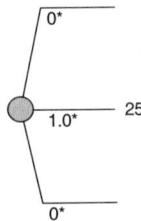

Once the samples have been allocated all the way to the endpoints, DPL calculates the endpoint value for each attribute and calculates the objective function value.

As DPL rolls back through the decision tree calculating expected values, it uses the number of samples on each branch to calculate the "simulated" probability distribution for each chance event. DPL marks these simulated probabilities with asterisks (*) on the policy tree. For example:

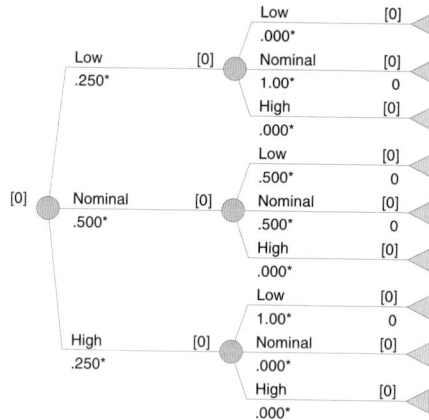

Note: For simplicity, we call this 3-technique combination 'distributed sampling,' and it is the default form of simulation. If you prefer to use Modified Monte Carlo at the beginning of the decision tree, you can specify the simulation method called 'Modified Monte Carlo.' If you would prefer to use the simple Monte Carlo approach throughout the tree, you can. Distributed Sampling is the most accurate and efficient simulation method.

Simulating Decisions

When there are decisions in the tree, simulation gets a bit trickier, because there are no probabilities to use to allocate samples. When DPL encounters a decision node with, say, 50 samples, it sends 50 down each decision branch. The good news is that the optimal policy will have been sampled with the full number of samples. The bad news is that runtime will be increased.

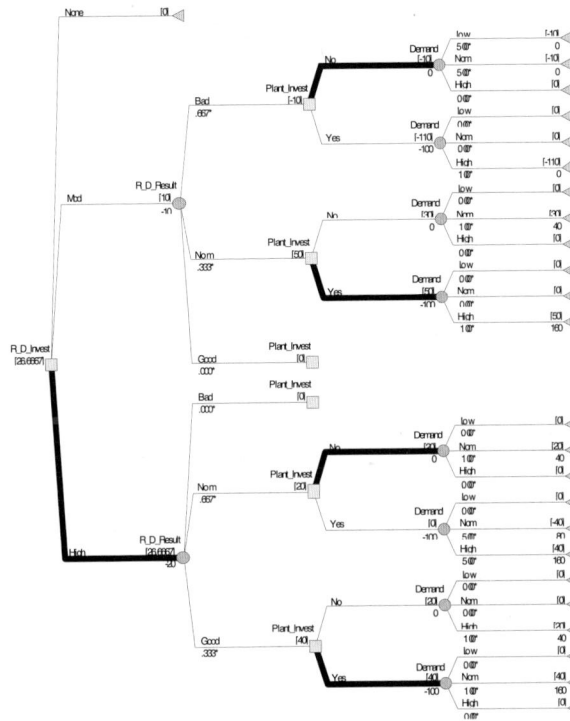

Simulating downstream decisions

If your decision tree has downstream decision nodes, DPL may encounter a decision on a path that only has one or two samples. Because decisions can be fairly discontinuous, you may not want the decision to be made on the basis of only a single sample. DPL offers a second simulation parameter called "decision restart" samples — this parameter allows you to specify a minimum number of samples to evaluate a decision with. These additional samples do not bias the probabilities (these new samples only get a fractional weight), but do increase run time, sometimes significantly.

How Many Samples?

The goal in choosing the number of samples is to find an acceptable balance between runtime and accuracy.

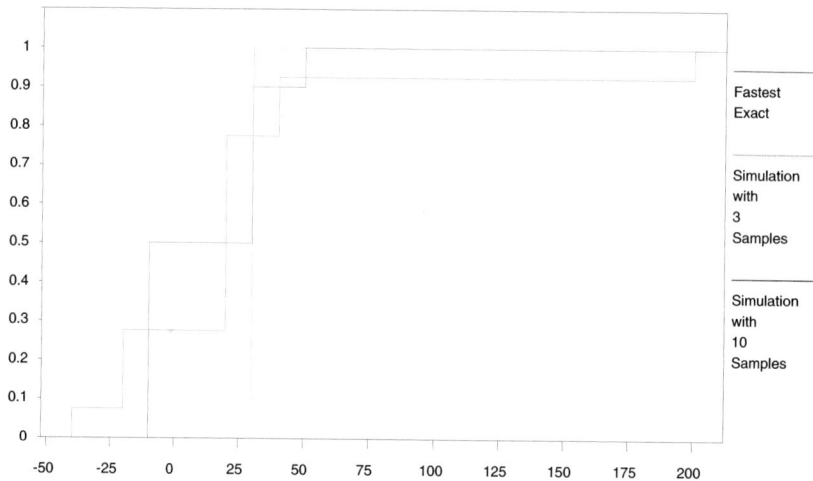

DPL's default is 1,000 samples, a number we have found to be a good starting point. Almost any model can run 1,000 samples in a minute or two, so you can get an estimate of how many samples you can run in, say, 5 minutes. Also, 1,000 samples is usually enough to verify that everything is calculating properly.

To estimate how accurate the simulation is, run it several times. If the expected values are significantly different for each run, the simulation has a lot of error. Try increasing the number of samples until the answer stabilizes (or the runtime gets too long). It is usually better to increase the total number of samples than to increase the decision restart samples, since the restart parameter can have an unpredictably large impact on runtime.

Tips for Successful Simulation

- If you are curious about how well a particular chance event is being simulated, look at the policy summary.
- Place important chance events, especially those with high consequence, low probability outcomes, as early in the decision tree as possible. Remember that at the beginning of the tree, DPL is better able to guarantee that these low probability paths get sampled.
- Try not to use simulation on models that have downstream decisions.
- If possible, try to avoid making a decision solely on the basis of simulation. Instead, use simulation to identify ways to reduce the size of the model (by eliminating obviously weak alternatives or chance events whose uncertainty doesn't significantly affect the results). Or, try to run a full decision analysis overnight or over a weekend. This will give you and your team a better handle on the robustness of the simulation results and more confidence when presenting your conclusions and recommendations to others.

Tutorial: Simulation

▸ Open simulation.da.

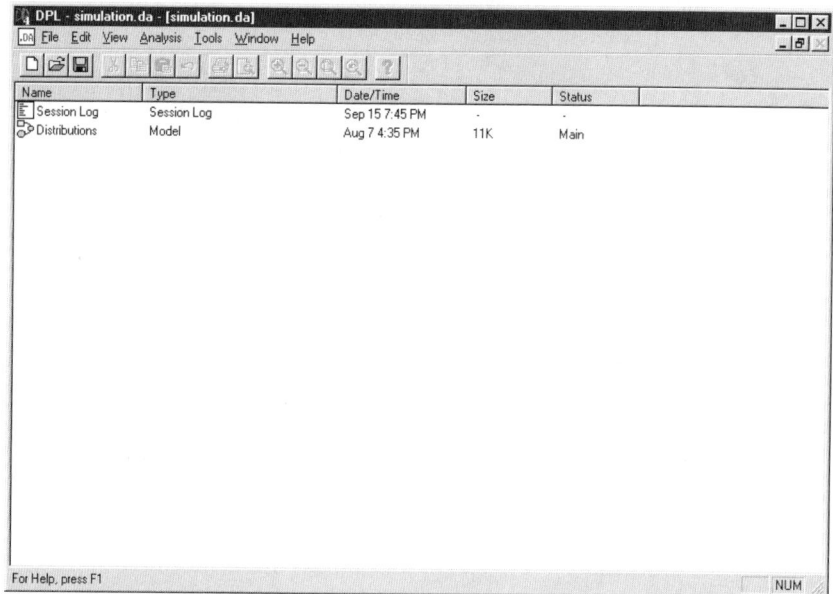

▸ Double-click on "Distributions".

With only 324 paths and a very simple spreadsheet model, you can run an analysis on this model in a few seconds. If the model had a greater number of paths or a larger spreadsheet, the run time would increase. To decrease the run time, you might choose to analyze the model using simulation. You will use simulation on this simple model for demonstration purposes.

▸ Select Analysis Decision Analysis
▸ If both Excel and the spreadsheet model are not open, DPL will open them.

DPL displays the Decision Analysis Options dialog box.

Evaluation Methods 387

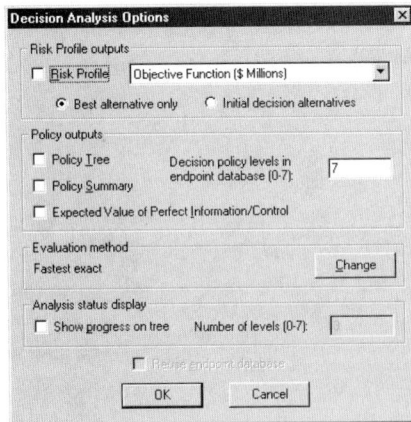

To run a simulation, you must change the evaluation method DPL uses to analyze the model.

▶ In the Evaluation method section, click the Change button.

DPL displays the Run tab of the Options dialog box. First, change the evaluation method to Monte Carlo simulation. Then, change the number of samples being used. Because this model only has 324 paths, you could fully evaluate the tree using the "Fastest exact" method to evaluate each path once. Thus, you want to choose sample numbers which will result in fewer than 324 paths (or else you are better off using Fastest Exact). Begin with 20 initial samples and 4 restart samples.

▶ In the Evaluation method section, click the radio button labeled "Monte Carlo simulation".
▶ In the Monte Carlo parameters section, change "Initial samples" to 20 and "Restart samples" to 4.

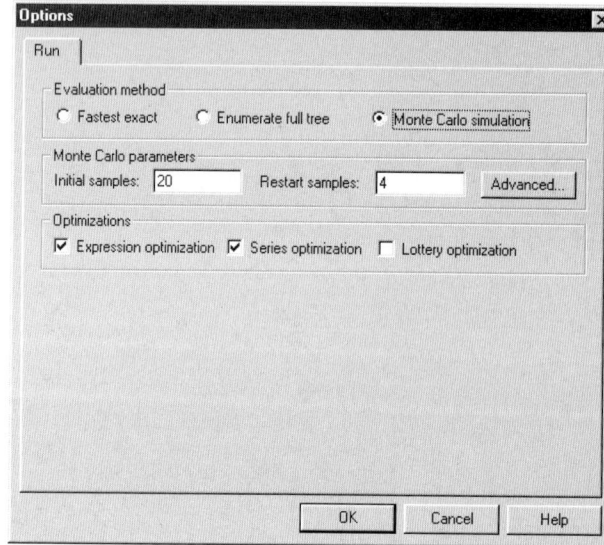

You can use the advanced Monte Carlo options to tell DPL how to handle multiple and single samples.

▶ Click the Advanced button.

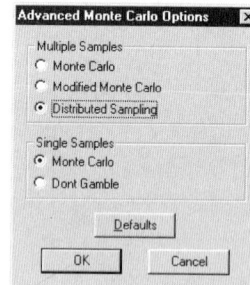

DPL displays the Advanced Monte Carlo Options dialog box. The default Multiple Samples evaluation method is Distributed Sampling, which provides the most accurate simulation results by using a combination of perfect sampling and stratified sampling techniques. The default Single Samples method is Monte Carlo, which will give the random results desired. Accept these defaults.

▶ Click OK.
▶ In the Run tab of the Options dialog box, click OK.

The Decision Analysis Options dialog box now shows that Monte Carlo simulation is the evaluation method.

▶ Click OK.

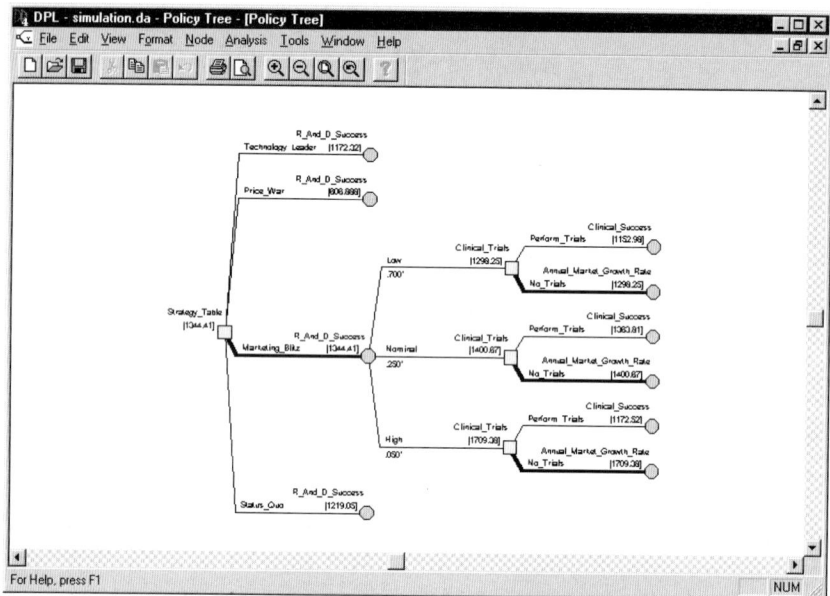

When DPL has finished running the analysis, it displays the Policy Tree. The Policy Tree indicates that the expected value is 1,530 (in $ Millions). From the previous example, you know the exact expected value of the model is 1,442 (in $ Millions). (**Note:** Due to the random nature of simulation, your answer will probably be different.) This simulation appears to give you a reasonable approximation.

Looking at the probability distribution on R And D Success, you see that the branch probabilities are given as 0.6, 0.3, and 0.1. These are the probabilities entered into the node when building the model. This means that when DPL reached R And D Success in the analysis, it had enough samples that it could use the exact probabilities rather than perform a Monte Carlo analysis. In this case it had 20 samples, so 12 went down the Low branch, 6 went down the Nominal, and 2 went down the High.

With only 2 samples entering the node Clinical Trials after the High outcome of R And D Success, DPL increases the number of samples to the Restart Sample value of

4 when it reached Clinical Trials. It therefore evaluated Clinical Success using 4 samples.

▶ Find the instance of Clinical Trials at the end of the High branch of R And D Success. Go to the end of the branch "Perform_Trials", and double-click on the green circle representing the chance node Clinical Success.

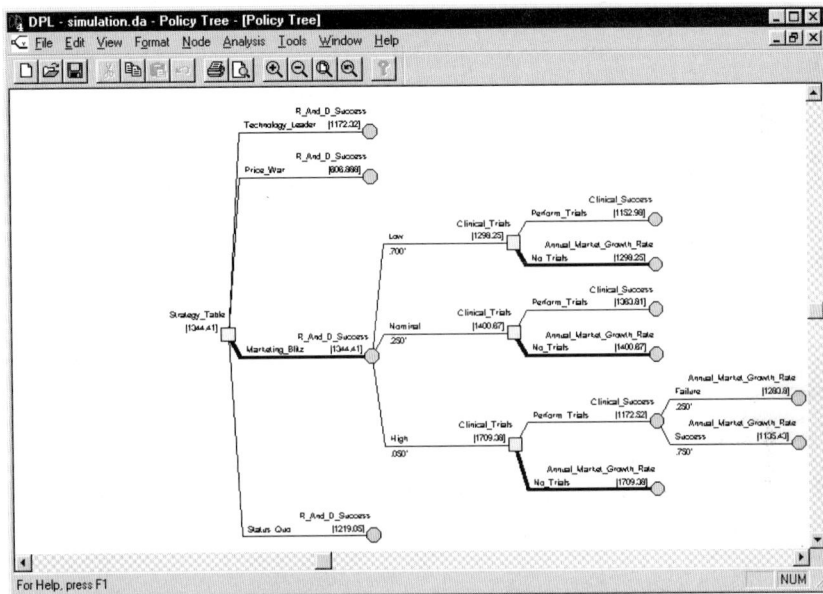

The probabilities of Clinical Success have asterisks, indicating that they are the probabilities determined through the simulation rather than the probabilities specified in the model. The Failure outcome occurred 25% of the time, while Success occurred 75% of the time. (In the model, the actual probabilities were specified as 20% and 80%.) You can deduce that 1 of the 4 samples was sent down the Failure branch and the remaining 3 samples were sent down the Success branch.

▶ Select Window Session Log.

```
DPL - simulation.da - Session Log - [Session Log]                    _ □ ×
File  Edit  View  Format  Analysis  Tools  Window  Help              _ 8 ×

Decision Programming Language (DPL)
Copyright (c) 1989-1998 Applied Decision Analysis
All Rights Reserved

Applied Decision Analysis, Menlo Park, California
1-888-926-9251 (U.S. and Canada), 1-650-926-9251

Professional Version
Release 4.00.06 (Alpha)


15:44:30  Compiling: (untitled)

15:44:46  Compiling: (edited) (untitled)
Number of paths = 324
15:44:47  Complete

15:49:21  Analyzing... Monte Carlo simulation
Initial samples 20    Restart samples 4


End points = 97

15:49:23  Complete

For Help, press F1                                                   NUM
```

The Session Log displays the number of paths in the full model (324), the number of
end points evaluated in the simulation (128), and the expected value calculated from
the simulation (1,452). (**Note:** The number of endpoints will also vary from run to
run.) The number of endpoints is much greater than the 20 samples we specified.
This is caused by the restart samples parameter.

You should run the model a few times with the parameters to verify that you
consistently get reasonable results. If you do not, you can increase the number of
samples and repeat the process.

How DPL Calculates Multiple Attribute Models

When evaluating a model with multiple attributes, DPL keeps track of each attribute separately during the roll forward, then combines them with the objective function at each endpoint. The rollback is done only with the objective function value. (For help building a multi-attribute model, refer to Chapter 7: Multiple Attribute Models.)

For example, consider a very simple model with two attributes, cost and solid waste reduction. The decision is a simple choice between an inexpensive, but wasteful, production process and a more expensive process that produces less waste.

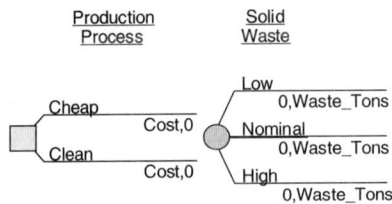

As DPL rolls forward through the decision tree, it keeps track of both cost and waste for each path. If you use a single get/pay expression, you will see the values for both attributes under each endpoint and in the decision policy.

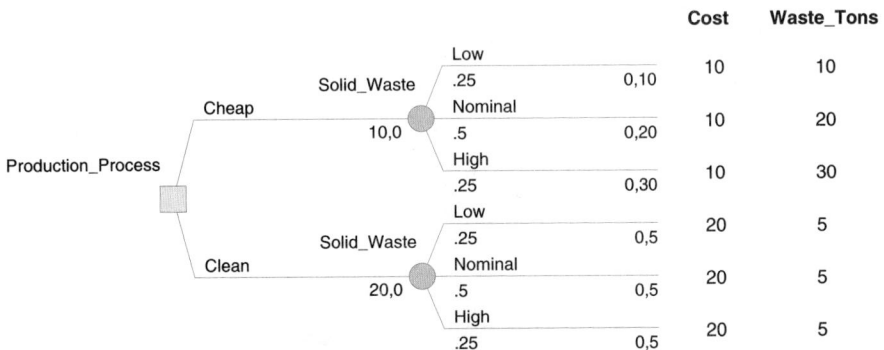

Calculating endpoint values for each attribute

Before DPL can roll back through the tree and calculate expected values, it must convert the attribute values into a single number for each endpoint, which it does with the objective function. The objective function value for each endpoint is shown above at the branches on the decision policy.

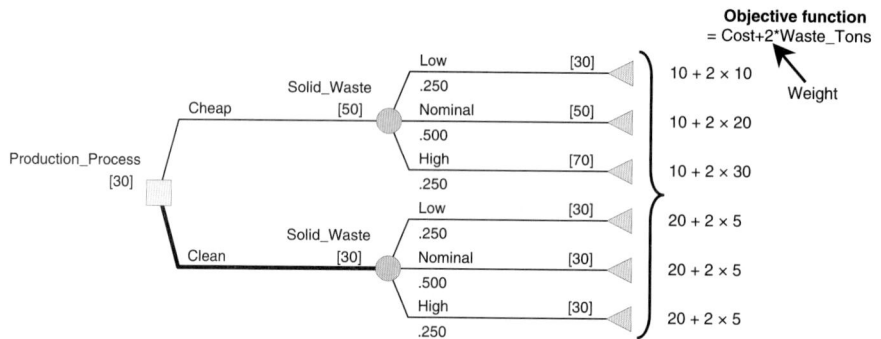

Calculating the objective function and expected values

For all sensitivity analyses, value of information, and other outputs, the values shown are objective function values.

You can graph risk profiles for any individual attribute or for the objective function values. This can be useful if, for example, you want to compare the cumulative risk profile for the least cost policy and the least waste policy.

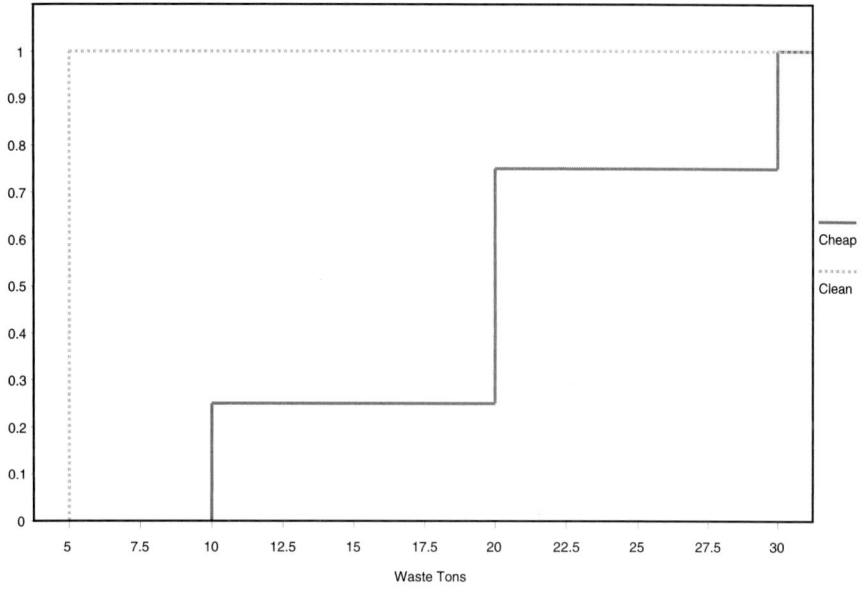

Tutorial: Analyzing Models with Multiple Objectives

This tutorial analyzes the model created in the tutorial on modeling decisions with multiple attributes.

▸ Open mua_eval.da.

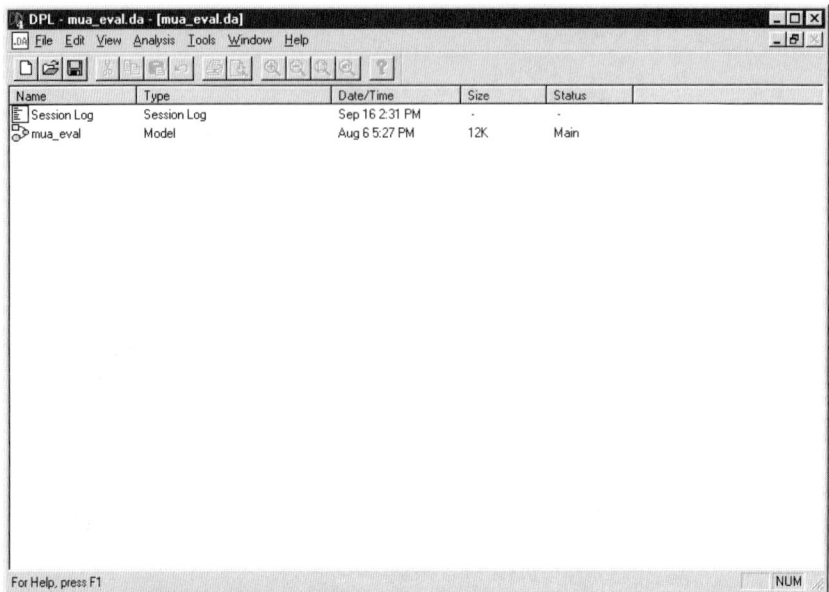

▸ Double-click on the model named "mua_eval".

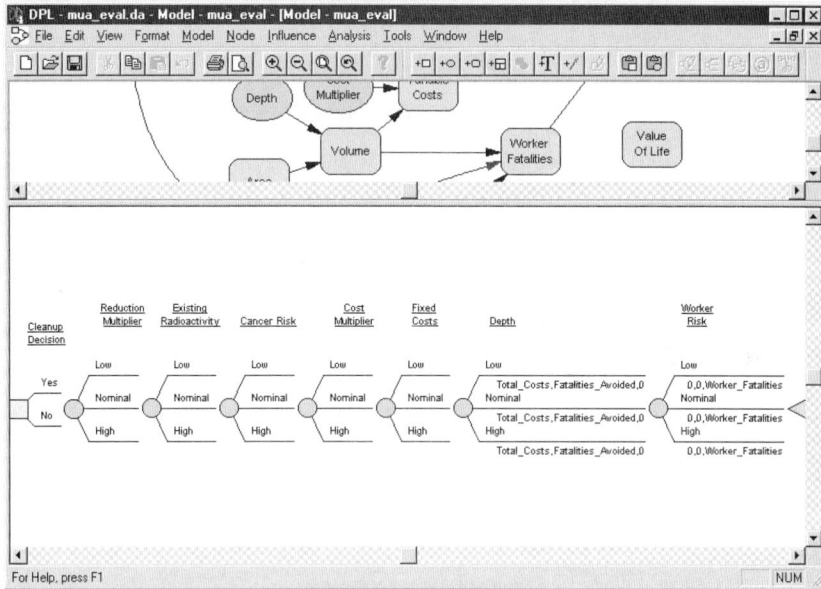

▸ Select Analysis Decision Analysis.

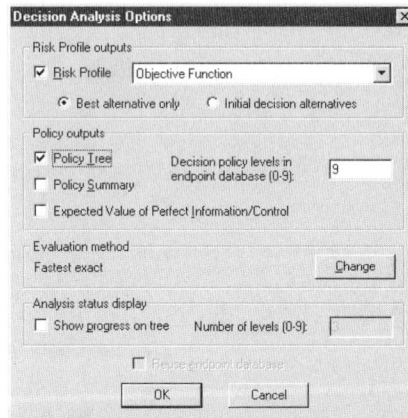

▸ Uncheck the box labeled "Risk Profile".

▶ Click OK.

▶ Select Yes in the DPL warning dialog box.

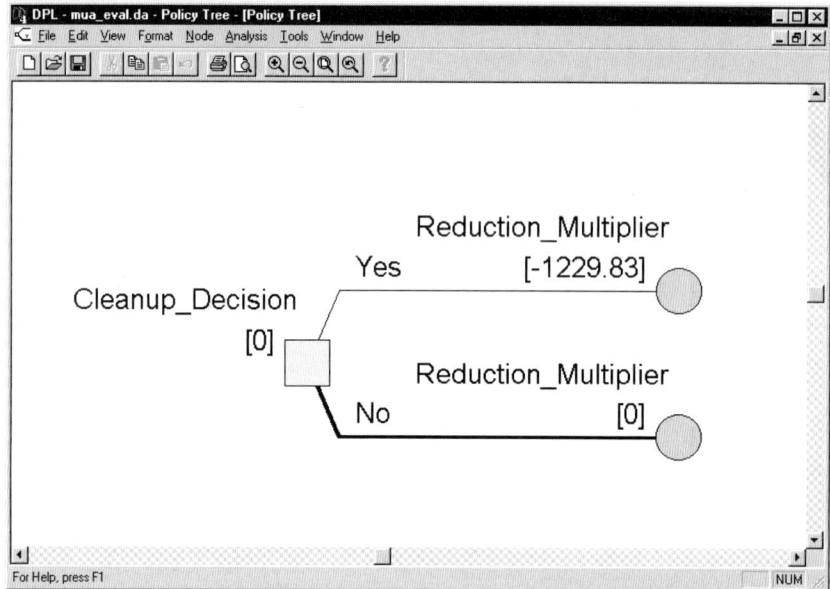

```
DPL - mua_eval.da - Policy Tree - [Policy Tree]                    _ □ ×
File  Edit  View  Format  Node  Analysis  Tools  Window  Help       _ 8 ×
```

```
                              Reduction_Multiplier
                        Yes          [-1229.83]
      Cleanup_Decision  ┌─────────────────────────○
          [0]        ┌──┐
                     └──┘
                        \    Reduction_Multiplier
                         No          [0]
                          \──────────────────────○
```

```
For Help, press F1                                          NUM
```

The Policy Tree reports that the value of cleaning up the site is negative, while doing nothing provides a (preferable) value of 0. It would be helpful to get more information about the "Yes" alternative.

▶ Select Window Model-mua_eval.

Investigating Non-Optimal Decision Alternatives

To get more information on what happens if you decide to clean up the site, you can run the model and force DPL to select the "Yes" alternative of Cleanup Decision.

▶ In the decision tree, click on the branches of Cleanup Decision.

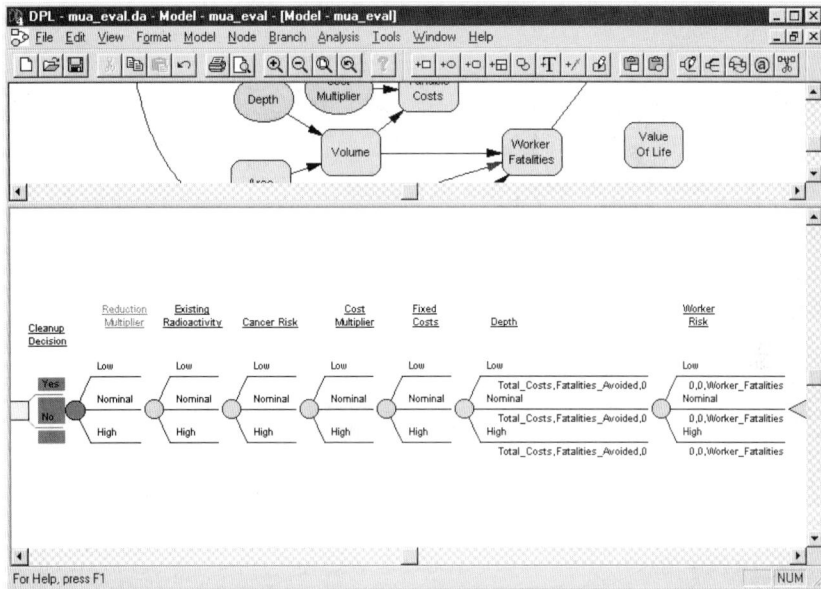

▸ Select Branch Control.

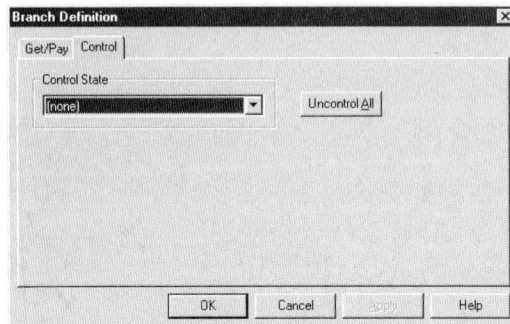

▸ Using the drop-down menu, set the Control State to Yes.
▸ Click OK.

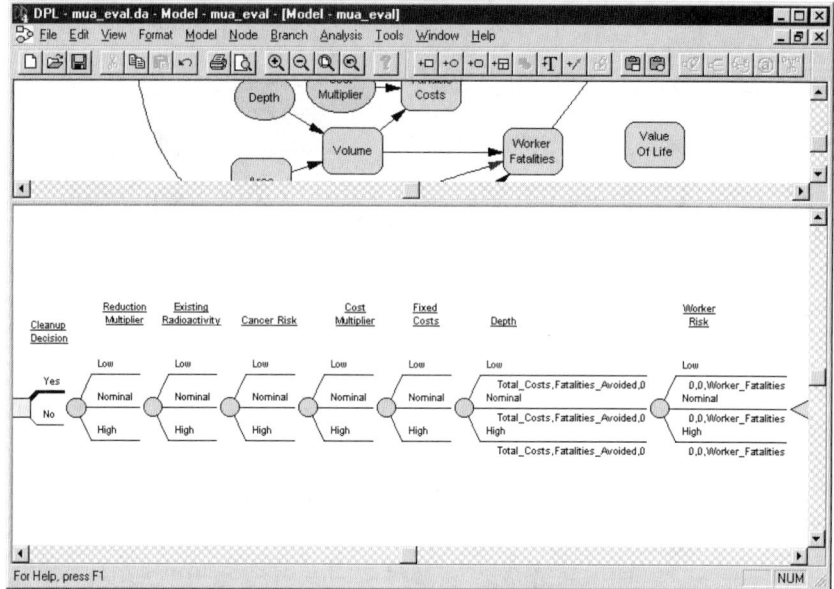

Now, re-run the model.

▶ Select Analysis Decision Analysis.
▶ Click OK in the warning dialog box.

Tell DPL to create a risk profile for the Objective Function.

▶ Uncheck the box labeled "Policy Tree".
▶ Check the box labeled "Risk Profile". (Notice that Objective Function is the default risk profile.)

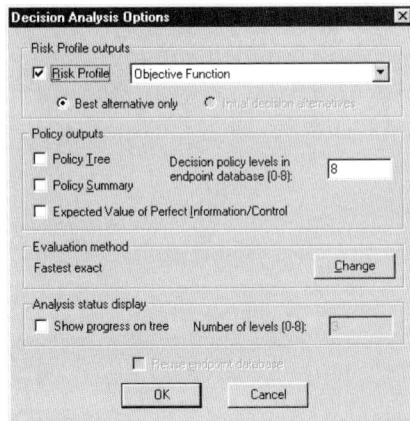

▸ Click OK.

▸ Select Yes in the DPL warning dialog box.

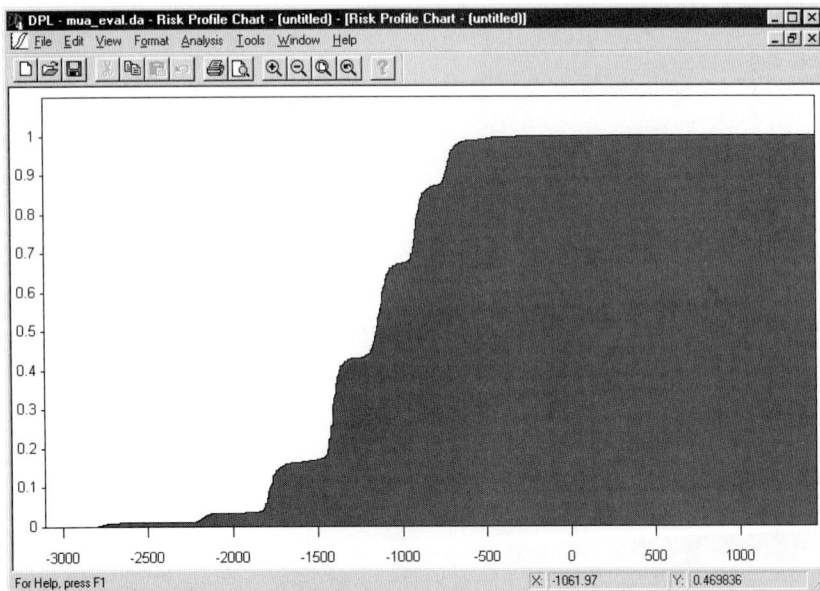

There are some scenarios which cause extremely negative values. You can investigate the individual attributes for more information.

Running Risk Profiles on Attributes

First, create risk profiles for the Cost attribute.

▸ Select Analysis Decision Analysis.
▸ In the Risk Profile outputs section of the dialog box, use the drop-down menu to change Objective Function to Public Safety.

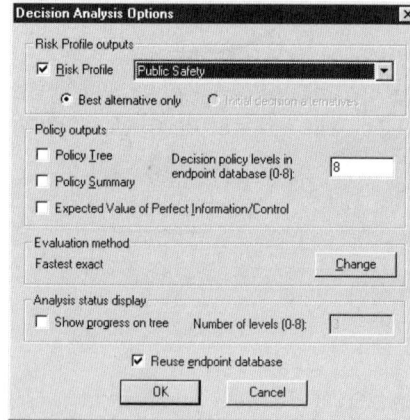

▸ Click OK.

The new risk profile is created immediately because DPL does not need to re-run the model.

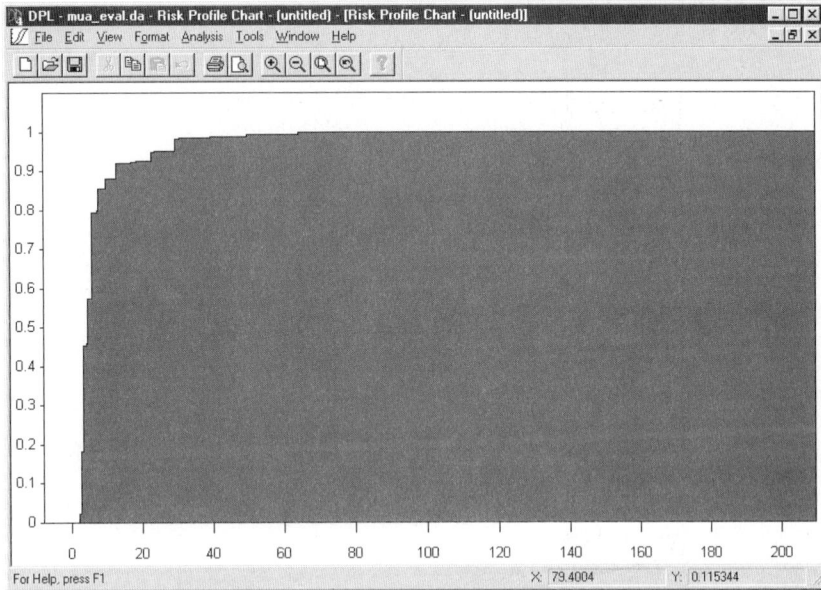

Rename the Public Safety risk profile in the Project Manager to save it.

▸ Select Window mua_eval.da.

▸ Right-click on the risk profile data item titled Expected Value.

▸ Select rename.

▸ Type "Public Safety".

Next, create a risk profile for Worker Safety.

▸ Select Analysis Decision Analysis.

▸ In the Risk Profiles outputs section of the dialog box, use the drop-down menu to change Objective Function to Worker Safety.

▸ Click OK.

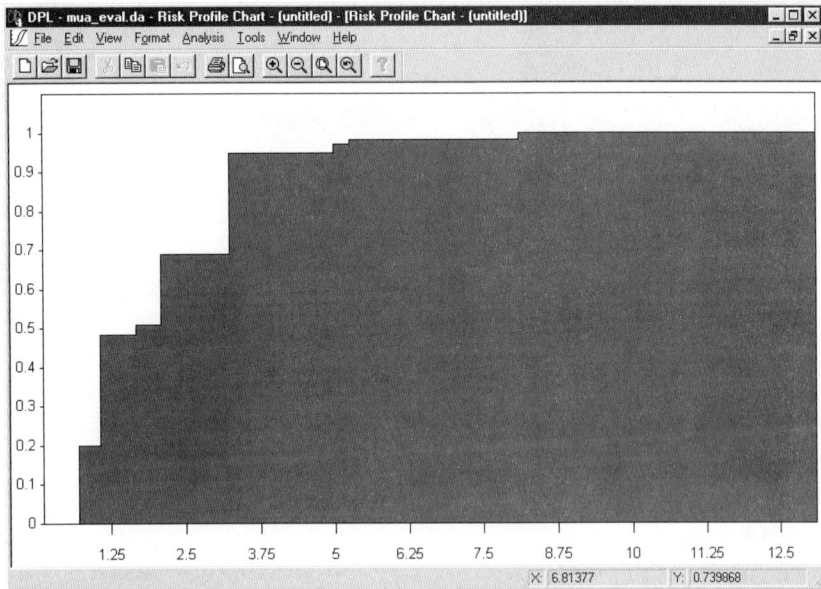

It would be interesting to compare the risk profile for Worker Safety with the risk profile for Public Safety.

▸ Select Format Series.
▸ Using the drop-down menu, change Series 2 from (none) to Public Safety.

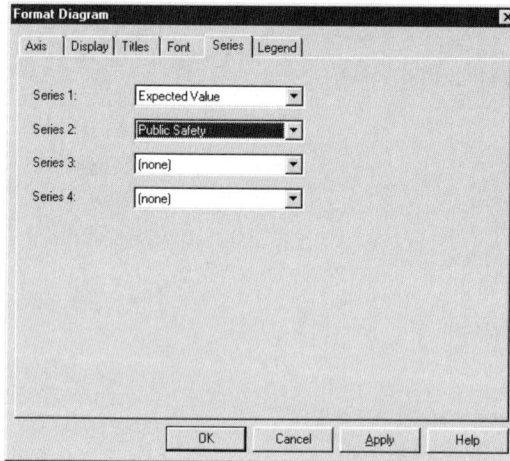

▶ Click on the Legend tab.

▶ In the input box for Series 1, type "Worker Safety".

▶ In the input box for Series 2, type "Public Safety".

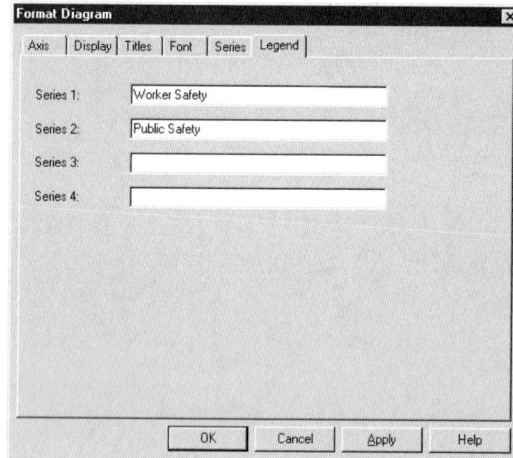

▶ Click OK.

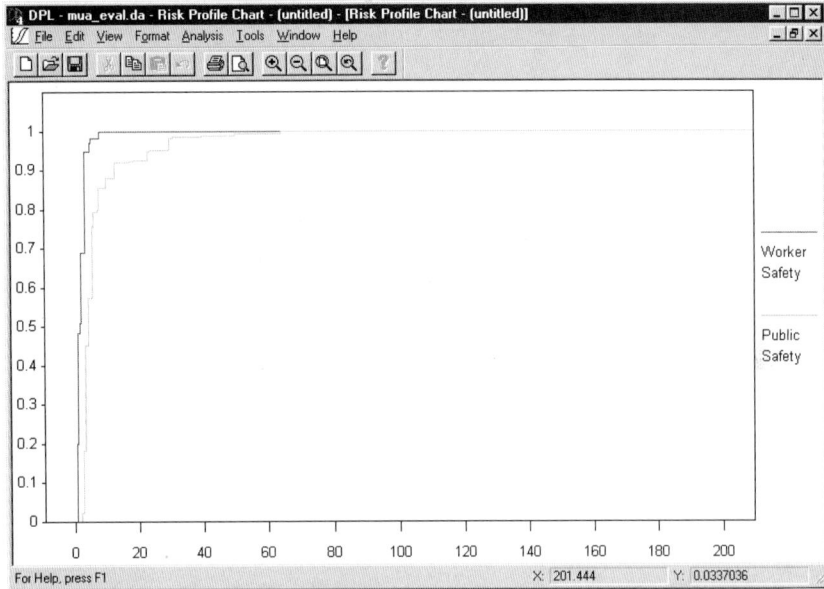

The units for both Worker Safety and Public Safety are number of fatalities. You can see that this alternative will have fewer expected worker fatalities than expected public fatalities.

Modeling Risk Attitude

Decision analysts, behavioral psychologists and insurance companies have known for years that people and companies have different attitudes regarding the amount of risk they are comfortable living with. The modeling concept of a "certain equivalent" gives us a useful way to represent the way an individual or company thinks about risk.

For example, imagine that you have received a ticket to play a game sponsored by a local radio station. If the local college basketball team wins its regional championship, you win a free dinner at a restaurant, valued at $100. If the team loses, you win nothing. You think there is a 60% chance they will win, so your expected value for this game is $60.

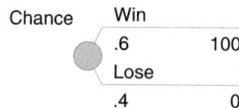

```
              Chance     Win
                         .6        100
                         Lose
                         .4         0
```

A co-worker, who is an avid college basketball fan, has offered you $50 for your ticket. Do you sell your ticket?

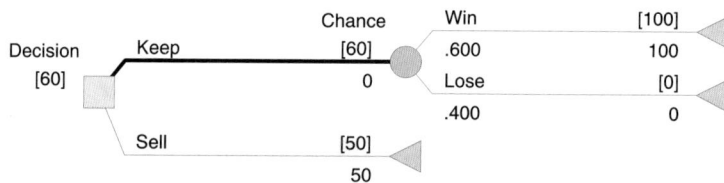

```
                         Chance    Win        [100]
     Decision    Keep     [60]     .600        100
       [60]                0       Lose        [0]
                                   .400         0

             Sell        [50]
                          50
```

On an expected value basis, you should keep this ticket, because the expected value of the lottery is greater than the selling price. But many people would take the $50, since it is a sure thing. This "a bird in the hand is worth two in the bush" argument illustrates a common phenomenon — frequently people are willing to give up expected value in return for certainty. The certain equivalent of a lottery is the minimum guaranteed amount of money you would be willing to sell the lottery for.

In general, there are three kinds of risk attitudes:

- **Risk-neutral.** A risk-neutral decision-maker would sell the lottery for no less than the expected value.
- **Risk-averse.** A risk-averse decision-maker would sell the lottery for less than the expected value.
- **Risk-preferring.** A risk-preferring decision-maker has a minimum selling price that is **higher** than the expected value. You might act in a risk-preferring way in the dinner game if you thought that dinner would taste better if you won it than if you bought it yourself.

There are two important things to remember about risk attitude:

- Risk attitude is a question of preferences. As you can see from the above example, risk attitude is determined by the decision-maker. It is **not** an "objective" property of the decision situation.
- Risk attitude is often a function of the amount that is at stake. People and companies usually become more risk averse as the stakes get higher. For example, you might not sell your dinner game ticket for $50. However, if the units were in the thousands of dollars, the lottery was the amount of money you might win in a jury trial, and the sell option was $50,000 you were being offered to settle the case for, you might think selling was an excellent idea.

When Should You Model Risk Attitude?

In general, most observers agree that most **business** decisions should be made on an expected value (risk-neutral) basis. The only decisions that really justify making in a risk-averse manner are those that have such high stakes that they present "bet the company" risks. Although this may seem to contradict our earlier statement that only the decision-maker can know what the appropriate risk attitude is, it really doesn't. It simply reflects the fact that most business decisions are evaluated by people who are acting as "agents" for the company at large, or its stockholders. These agents, frequently project, department or business group managers, may forget to place the risk of an individual decision in the context of the whole company, rather than in the context of their group or department. When considered as part of the whole company, most decisions are small enough to be evaluated on an expected value basis.

How Does Risk Aversion Get Used in Evaluating a Decision Analysis Model?

To reflect any risk attitude other than risk neutrality when evaluating a decision tree, DPL needs a way to look at any lottery and convert it to the decision-maker's certain equivalent (the expected value is a risk-neutral decision-maker's certain equivalent). It needs a "utility function" specified as part of the model.

Creating a utility function

First, define the x-axis of the graph. This axis represents the range of objective function values. It usually starts at the lowest possible objective function value or zero and ends a little above the highest possible objective function value. For our dinner decision, we might choose a range of $0 to $100.

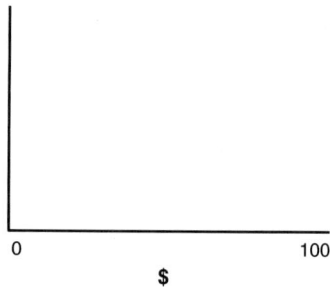

Second, define the y-axis of the graph. This axis represents utility and usually starts at zero and goes to a nice round number like 1, 100 or 1,000. It doesn't have any natural units, but is sometimes called "utiles."

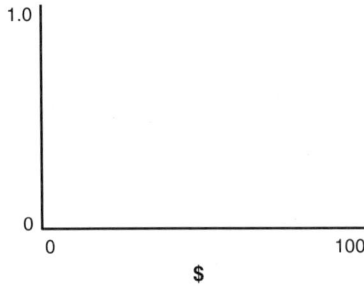

Third, define the lowest point on the x-axis to be equal to zero utiles and the highest to be 1 (or 100 or 1,000).

Fourth, assess the third point on the graph. Ask the decision-maker for a value, *n*, such that he would be indifferent between choosing between *n* or a lottery with a 50% chance of winning the lowest x-axis value and a 50% chance of winning the highest.

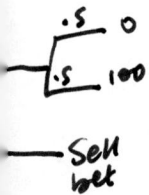

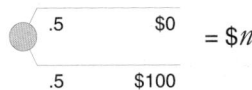

Say, for example, the decision maker says that he would be indifferent at $30. This $30 must have the same utility as the expected utility of the lottery, which is (.5 × the utility of $0) + (.5 × the utility of $100) = .5 × 0 + .5 × 1 = .5

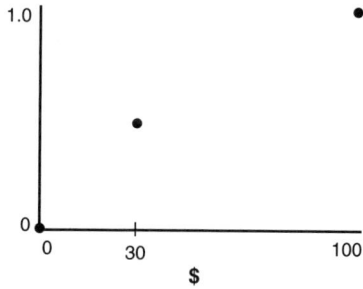

Repeat, using lotteries such as

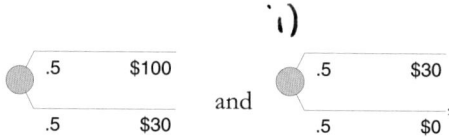

'ı)

	.5	$100			.5	$30
⬤			and	⬤		
	.5	$30			.5	$0

,

until you have several points.

Fifth, draw or fit a curve to the points. This is your utility function.

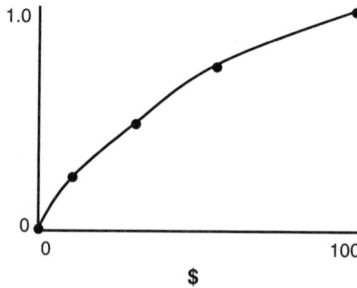

(handwritten margin note:) how get Prob. ı) would be 0.25 only?

Many people and organizations find that their utility functions are exponential. This corresponds to a marginally decreasing utility for something (usually $). Most people find this idea that the more money you have, the less you value your next dollar, fairly reasonable and intuitive.

Using a Utility Function

First, build the full decision tree and calculate the endpoint and objective function values.

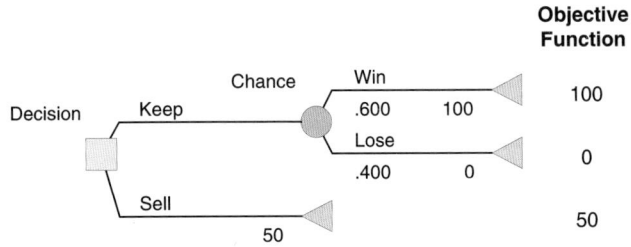

Second, use the utility function to convert the objective function values to utiles.

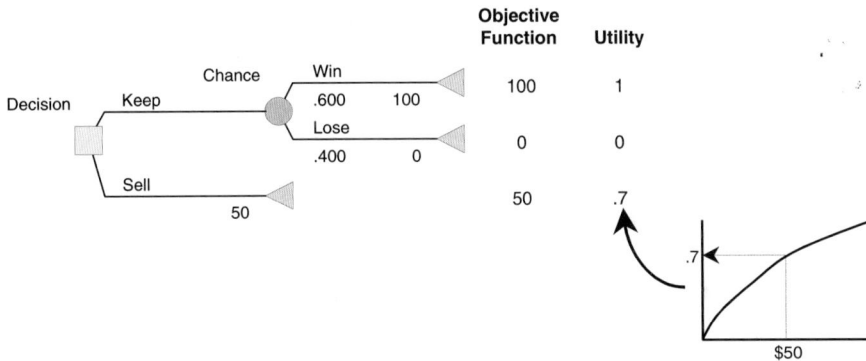

Third, roll back the tree using the utiles to calculate "expected utility."

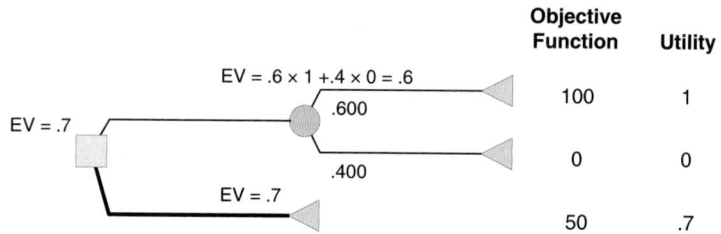

	Objective Function	Utility
	100	1
	0	0
	50	.7

EV = .6 × 1 + .4 × 0 = .6

EV = .7

EV = .7

.600

.400

Fourth, use the utility function to convert the expected utility into the certain equivalent, in $.

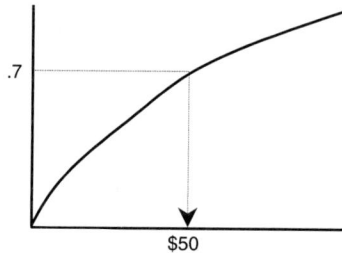

The certain equivalent of this decision is $50.

Defining a Utility Function in DPL

DPL offers a built-in exponential utility function. To use it, you simply specify a risk tolerance coefficient, R, which gets used in the formula $-e^{(-x/R)}$ where x is the objective function value that is being converted to utility. The built-in utility function should only be used in models that contain only maximizing decisions.

To assess R, the risk tolerance coefficient, ask the decision-maker the following questions:

If you are offered the following risky opportunity, where you have a 50% chance of doubling your investment and a 50% chance of losing half of your investment, how much are you willing to invest?

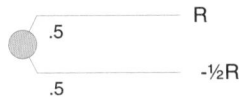

The amount you are willing to invest is the risk tolerance coefficient. Be sure to specify it in the same units as your objective function — don't define R in $ if your endpoint values are in $millions!

Note: the expected value = .5 × R + .5 × -½R = .25R

If you define R as a named value in your model, you can run a rainbow diagram sensitivity analysis and see how sensitive your model results are to the assessment of R.

Getting Fancy

Although DPL's built-in utility function is simple to use, it does limit you to using a single, exponential function for the entire model. If you prefer, you can define your own utility function. You can also specify multiple utility functions to apply in different parts of the decision tree. These functions can be the built-in functions with different risk tolerance coefficients, or user-defined functions, or a combination. Refer to the On-Line Help system or contact technical support for more information on these features.

Tutorial: Risk Tolerance/Certain Equivalent

In the first tutorial in this chapter (on Expected Value), you calculated an expected value of 1,442 ($ Millions). However, this assumes that you will select the decision policy with the highest expected value, regardless of the associated risks. Re-run this model using a Risk Tolerance to see if it affects the decision policy.

▶ Open risk.da.

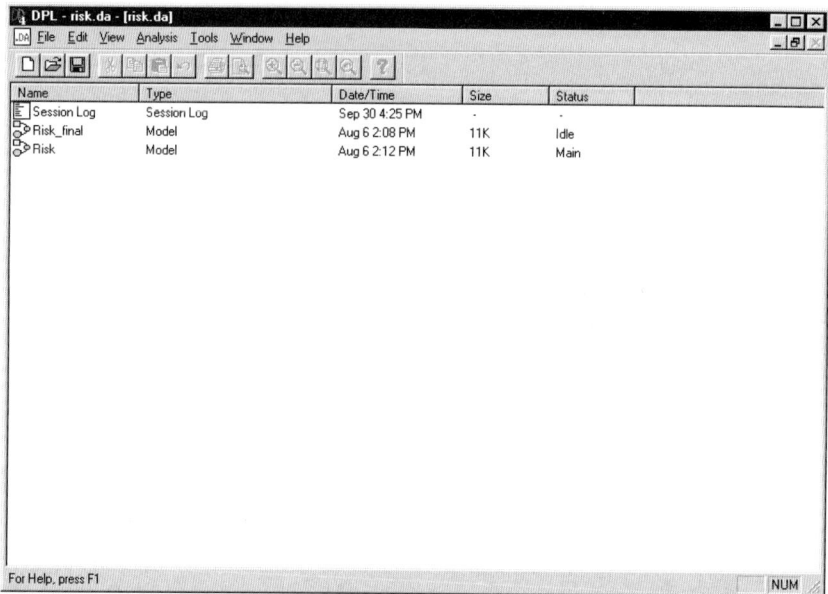

Specifying a risk tolerance is done in the Model Window.

▶ Double-click on the Model named "Risk".

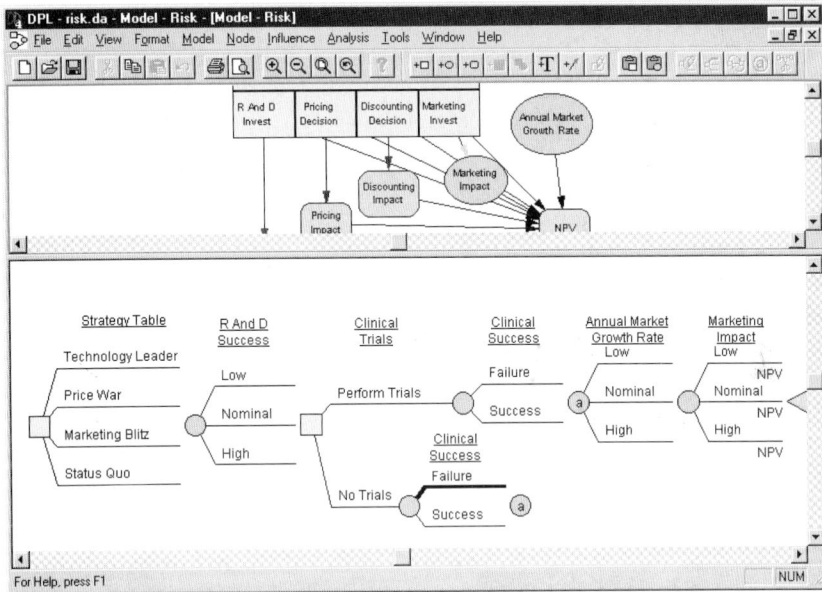

Adding Risk Tolerance to the Model

▶ Select Model Risk Tolerance.

You can model your risk tolerance using the built-in exponential utility function with a Risk Tolerance Coefficient of 500. (Remember that your units are $ Millions).

▶ Type "500" in the Risk Tolerance Coefficient entry bar.

▸ Click OK.

Analyzing the Risk-Adjusted Model

Now, run a decision analysis to see if the risk tolerance affects the decision policy.

▸ Select Analysis Decision Analysis.

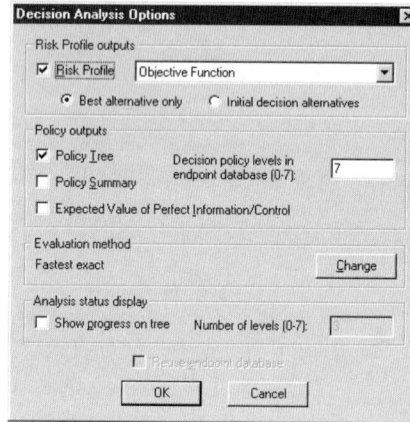

▸ Click OK.

DPL displays a decision policy based on the expected value of the model. This is not changed by the inclusion of risk tolerance.

The expected value incorporating the risk tolerance is called the Certain Equivalent, generally abbreviated in DPL as CE. You can view the decision policy based on the Certain Equivalent.

▶ Select View CE.

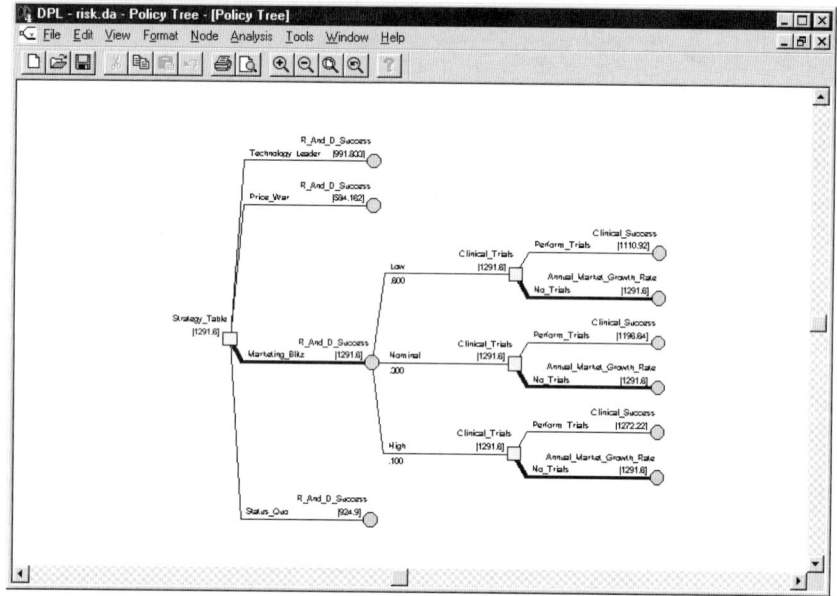

The CE for our model is 1,291, which is lower than the expected value of 1,442. While you continue to choose the strategy Marketing Blitz, the Clinical Trials decision policy has changed. If the outcome of R And D Success is High, you will choose the No Trials alternative. (The expected value decision policy chooses the Perform Trials alternative.)

You can also see the impact of our risk tolerance in the Risk Profile Chart.

▶ Select Window Risk Profile Chart.

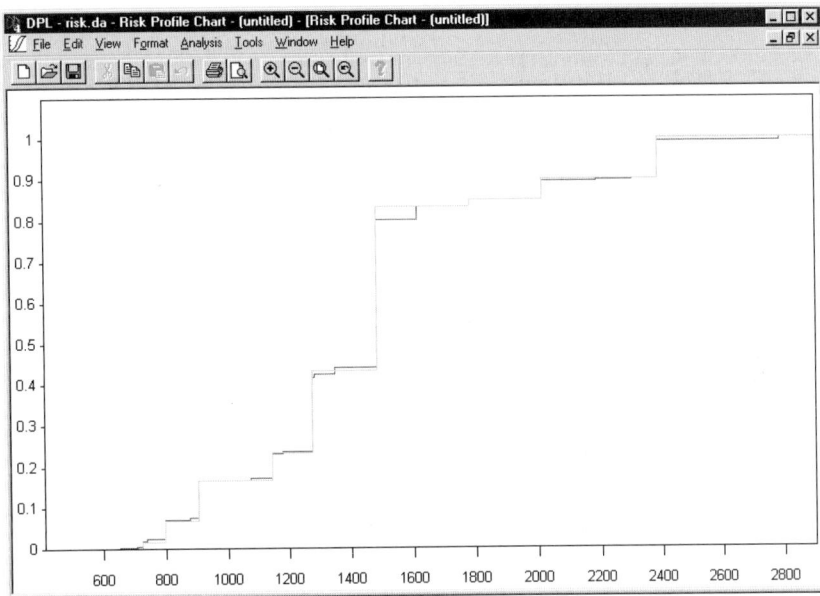

DPL displays risk profiles for both the Expected Value and Certain Equivalent decision policies. The values plotted for the Certain Equivalent series are not risk-adjusted. They are simply the endpoint values of the model under the Certain Equivalent decision policy. The two risk profiles are only slightly different because the Expected Value and Certain Equivalent decision policies only differed on one aspect of the Clinical Trials decision.

You can display the Expected Value and Certain Equivalent as vertical lines on the chart.

▶ Select Format Display.

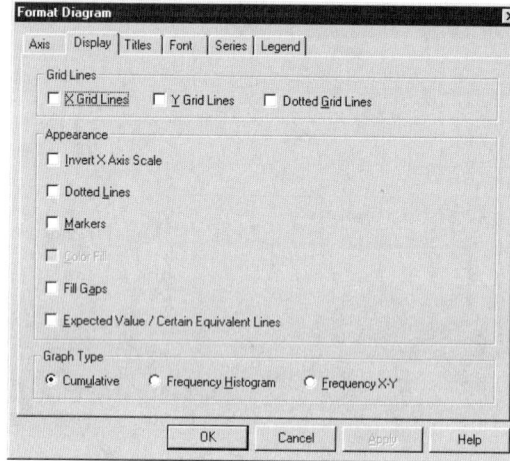

▸ Check the box labeled "Expected Value/Certain Equivalent Lines".
▸ Click OK.

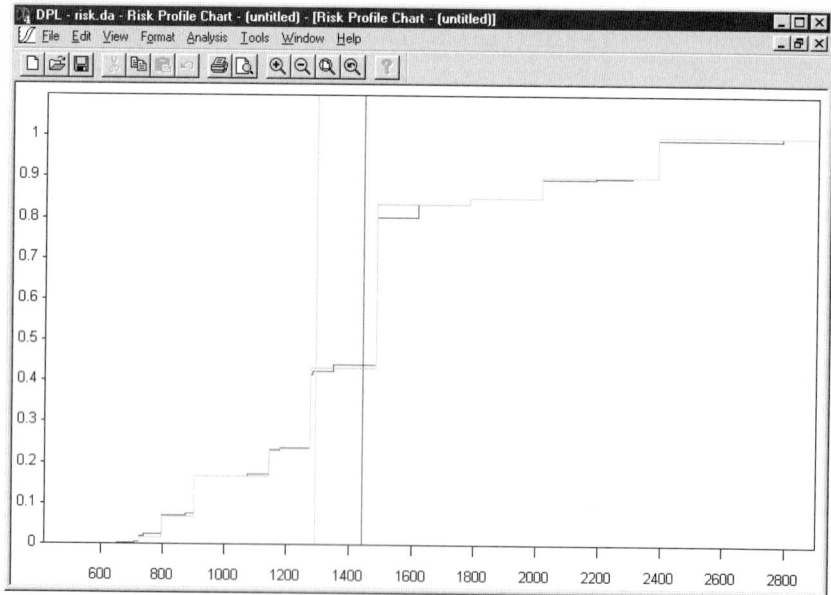

Running a Sensitivity Analysis on the Risk Tolerance Coefficient

It is possible that the model is very sensitive to the Risk Tolerance Coefficient of 500. You can investigate this by modeling the Risk Tolerance Coefficient as a value node and running a Rainbow diagram.

▸ Select Window Model-Risk.
▸ Slide the splitter bar down to the bottom of the window so the Influence Diagram is maximized.

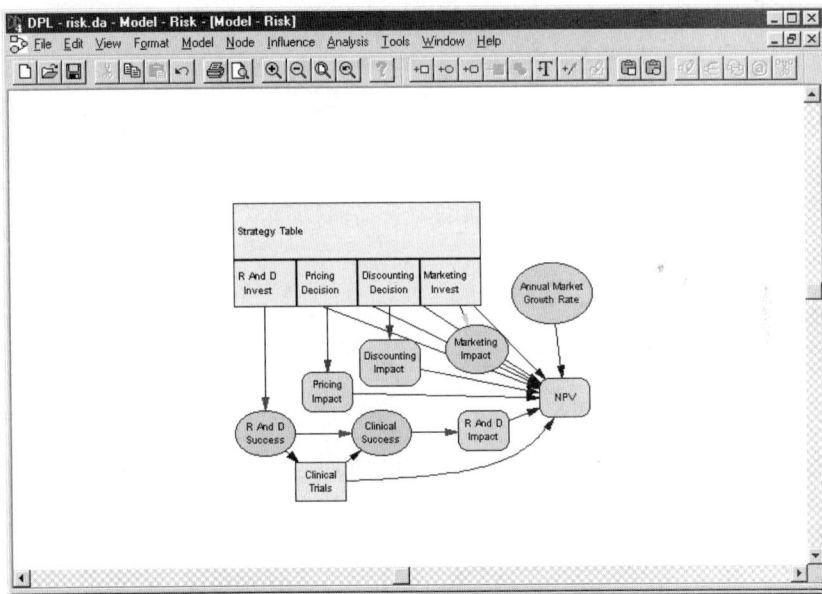

▸ Click the Create Value icon to add a value node to the diagram.

▸ Place the node in the bottom right-hand corner of the diagram.
▸ Name the node Risk Tolerance.

▸ Click on the Data tab.

▸ Type "500" in the entry bar.

▸ Click OK.

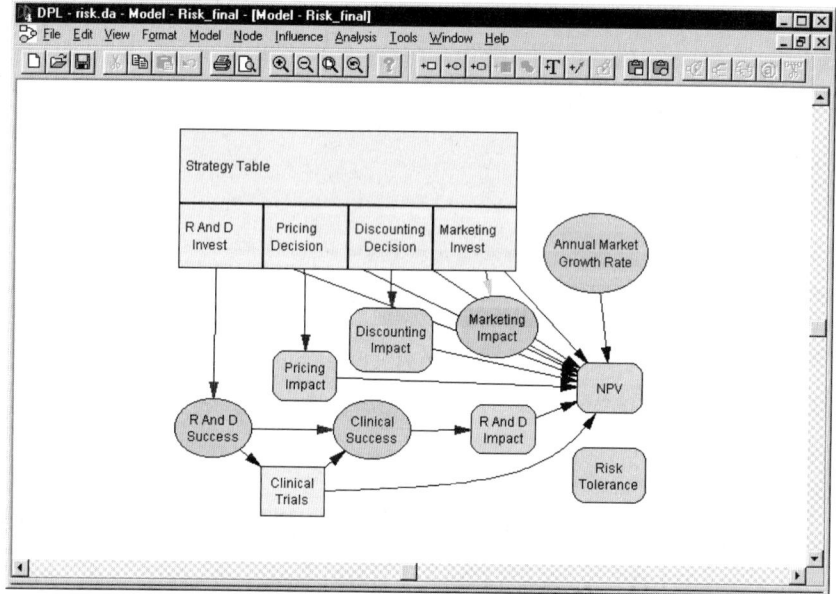

▸ Select Model Risk Tolerance.

▸ Replace the "500" with "Risk_Tolerance", using the Variable icon to insert the variable name.

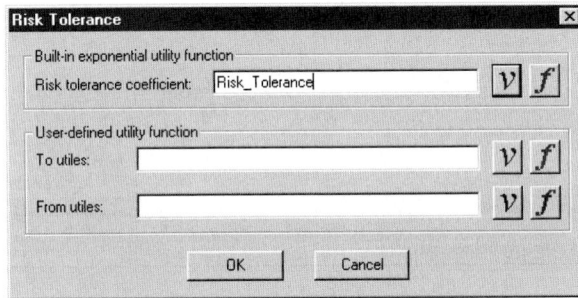

▶ Click OK.

Now, you can run a Rainbow Diagram on Risk Tolerance.

▶ Run a Rainbow Diagram on Risk Tolerance.
▶ Click OK.
▶ Use a range from 100 to 1000, with a Step size of 100.
▶ Click OK.

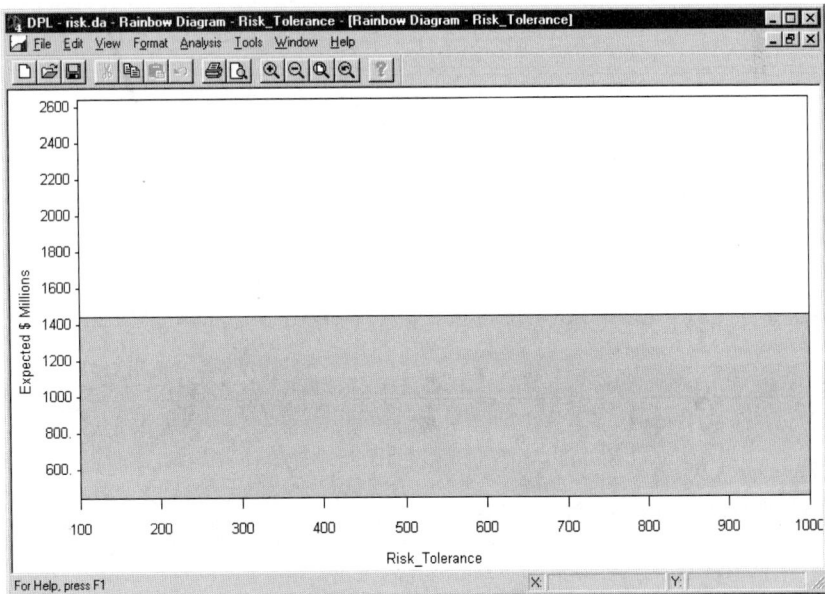

As you would expect, varying the risk tolerance coefficient has no impact on the expected value of the model. You're interested in the impact on the Certain Equivalent.

▸ Select View CE.

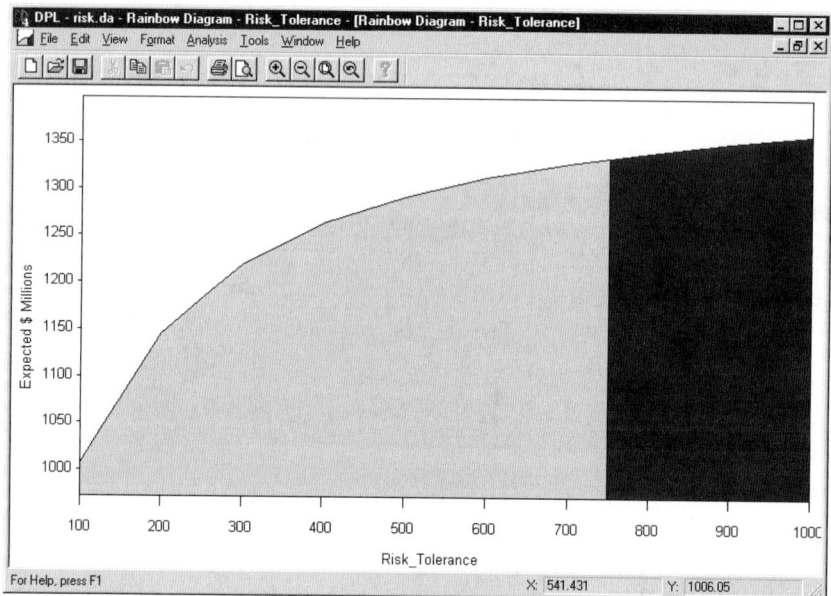

You can see that as the Risk Tolerance Coefficient increases from 100 to 1000, the Certain Equivalent also increases. The Rainbow Diagram also shows that as Risk Tolerance changes from 700 and 800, the decision policy changes as a result. You may wish to re-assess your Risk Tolerance to determine if you might be in this range.

A policy tree is a graphical display of an evaluated DPL model. It explicitly displays every path, or scenario, in the model, along with calculated values, probabilities, and expected values or certain equivalents for every node.

Chapter **11**

POLICY TREES

Chapter 11: Policy Trees

How DPL Generates Policy Trees

Policy trees are generated during regular decision analysis (not during sensitivity analysis).

For example, consider the plant investment model, which has an influence diagram and a decision tree.

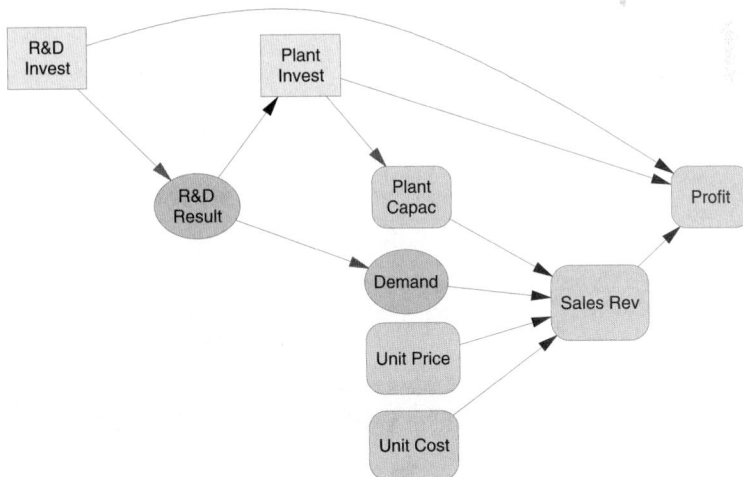

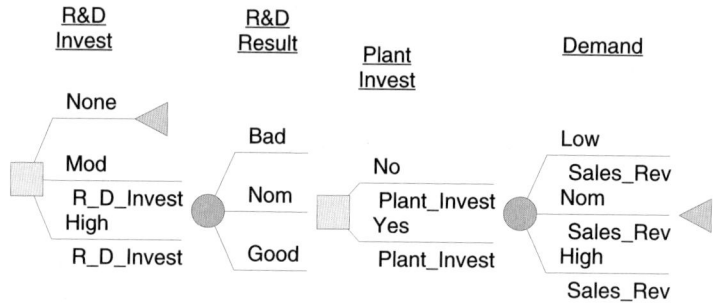

First, DPL expands the influence diagram and/or decision tree into a tree that explicitly displays every path.

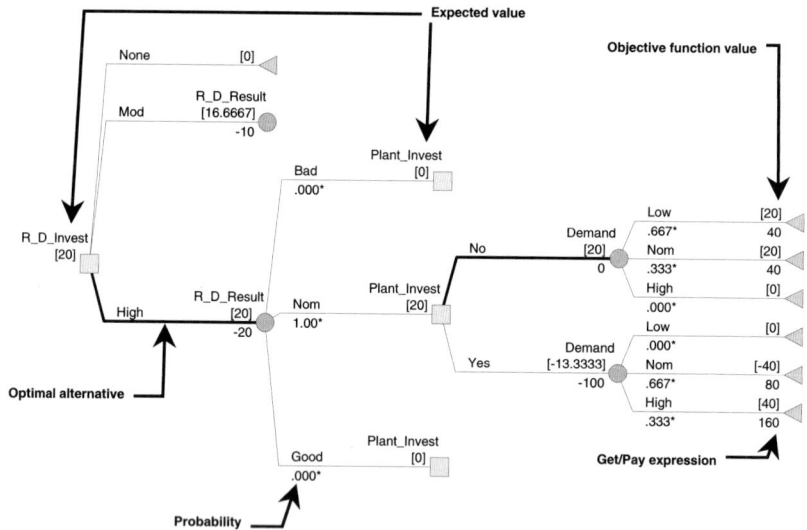

Second, DPL fills in probabilities for each chance event. If the probabilities are constants (0.5), DPL simply writes them in under the branch, to the left. If the probability is defined with a formula or named distribution, DPL calculates it, then writes it under the branch. If the probability is the result of a simulation, DPL marks it with an asterisk (0.5*)

Third, DPL replaces each get/pay expression with its calculated value. These are placed under the branches, to the right.

Fourth, DPL places the objective function value for each endpoint above the endpoint, in brackets [ ].

Fifth, DPL labels each node with its expected value (or certain equivalent, if appropriate). These are placed just to the left of the node, in square brackets [ ].

Sixth, DPL indicates the optimal alternative for each decision node by drawing it with a heavier line.

Reading a Policy Tree

A policy tree is the most detailed DPL output, packed densely with a lot of information. If your model is relatively small, just glancing through it will probably give you the information you need. If your model is large, it helps to have a game plan.

Reading the Endpoints

First, spot-check the endpoint values. Are they all the same? If every endpoint has the same endpoint value, your calculations aren't working — this usually means that a spreadsheet link has broken. Are they all the same for some portion of the tree? This may indicate that one or more events is not having an impact on the endpoint value. This may be an error caused by leaving the variable out of a calculation. It could also be that the model is correctly reflecting the impact of a minimum or a maximum, or of a decision to exit or sell the business before the event becomes relevant.

Are the numbers in the right ballpark? Do they move in the right direction? For example, if the only difference between two endpoints is the state of the Sales chance event, then you would expect the Profit endpoint value on the high state branch to be higher than the one on the low branch.

Reading the Probabilities

Second, look at the probabilities on the branches, especially any conditional or calculated probabilities. Are they correct?

If you used simulation, look to see the effect of simulation on the probabilities. If the approximation is too rough, run the analysis again, using a larger number of samples

Reading the Optimal Policy

Third, take a good look at the optimal policy. Is it a surprise? Scan the endpoints for just the optimal policy — is there much of a range? If there is, a risk profile may give you a better picture of the range. The policy summary may help you discover patterns that show the events that contribute to good or bad performance.

How does the optimal policy compare to the alternatives in expected value? If they are close, perhaps you should compare their risk profiles to see if the optimal policy is significantly more or less risky than the others with a similar expected value.

If there is one policy that has dramatically different endpoint and expected values (say by an order of magnitude), you might want to verify that its calculations are correct.

Tutorial: Policy Tree

Generating a Policy Tree

You can generate a Policy Tree by running a Decision Analysis.

▶ Open Policy.da.

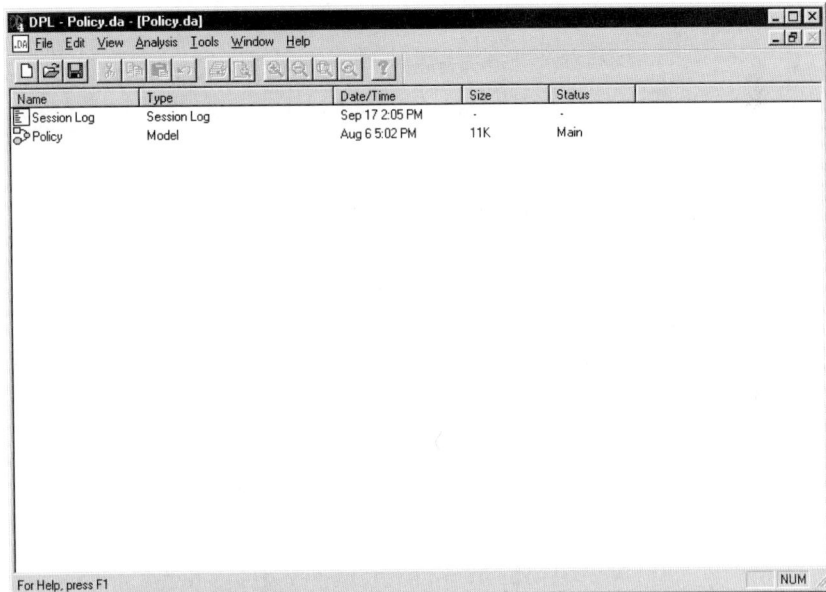

▶ Select Analysis Decision Analysis.

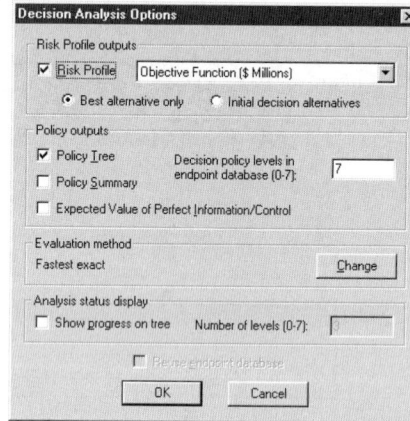

▶ Uncheck the box labeled Risk Profile so that Policy Tree is the only output you'll get from this run.

▶ Click OK.

Reading a Policy Tree

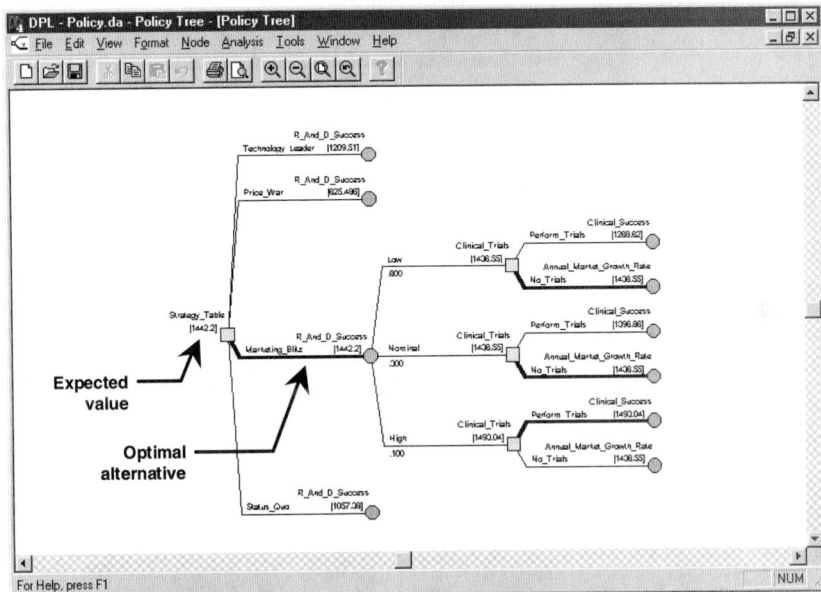

The rollback values are displayed in brackets on the branches of the Policy Tree. The expected value of the model is the left-most rollback value, displayed to the left of the Strategy Table node. The expected value is 1,442 (units are $ Millions).

DPL has made the Marketing Blitz branch of the Strategy Table darker than the other branches, indicating that it is the optimal decision alternative. It has also expanded the tree beyond the Marketing Blitz branch through the downstream decision, Clinical Trials. In each of the three occurrences of Clinical Trials, DPL has darkened the optimal alternative. You can see that only if the outcome of R And D Success is high will you choose the Perform Trials alternative of Clinical Trials.

Viewing the Policy Tree in Detail

You can expand or contract the tree to adjust the number of nodes and branches in view. Do this to investigate the uncertainties which occur after Clinical Trials.

▸ Find the bottom instance of Clinical Trials. Go to the end of the highlighted branch "Perform_Trials", and double-click on the green circle representing the chance node Clinical Success.

▸ Click on the Zoom Full icon.

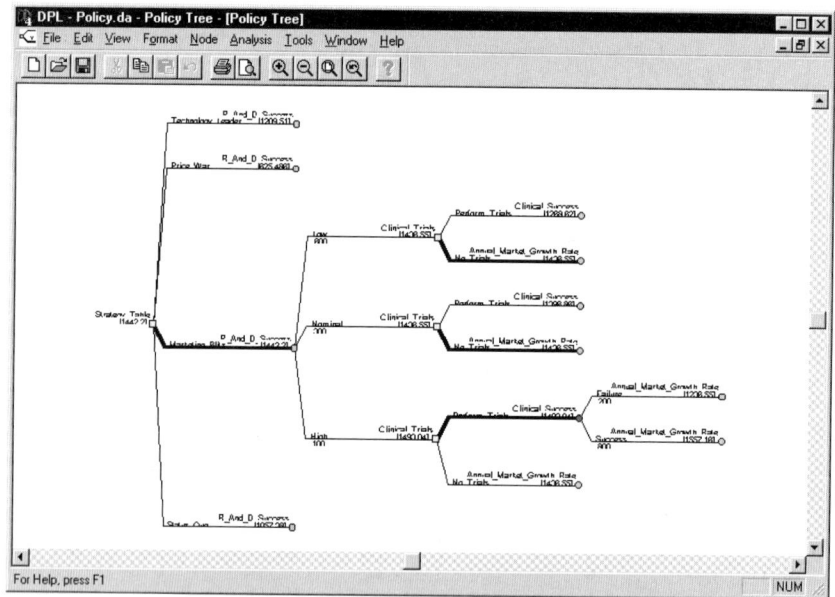

You can use the mouse to zoom in on the node you just expanded.

▸ Using the mouse, position the cursor above and to the left of the instance of Clinical Success you just expanded. Click and hold down the left mouse button.

Drag the cursor until both branches are enclosed in the zoom box. Release the mouse button.

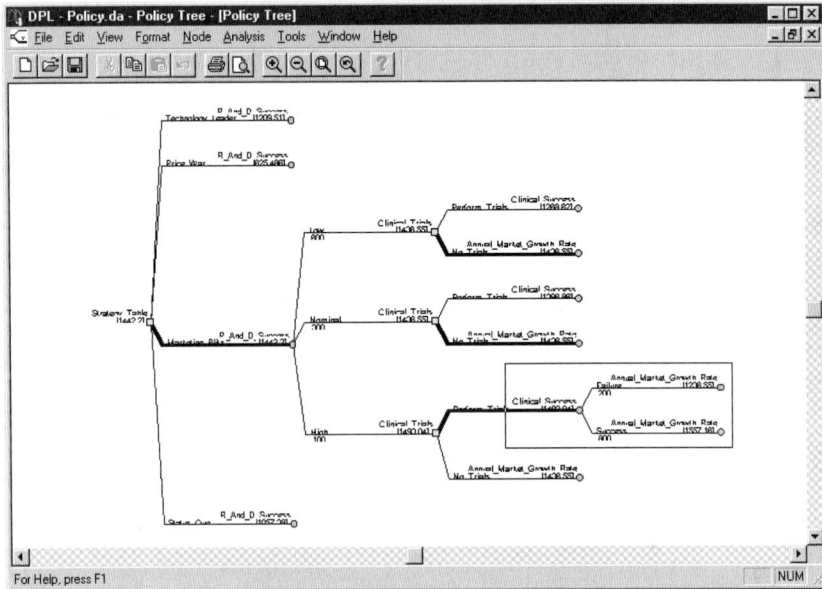

You now see a close-up view of the Clinical Success instance. You can see that if you have a success, the expected value is over $300 million higher than if you have a failure. You can also see that you had a 20% probability of the Failure outcome occurring and an 80% probability of the Success outcome occurring.

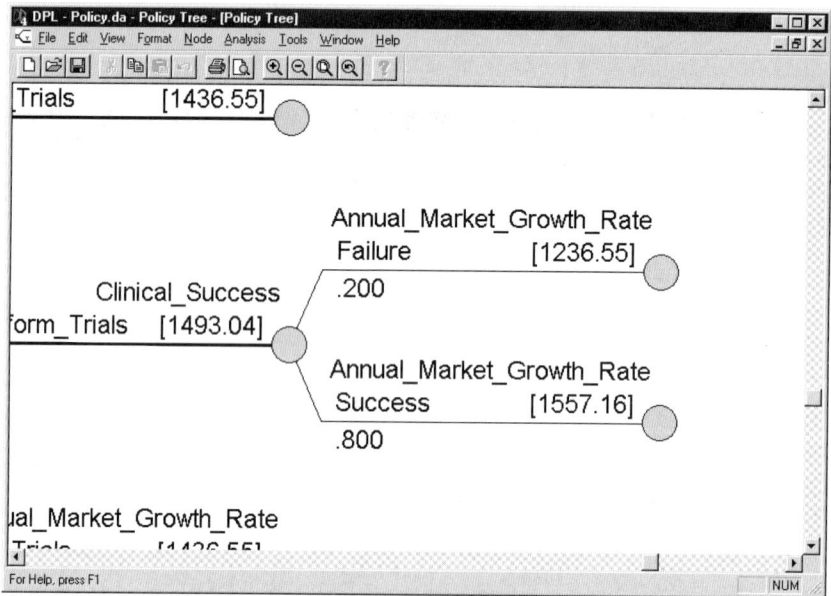

▸ Double-click on the Annual Market Growth Rate instance on the end of the Success branch of Clinical Success.

▸ Scroll to the right to bring the newly-expanded node into view.

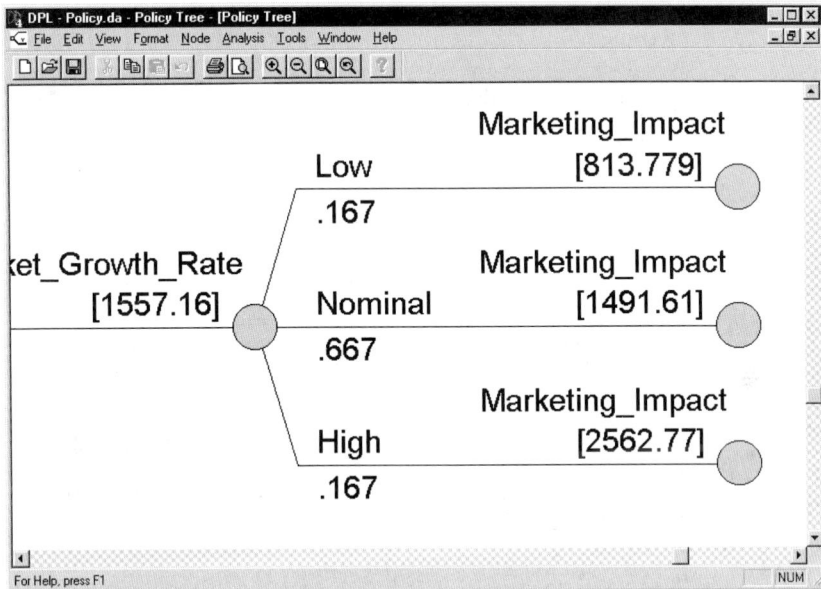

You can see that the outcome of Annual Market Growth Rate had a very large impact on the expected value, ranging from $813 million in the Low case to $2,562 million in the High case.

▸ Double-click on the Marketing Impact instance on the end of the High branch of Annual Market Growth Rate.

▸ Scroll to the right and down to bring the newly-expanded node into view.

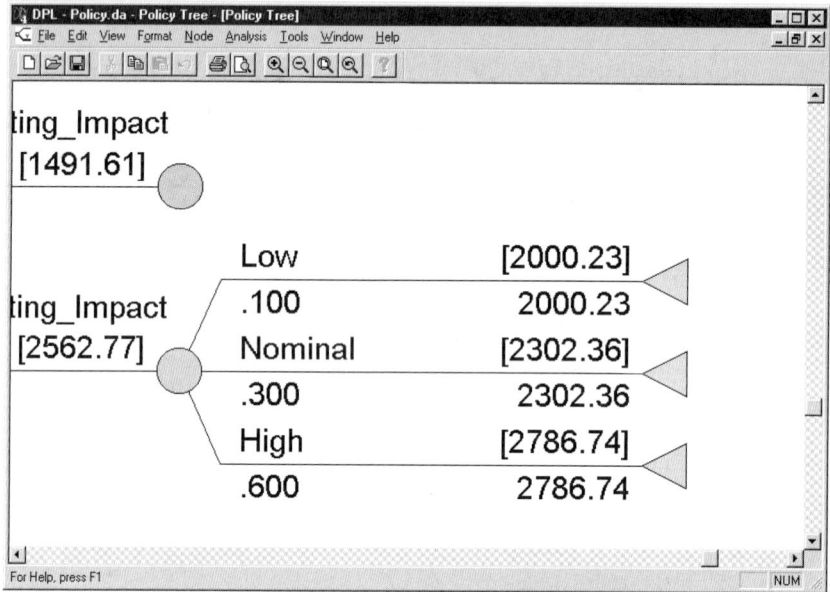

You can see that Marketing Impact also had a significant impact on the expected value of the model. The blue triangles on the end of the branches are endpoints, indicating that Marketing Impact is the final node in the Policy Tree.

▶ Click the Zoom Full icon.

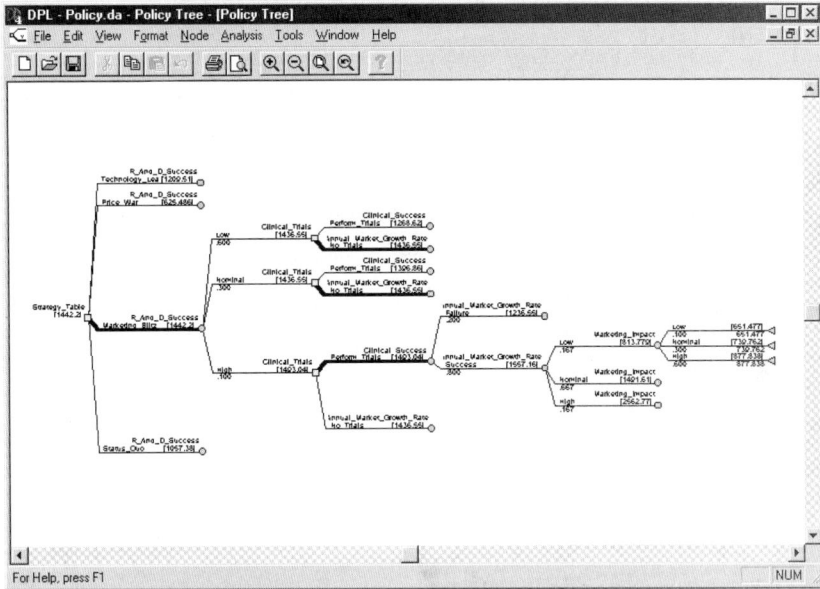

Looking at a full view of the Policy Tree, you can see that you have only expanded a small portion. Most Policy Trees present a great deal of information. This chapter offers some suggestions for making sense of it. You will also find that Risk Profiles and Policy Summaries can help condense the information.

Saving a Policy Tree

The Policy Tree is saved when you save the project. However, you cannot have more than one Policy Tree in your project. The Policy Tree is generated from data in the Endpoint Database, which is overwritten each time you run the model. Therefore, the Policy Tree will be overwritten each time you run the model.

Be careful about saving your model when you have a Policy Tree. When you open a project, DPL uses the Endpoint Database to regenerate the Policy Tree. If the Main model in the project is not the model which was used to generate the Endpoint Database (if you changed the model after running it, for example), DPL will not be able to reconstruct the Policy Tree.

A risk profile is a graphical display of the range of outcomes possible under a particular alternative or strategy. A single profile is useful to show how risky a particular course of action is. Multiple profiles on the same graph are useful to compare the riskiness of several alternatives or policies.

Chapter 12

RISK PROFILES

Chapter 12: Risk Profiles

How DPL Generates Risk Profiles

Risk profiles are generated during a regular decision analysis (they may be generated, but are not displayed, during sensitivity analysis).

First, for any particular decision policy (most often the optimal policy), DPL identifies all the endpoints that can be reached by following that policy.

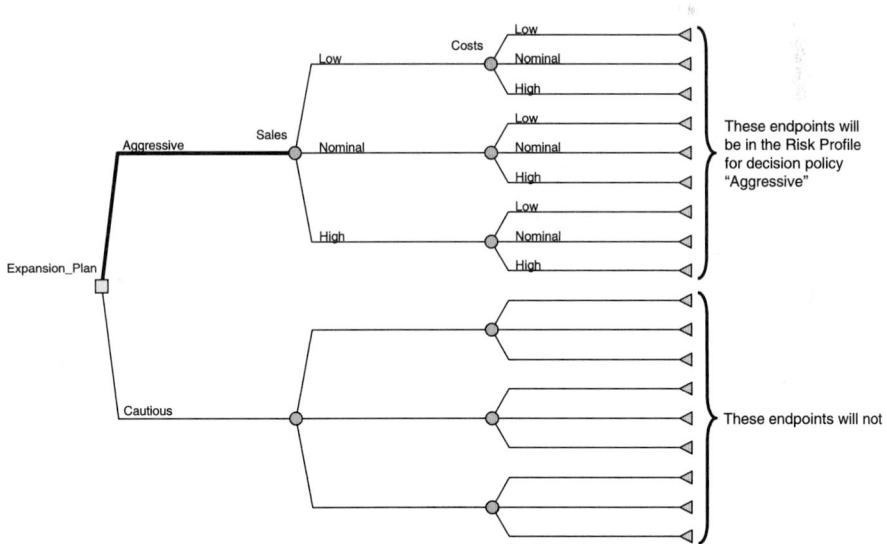

Second, DPL calculates the endpoint value for each endpoint. The endpoint value can be the objective function value (such as Profit) or the value of a particular attribute (such as worker injuries).

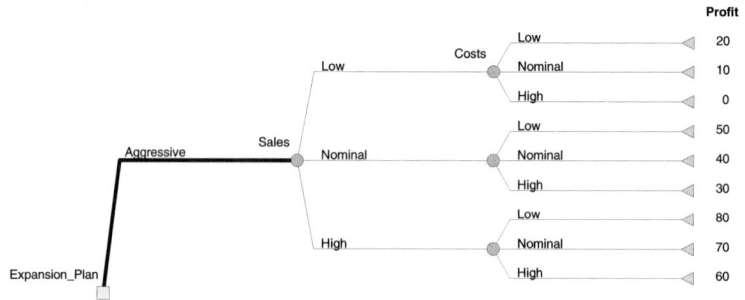

Third, DPL calculates the joint probability of each endpoint. The joint probability of an endpoint is the product of the probability of each chance event state that leads to the endpoint.

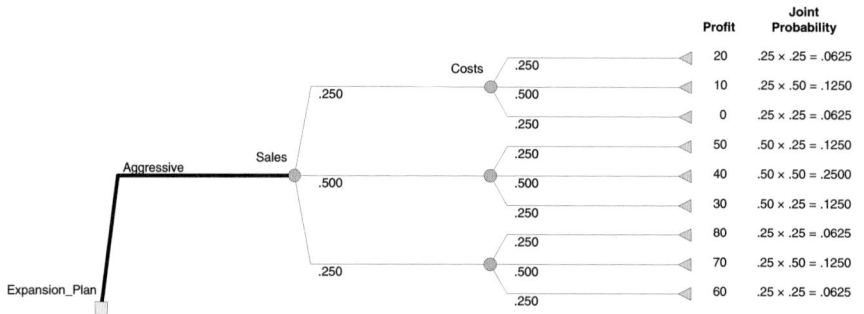

Fourth, DPL sorts the endpoints from lowest endpoint value to highest. If there are two endpoints with the same endpoint value, their probabilities are added together. DPL also calculates the cumulative probability for each endpoint, which is the sum of the probabilities of the endpoint and all the probabilities for the endpoints with lower endpoint values.

Profit	Joint Probability	Cumulative Probability
0	.0625	.0625
10	.1250	.1875
20	.0625	.2500
30	.1250	.3750
40	.2500	.6250
50	.1250	.7500
60	.0625	.8125
70	.1250	.9375
80	.0625	1.000

Finally, DPL graphs the distribution. There are two formats for the graph. Cumulative is the most common, and usually the most useful format. See the next section for tips on reading cumulative risk profiles.

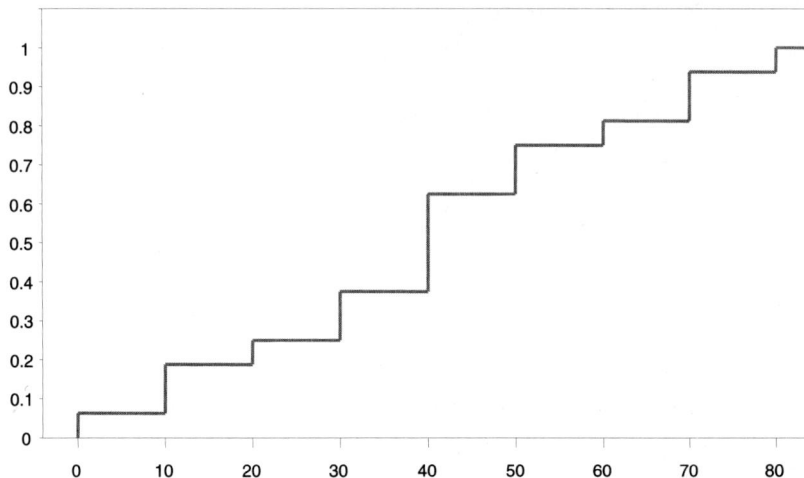

Frequency shows the probability of any particular endpoint value.

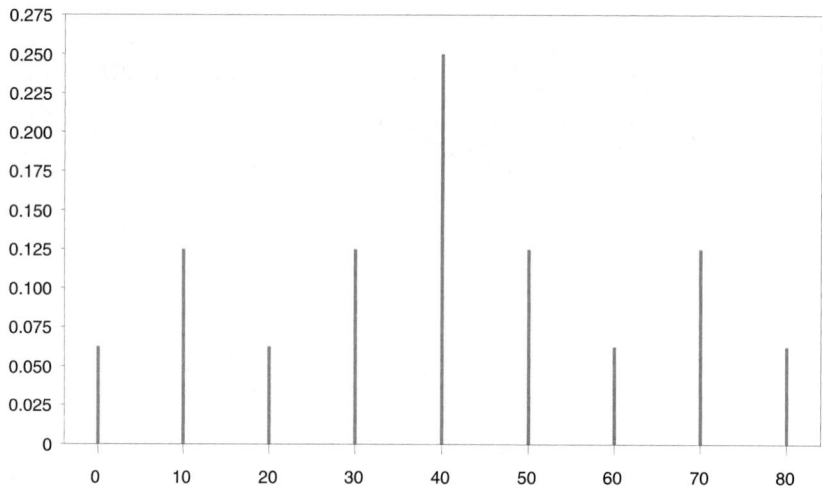

Reading Risk Profiles

Reading a Single Risk Profile

A single risk profile displays the range of outcomes that are possible for a single decision policy. It shows the amount of risk that is still present once you have chosen a particular course of action.

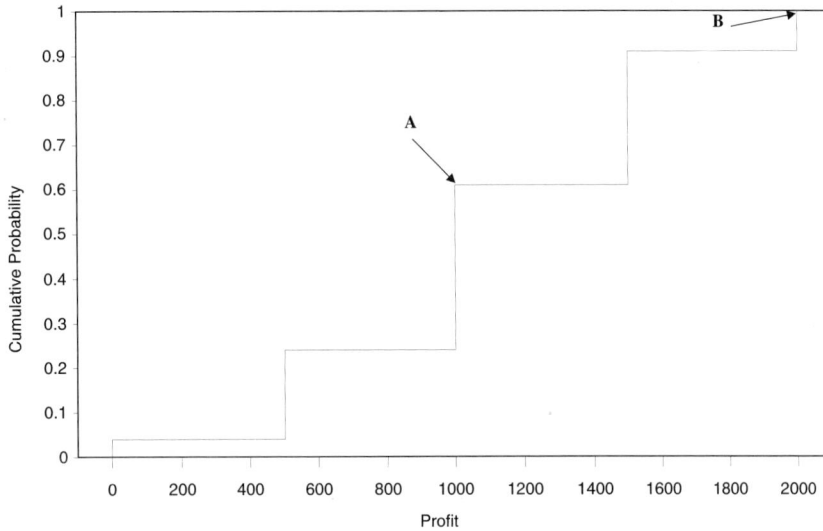

The horizontal axis shows the range over which the variable of interest varies. This variable may be the objective, such as Profit (or Total Cost) or a single component of the objective (attribute) such as Cost (or Acres of Wetlands).

The vertical axis shows the cumulative probability.

Point A is read as "there is a 60% chance that Profit will be 1000 **or less**."

Point B is interpreted as "the probability that Profit will be 2000 or less is 1.0; therefore, 2000 is the maximum profit we can achieve with this decision policy."

The probability that Profit will be between 1,000 and 2,000 is 0.4 (40%). This is calculated by taking the probability that Profit will be 2000 or less (1.0) and subtracting the probability that Profit will be 1000 or less (0.6).

Comparing Risk Profiles

If the variable of interest is something you wish to maximize, such as Profit, then, in general, one distribution is preferred if it is below and to the right of another distribution. The best possible distribution would be one that lies along the bottom and right axis, since it has probability 1.0 of the highest possible value and probability 0 of the lower values.

If the variable of interest is something you wish to minimize, like Total Cost, then in general, one distribution is preferred if it is above and to the left of another.

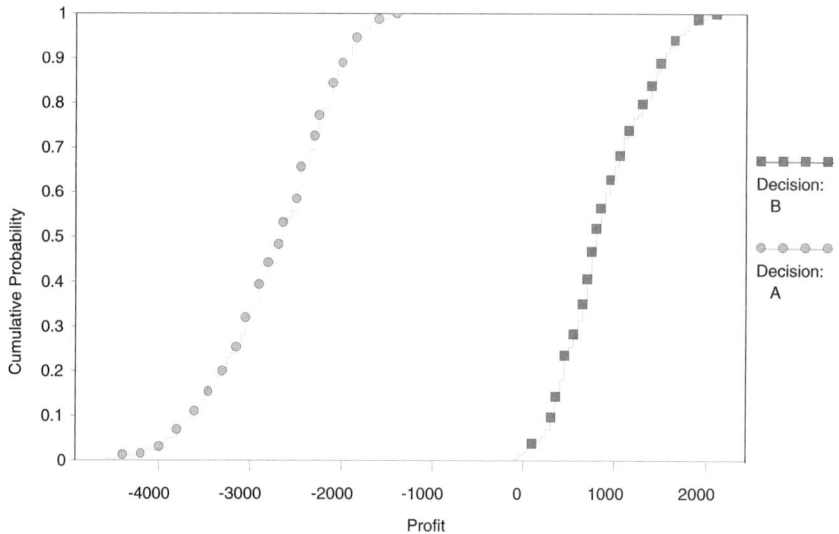

In the above graph, distribution B **deterministically dominates** distribution A because the lowest possible value for B is higher than the highest possible value for A. We would always prefer B to A because there is no scenario in which A has a better outcome than B. The really technical term for this is "no-brainer."

Tip: Truly dominated alternatives are rare—if they're that bad, they would have been screened out before this point in an analysis. Before ruling this alternative out, be sure your model is right!

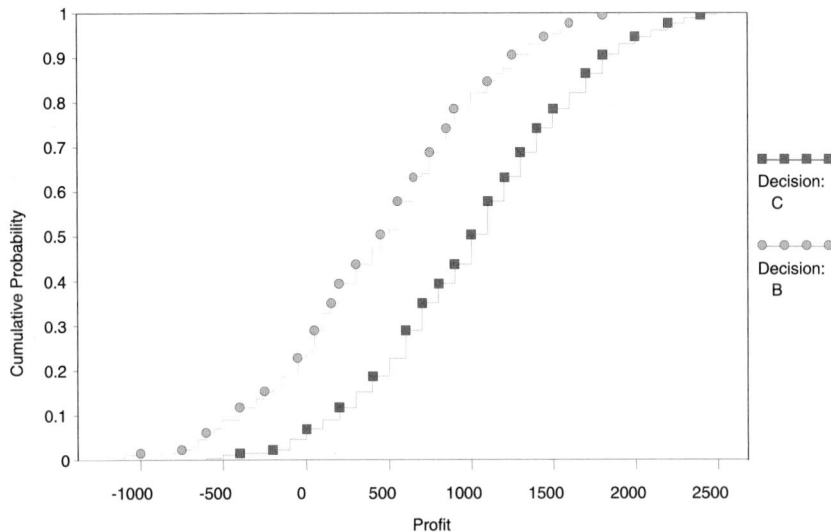

Distribution C probabilistically dominates distribution B because it lies entirely below and to the right, without crossing. Probabilistic dominance means that, for any Profit value, distribution C will have a lower probability than distribution B—and that is good because the probability is the likelihood of being **at or below** that Profit—so policy C has a greater probability of higher values than policy B does. Another way to look at it is to say that for any cumulative probability, distribution C has a higher cap value than distribution B—that is, for example at the .50 point we can say that distribution B will be at or below 500, but distribution C will be at or below 1000, which is better.

Although probabilistic dominance is good, it does not guarantee that policy C is better **in every scenario** than policy B. For any particular scenario (i.e. combination of chance event outcomes), it is possible for policy B to have a better outcome than policy C. On the whole, however, policy C is better. The best way to identify scenarios where policy B would be better than policy C is to look at the rolled back decision tree.

When two distributions cross, it becomes much harder to make strong statements about which policy is better. For example, distributions D and E show a very common situation—policy D has much less downside potential, but also less upside. In this case, we would say that policy E is riskier than policy D.

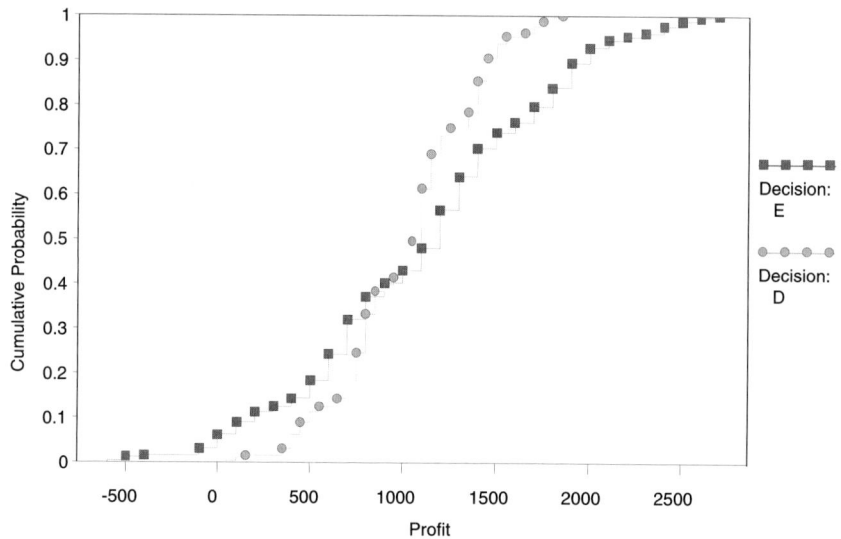

In such a situation, your first step is to look at the expected values of the two policies. If one policy has a much higher expected value, you may wish to choose it. If the expected values are very similar, you should consider whether your organization would prefer to choose the policy with less risk or the policy that offers the possibility of greater upside outcomes.

Tip: In each of these cases, and especially the last, don't be in too great a hurry to recommend one policy over another. First, spend some time trying to think of ways to combine alternatives or create new ones that capture the good outcomes and eliminate the bad ones. Policy Summary comparisons and the value of information and control graphs can help get you started.

Tutorial: Risk Profile Chart

Generating a Risk Profile

You can generate a Risk Profile by running a Decision Analysis.

▸ Open Risk_profiles.da.

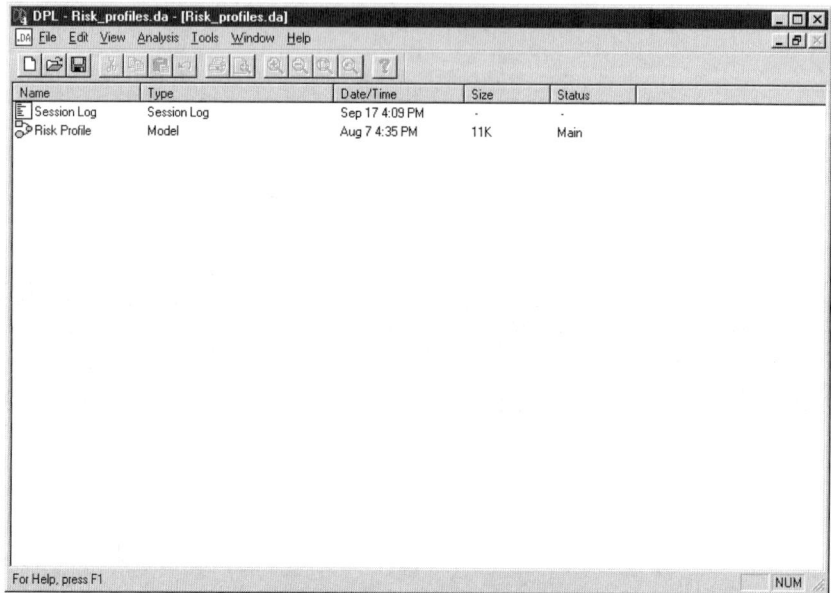

▸ Select Analysis Decision Analysis.

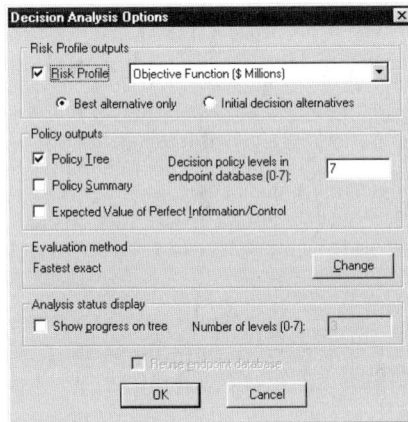

▸ Click OK.

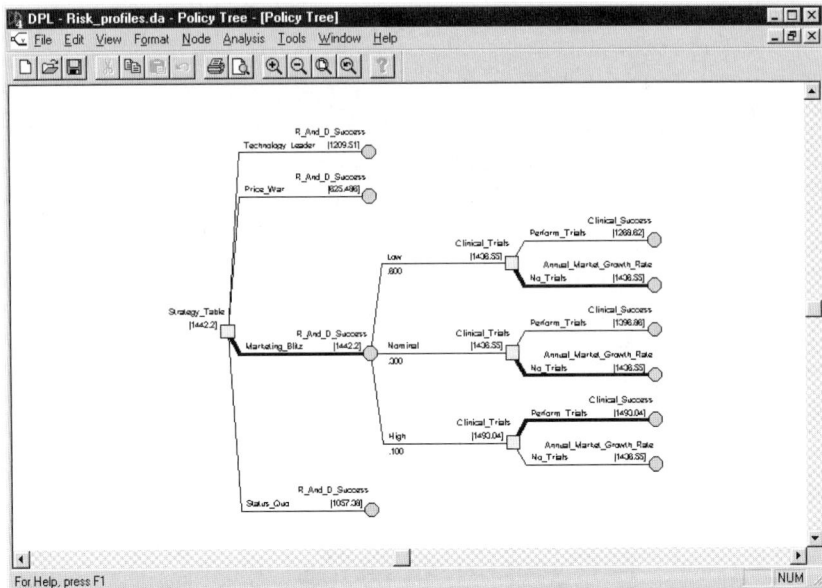

DPL displays the Policy Tree. The highlighted branches report the optimal decision policy: Choose the strategy Marketing Blitz and then decide which alternative of Clinical Trials to select based on the outcome of R And D Success. The expected value is 1,442 (in $ Millions).

▸ Select Window Risk Profile Chart-(untitled).

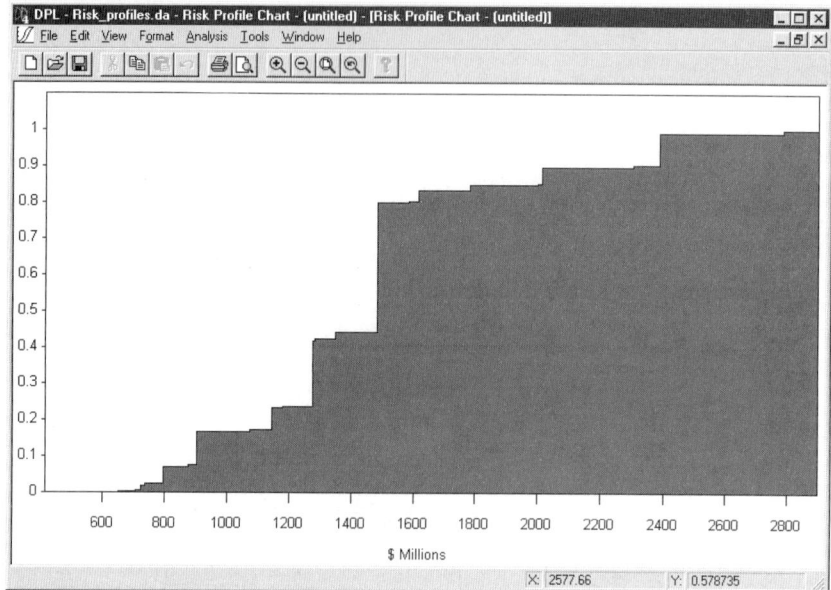

Reading the Risk Profile Chart

The Risk Profile Chart displays a cumulative probability distribution based on the optimal decision policy you viewed in the Policy Tree. The x-axis is a range of value for the objective, Profit. The y-axis represents the probability of achieving a Profit equal to or less than the corresponding x-value. You can see that the lowest value for Profit is about $650 million, and that the highest value is around $2.8 billion.

You can also display the expected value on the Risk Profile Chart. Sometimes the Risk Profile Chart is easier to read if you remove the color fill. You can do both at the same time.

▸ Select Format Display.

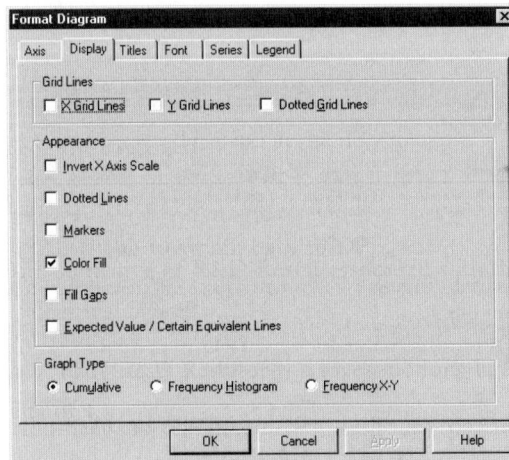

▸ Check the box labeled "Expected Value/Certain Equivalent Lines".
▸ Uncheck the box labeled "Color Fill".
▸ Click OK.

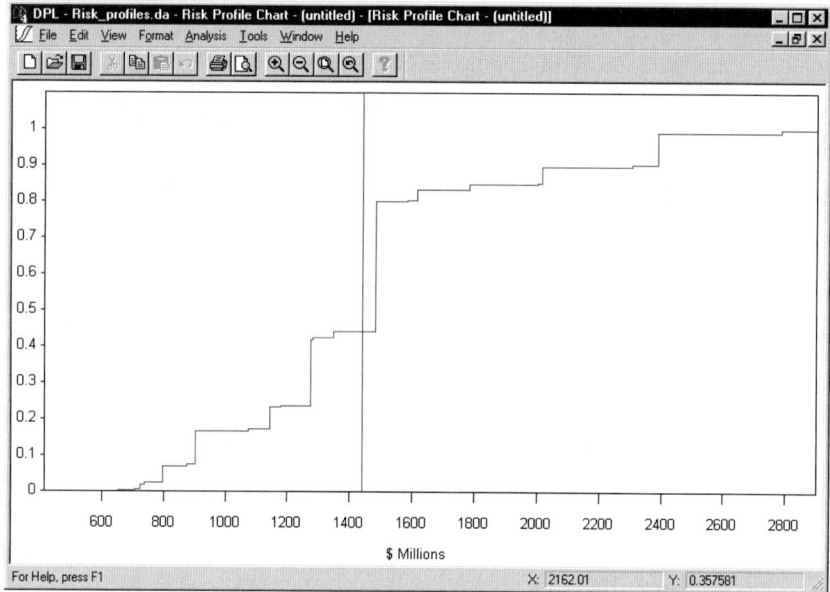

Zooming

You can zoom in on a section of the Risk Profile Chart to view it in more detail. Zoom in on the intersection of the Expected Value line and the Risk Profile line.

▸ Position the cursor above and to the left of the intersection of the two series.
▸ Click and hold down the left mouse button. Drag the zoom box until it encloses the intersection.
▸ Release the mouse button.

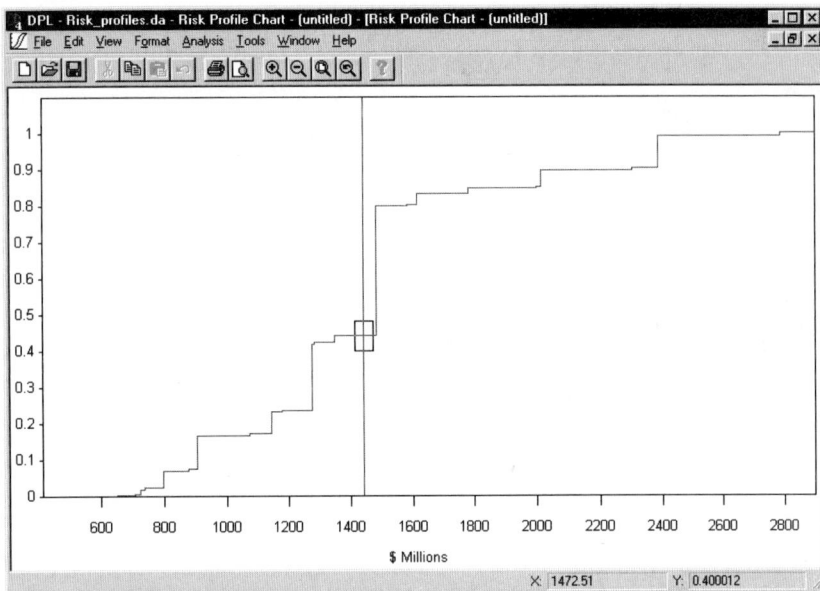

DPL now displays a close-up of this area.

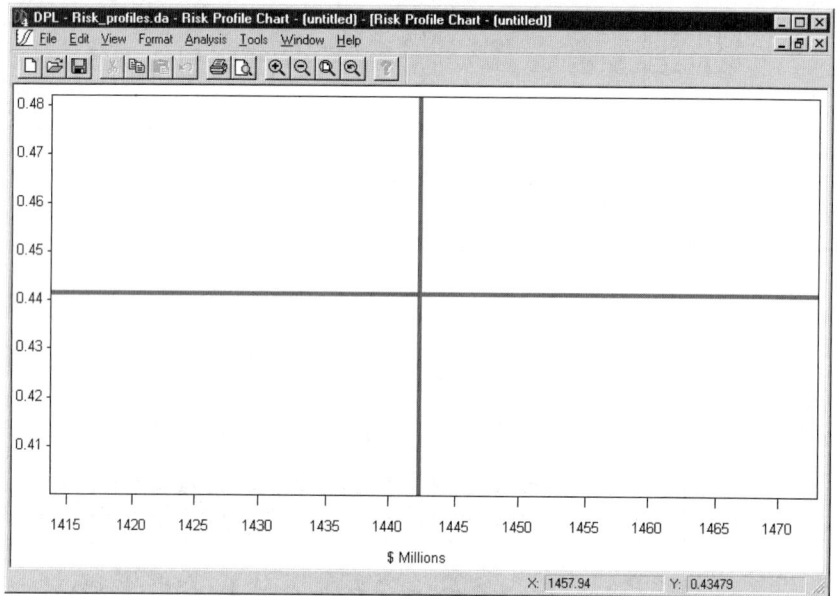

This view shows that there is roughly a 44% probability the outcome will be less than or equal to the expected value.

Coordinates

You can get a more accurate readings of points by placing the cursor over the point you are interested in.

▸ Position the cursor over the intersection of the two series.

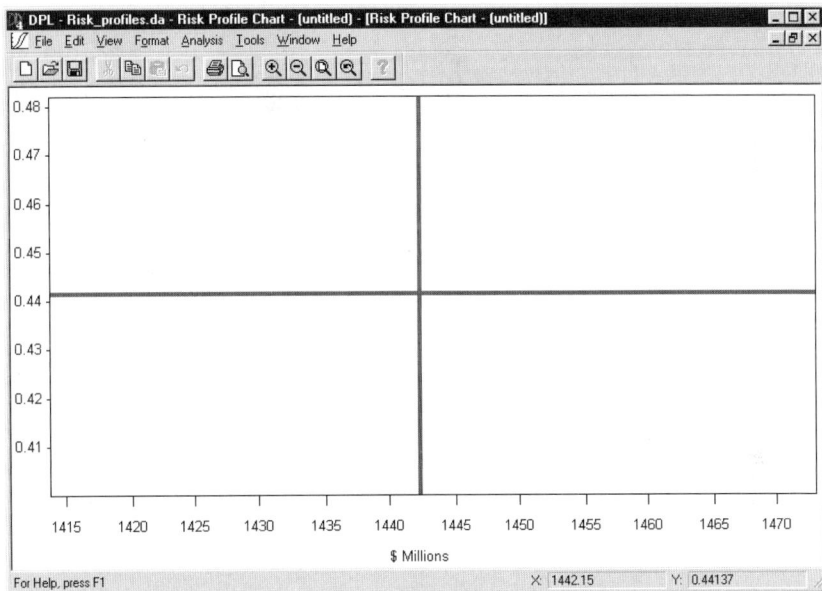

The x and y-coordinates of the cursor's location are displayed in the status bar in the bottom right-hand corner of the window.

▸ Click the Zoom Full icon.

Note: The accuracy of the coordinates display is limited somewhat by the resolution of your monitor.

Changing the Graph Type

You can also view the Risk Profile as a frequency histogram or a frequency x-y chart.

▸ Select Format Display.
▸ In the Graph Type section of the dialog box, click "Frequency Histogram".
▸ Click OK.

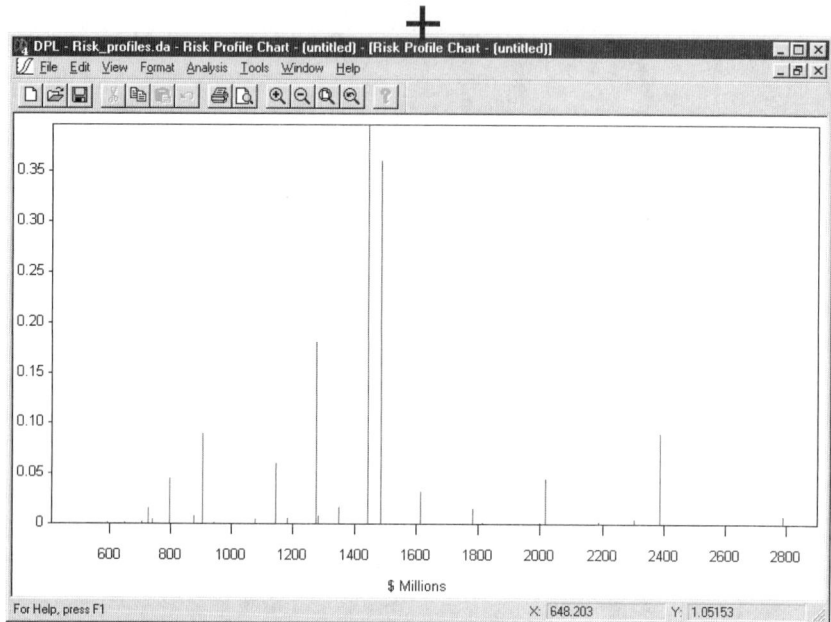

▸ Repeat, changing the Graph Type back to "Cumulative".

Percentiles in the Session Log

When you run a Risk Profile, DPL automatically writes the 10th, 50th, and 90th percentiles of the Risk Profile to the Session Log.

▸ Select Window Session Log.

```
DPL - Risk_profiles.da - Session Log - [Session Log]                    _ □ x
File  Edit  View  Format  Analysis  Tools  Window  Help                 _ 8 x

Decision Programming Language (DPL)
Copyright (c) 1989-1998 Applied Decision Analysis -- All Rights Reserved
Applied Decision Analysis, Menlo Park, Calif.  (888) 926-9251

Professional Version
Release 4.00.04 (Alpha)


16:09:58  Compiling: (untitled)

16:10:21  Compiling: (edited) (untitled)
Number of paths = 324
16:10:23  Complete

16:11:09  Analyzing...
16:11:11  Complete

Expected value = 1442.2
Percentiles 10/50/90:
906.015657224278,1483.79772527333,2302.36159709784
                                                                        NUM
```

You can see that the 10th, 50th, and 90th percentiles are 906, 1484, and 2302, respectively.

Saving Risk Profiles

Because you often run a model a number of times before generating a Risk Profile you choose to keep, each time you run a new Risk Profile DPL overwrites the old Risk Profile data. If you would like to keep a Risk Profile you generate, you must save it in the Project Manager.

▸ Select Window Risk_profiles.da.

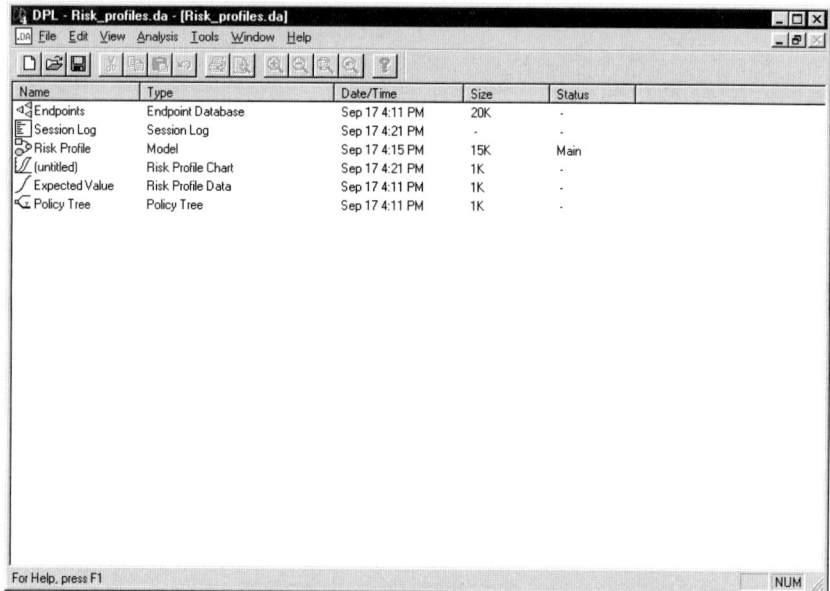

Two items were created by running a Risk Profile: An unnamed Risk Profile Chart and Risk Profile Data named "Expected Value." The Risk Profile Data is the set of points which defines the Risk Profile for the optimal policy. You can save it simply by renaming it.

▸ Right-click on the Risk Profile Data named "Expected Value."
▸ Select Rename.
▸ Type "Optimal Policy."

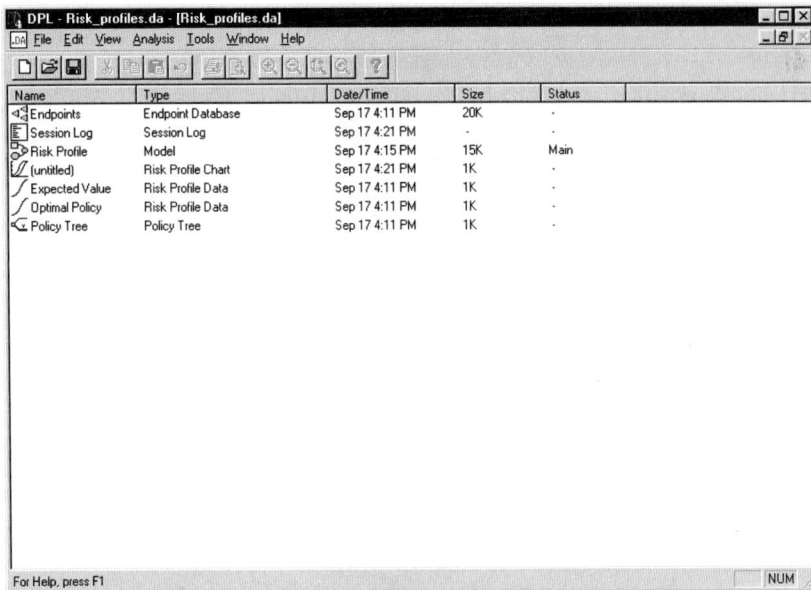

DPL adds an item to the list representing the saved Risk Profile Data with the name Optimal Policy. Notice that the Risk Profile Data Expected Value was not deleted. It will not be deleted until we run another Risk Profile or choose to delete it.

Risk Profile Charts

Running a Risk Profile has also created an untitled item in the Project Manager of the type Risk Profile Chart. Risk Profile Charts are separate from Risk Profile Data. The Risk Profile Chart contains chart formatting, such as titles and legends, and links to the Risk Profiles to be graphed. DPL creates a new Risk Profile Chart each time a Risk Profile is run. Unlike Risk Profile Data, Risk Profile Charts are not overwritten, and are instead automatically saved. To remove a Risk Profile Chart from the project you must delete it.

  ‣ Right-click on the Risk Profile Chart.
  ‣ Select Delete.
  ‣ Click OK in the DPL warning dialog box.

Risk Profiles for Initial Decision Alternatives

The Risk Profile you created is for Marketing Blitz, the optimal alternative of the initial decision in the model. (In this case Marketing Blitz was the optimal strategy, since the initial node is a strategy table). You can also have DPL generate Risk Profiles for the non-optimal alternatives of the initial decision. (Again, in this case they are for the non-optimal strategies.)

  ‣ Select Analysis Decision Analysis.
  ‣ Click OK in the DPL warning dialog box.
  ‣ In the Risk Profile Outputs section of the dialog box, click on the radio button labeled "Initial decision alternatives."

You do not need to run the Policy Tree again.

  ‣ Uncheck the box labeled "Policy Tree".
  ‣ Click OK.

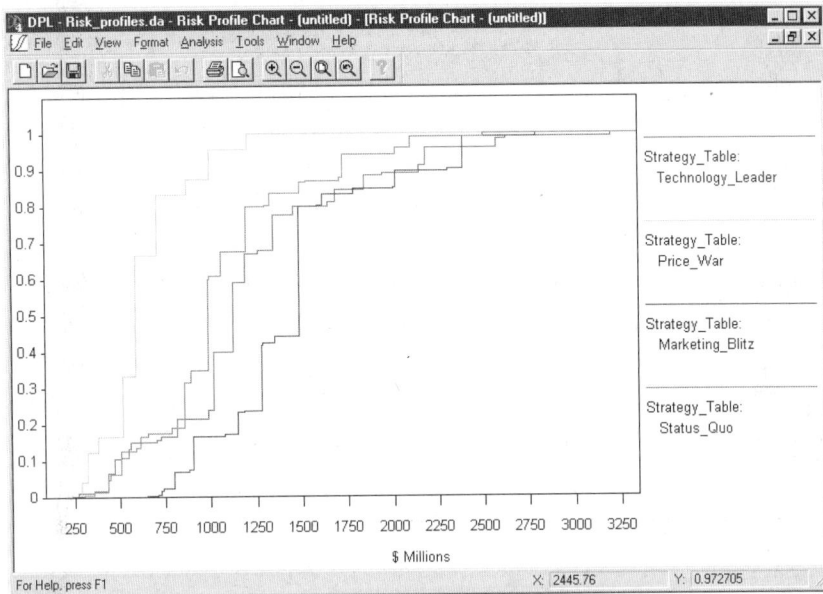

You can control which series are displayed. Since Price War appears to be an obvious loser, remove it from the chart.

▸ Select Format Series.

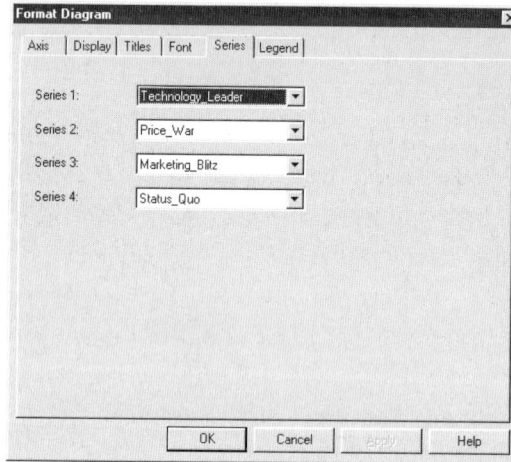

DPL can display up to four series simultaneously.

▸ In the drop-down menu for Series 2, replace "Price War" with "(none)".
▸ Click OK.

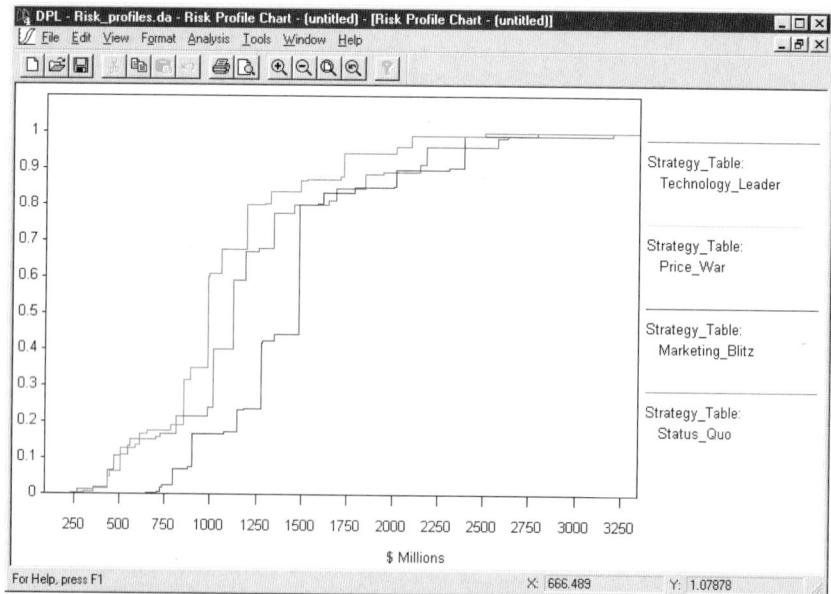

Using Reduction to Convert a Risk Profile Into a Chance Event

A Risk Profile is a probability distribution, just like any other probability distribution. You may occasionally need to use the Risk Profile from one model as an input in another model. Rather than duplicating the entire first model within the second, you can use DPL's reduction feature to approximate the Risk Profile with a single chance event.

DPL will create the new chance node using a general form of the same Gaussian Quadrature method it uses for named distributions. The new chance event will have the same moments (such as mean and variance) as the original Risk Profile.

You can also use Reduction to replace several not very significant chance events in a model with one chance event, which can reduce the number of paths and make a model run faster. You can also use reduction to define a chance event to represent an uncertainty for which you have data but haven't fit a named distribution.

Tutorial: Reduction

▸ Select Analysis Reduction.

▸ Click OK in the DPL warning dialog box.

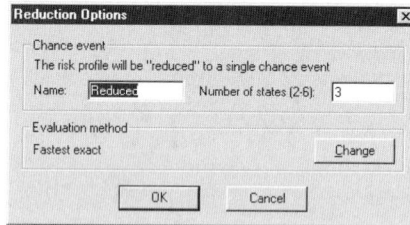

DPL gives you a number of Reduction Options. You can select the number of states in the reduction (between 2 and 6). You can also choose which evaluation method to use when performing the reduction.

▸ In the box labeled Chance event name, type "Optimal".

▸ Click OK.

DPL runs the reduction, and writes the results to both the Clipboard and Session Log.

▸ Click OK.

The {3} indicates that the resultant node has three states. The first two numbers are the probabilities associated with the first two states, approximately 0.26 and 0.62. (The probability of the third state is simply 1 minus the sum of the first two probabilities.) DPL also lists a set of three numbers. These represent the values associated with the three states: 903, 1486, and 2421.

A n expected value tornado diagram is one form of sensitivity analysis that compares the effects of several variables. This form of tornado has several modes, all of which are used to prioritize the modeling of additional sources of uncertainty. Because of this focus on prioritization, the expected value tornado is used extensively during an analysis, but may play a less important role during the presentation and explanation of results at the end of a project.

Expected value tornado diagrams are used to answer three important questions:

1) If I have a model that is currently deterministic, which variables are the most critical for distinguishing between candidate decision policies?
2) If I have a deterministic model, which variables create the most risk for any particular policy?
3) Once I have started to model some variables probabilistically, how sensitive is the model to uncertainty in the remaining deterministic variables?

Fundamentally, the expected value tornado diagram helps answer the question "If I can only model one (more) uncertainty, which one will contribute the most to my understanding of the problem and the selection of the best alternative?"

Chapter **13**

EXPECTED VALUE TORNADO DIAGRAMS

Chapter 13: Expected Value Tornado Diagrams

How DPL Generates an Expected Value Tornado Diagram

First, DPL runs a basic decision analysis on the current model. From the results (which are not displayed), DPL identifies the optimal policy and its expected value. This policy will be used as the nominal or base case policy in the tornado; the expected value will be used as a vertical axis for the tornado.

Second, DPL identifies the variables available for sensitivity analysis. You can do sensitivity analysis on variables that have been defined as named variables and which have been initialized with constants. For example:

Variable	Description	Data	Available for Sensitivity Analysis?
Price	the value Price	10	Yes
Sales	the chance event Sales	10 / 15 / 20	No
Sales	the Low state of the chance event Sales	Low 10 / 15 / 20	Yes

Variable	Description	Data	Available for Sensitivity Analysis?
Cost	the value Cost	Sales × 1.5	No
Marketing Budget	the High state of the decision Marketing Budget	High — 20 / 40	Yes

Third, DPL asks you which variables to use. As you select them, DPL displays their current values. For each variable, DPL needs a low and high value; for best results, the low and high values should define a range that includes the current value. For example:

	Current Value	Low	High
Price	10	5	20
Sales (10 / 15 / 20)	10	2	12
Marketing Budget (20 / 40)	40	30	60

Tip: When choosing ranges for sensitivity analysis, remember that you are comparing the effects of uncertainty in the variable. It is important to use ranges of equal uncertainty, not equal percentage or absolute change. A good rule of thumb is to put in high and low values that capture 80% of your range of uncertainty—there is about a 10% chance the true value could be lower and a 10% chance it could be higher.

Fourth, DPL runs the entire model two times for each variable. During the first run, DPL substitutes the low value wherever the sensitivity variable is called for, and during the second run it substitutes the high value. For example:

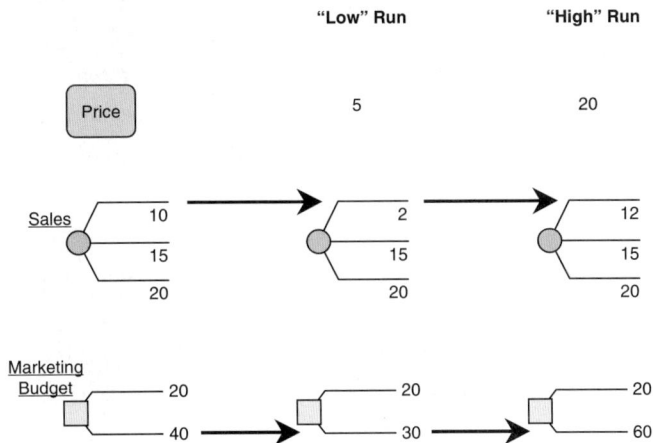

For each run, DPL remembers the optimal policy and expected value.

Fifth, DPL creates a bar on the tornado diagram for each variable. The ends of the bars display the expected values of the two runs. The color of the bar indicates whether the optimal policy changes or not — if the bar has a color change between the vertical base line and the end, the high or low value results in an optimal policy different from the nominal; if the bar has no color change, the optimal policy is the same as the nominal. (The tornado only has policy information for the base case and the end of the bar — it does not show the exact value of the sensitivity variable where the policy changes. If you want to know where the policy changes, run a rainbow diagram.)

Finally, DPL sorts the bars from widest to narrowest. This gives the diagram its characteristic tornado shape, which helps you identify the most (and least) sensitive variables at a glance.

Reading an Expected Value Tornado Diagram

The expected value tornado is an extremely versatile form of sensitivity analysis. Depending on the phase of the project, it can be used to investigate a number of things. This section will discuss three of the most common ways to use an expected value tornado. Despite their differences, they all come down to answering the question "Given the current state of the model, are there any variables currently modeled as constants that would add value if they were modeled in more detail?" It can be useful to think of the expected value tornado diagram as helping quantify the benefit side of a cost/benefit analysis on continuing to add complexity to your model.

Tip: This can be a very valuable tool when you are working on a high stakes, high profile project, where the natural tendency is to delay having to make a decision by continuing to add to the model. An expected value tornado diagram can help you know, and explain to others, when it's time to stop analyzing and start acting.

Using an Expected Value Tornado to Identify Which Variables Help Choose Between Policies

The most common application of the expected value tornado diagram is near the beginning of a decision analysis, right after the value model has been constructed and before data collection, especially probability assessment, begins. At this point, you have an influence diagram that has all the decisions plus value nodes for the model variables. This is a deterministic model, since there are no chance events. Depending on the size of your model, you may have as few as three or four, or as many as thirty or forty variables that are uncertain and could potentially be modeled probabilistically.

If you create an expected value tornado diagram for this deterministic model, you can sort these variables into four categories:

Decision Sensitive Variables

These are the variables whose bars show a color change. These are the tricky ones, the variables for which the best policy depends on the outcome of the uncertainty, but you have to decide on a policy before the uncertainty is resolved. You will want to model probability distributions for as many of these as you have time for, regardless of the width of their bars.

Risky Variables

These are the variables with the widest solid-color bar. They are not decision sensitive, but they may contribute significantly to the uncertainty of the final outcome. Their solid-color bars indicate that the optimal policy remains the same, no matter where in the range this variable falls, so going to the trouble of assessing a probability distribution probably won't add much to the model's ability to help you choose among alternative decision policies. On the other hand, it could help paint a better picture of the riskiness of the optimal policy, which may help build insight into the problem. Whether you model these variables further depends on how hard it will be to gather the data, how big the model is getting, and how much time you have.

Insensitive Variables

These are the variables with narrow solid-color bars. Modeling them further will add neither to your ability to choose between alternative policies, nor to your understanding of the total risk of these policies. Most likely, these can be left as constants while you concentrate your efforts elsewhere.

Variables with No Bar

Sometimes, the expected values of the nominal, low and high value runs for a variable are all the same, which DPL indicates by placing the variable name on the diagram, but not drawing a bar (or drawing a bar with zero width, if you prefer to think of it that way). There are three causes for this result:

- **Modeling Error.** The first thing to do with one of these variables is to verify that it is modeled correctly. If this is a variable used in a formula, is the formula entered correctly? If the variable is supposed to be linked to a spreadsheet, is it linked to the correct cell? Are the numbers in the correct range or order of magnitude? Obviously, if you find an error, fix it and update the entire tornado diagram.

- **No Impact on Nominal Policy.** Remember, an expected value tornado is showing the impact of variables relative to the nominal policy and expected value. This variable may not have an impact on the expected value of the nominal policy and not have enough impact on another policy to ever make it more attractive than the nominal policy. You can leave it set to a constant for now, but if the optimal policy changes as you add probability distributions for other variables, remember to revisit this variable.

- **Not Sensitive.** Finally, some variables simply don't contribute much uncertainty to a model. This does not necessarily mean that the variable is not significant—a $100 million cost may be a very significant part of the total cost of a project, but if there isn't much uncertainty about it, it isn't a sensitive variable. Set these variables to constants and focus your probabilistic modeling efforts elsewhere.

Using an Expected Value Tornado to Identify which Variables Contribute Most to the Riskiness of a Particular Policy

Another common application of this type of tornado diagram is to look at the sources of risk within a particular policy. It, too, is run on a deterministic influence diagram, but instead of allowing DPL to optimize the choice of policies, you use Branch Control on the decision tree to select the policy you wish to examine.

This kind of tornado differs from the first:

- **Decision-Sensitive Variables.** There are none, because you have limited your view to a single decision policy.
- **Risky Variables** (the variables with the widest green bars). These are the variables that contribute the most to the riskiness of the selected policy. They may or may not be decision sensitive.
- **Insensitive Variables.** These variables contribute relatively little risk to the selected policy. They may or may not be decision sensitive (but it's less likely).
- **Variables with No Bar.** Again, these may be modeled incorrectly, they may not have any impact on the selected policy, or they just may not be very sensitive.

There are several times when this type of tornado is helpful.

If the regular expected value tornado is using one policy as the base, this type of tornado can give you more information about the others. This is especially useful when the policy chosen by DPL is unexpected, not the "business as usual" case, or not management's favorite.

Although not as useful for prioritizing modeling activities, this type of tornado may be easier to explain when presenting your results.

This type of tornado can be useful even for the same policy used as the nominal one by the decision-optimizing run, especially if there are many decision-sensitive variables. If DPL is showing decision sensitivity, the end of a bar shows the value for the optimal policy, not the nominal one. A single-policy tornado shows how bad things can get under the nominal case.

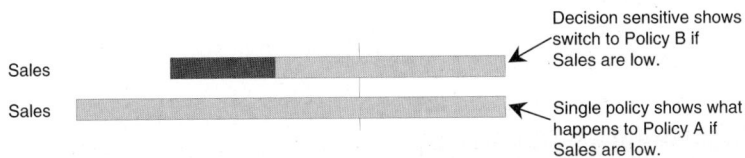

Sales — Decision sensitive shows switch to Policy B if Sales are low.

Sales — Single policy shows what happens to Policy A if Sales are low.

Using an Expected Value Tornado to Know When to Stop Adding Uncertainty to a Model

While the first two uses of the expected value tornado are useful for knowing where to start modeling uncertainty, the third helps you know when to stop.

This type of expected value tornado is run on a model that has some variables modeled as uncertainties already. The "nominal" policy and the vertical line are calculated using the full, probabilistic model, and the vertical line is the expected value.

When you add bars, you should only use variables that are still set to constants. The bar will show the change in the optimal policy and expected value of the whole decision tree. It doesn't (usually) make sense to add a bar for a variable already modeled as a chance event, because the effect of that variable's uncertainty is already included in the vertical line expected value and the nominal policy.

Tip: If you want to summarize the relative contribution to risk the uncertainties in a model, use a base case tornado diagram or an event tornado diagram.

The interpretation of some of the bars is a little different in this case:

- **Decision Sensitive** (with color change). If you have already entered probability distributions for all the decision sensitive variables from your deterministic tornado, you may be surprised to see any decision sensitive variables on this diagram, but it is possible. There are two common causes. The first is that adding uncertainty to a model may cause a change in the optimal policy from the deterministic case. This variable may not have been very important to the deterministic nominal policy but is important to the probabilistic nominal policy. The second common cause is "second order" effects, where a variable that isn't critical when others are at their deterministic values becomes critical when others are at their high or low values. As in the deterministic case, these variables merit further modeling, if time permits.

- **Risky Variables** (wide single-color bars). These are interpreted as they are in deterministic case — modeling them further may add to the picture of the riskiness of the optimal policy, but won't affect the choice between policies. Model them if you have the time.

- **Insensitive Variables** (narrow single-color bars). The evidence is mounting that the uncertainty in these variables doesn't have much impact. Don't model further.

- **Variables with No Bar.** In the deterministic case, there were three causes for this. By now, however, you should have already caught any modeling errors, so that shouldn't be a factor. Variables that affected policies other than the deterministic nominal policy should have moved up to decision sensitive, risky, or insensitive in this diagram. What's left are the really, truly insensitive variables.

Tutorial: Expected Value Tornado Diagrams

You can generate a tornado diagram by running a sensitivity analysis.

▸ Open EVTornado.da.

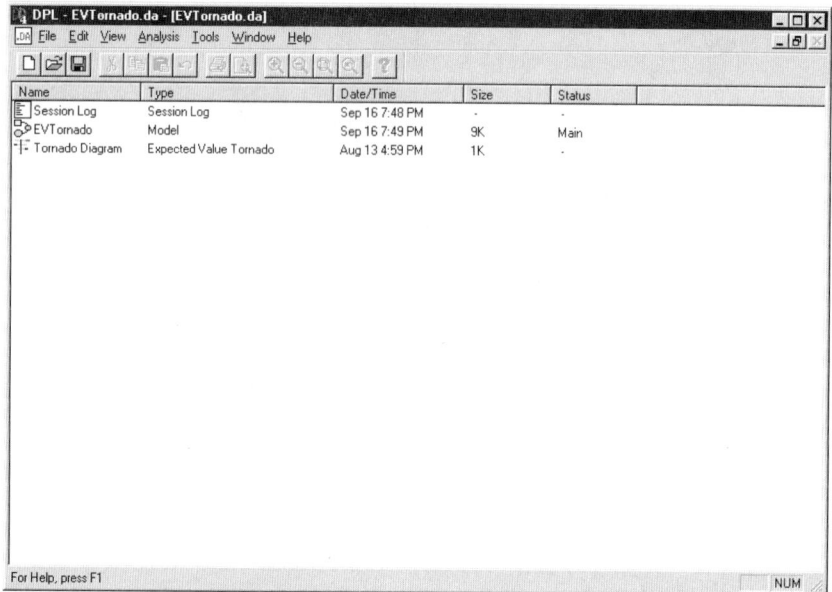

▸ Select Window Model-EVTornado.

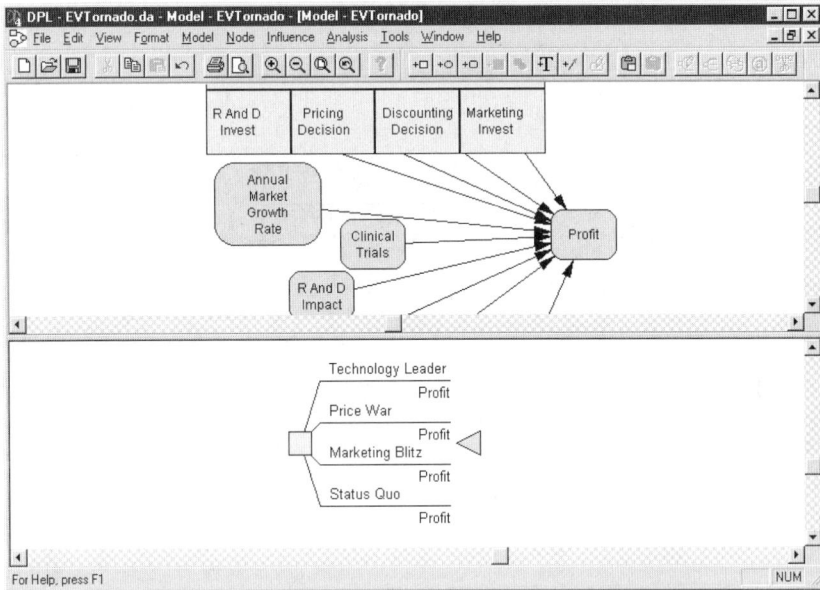

▶ Drag the splitter bar down to maximize the influence diagram pane.

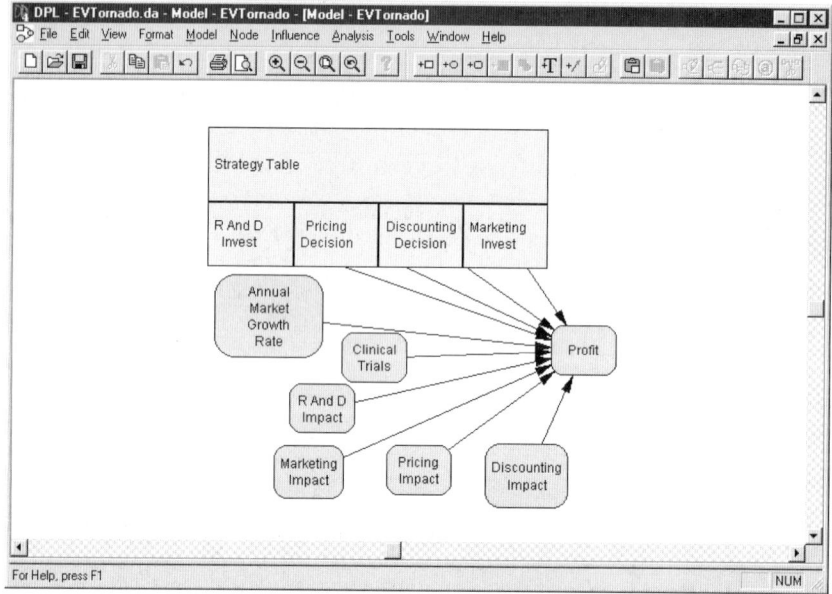

Running an Expected Value Tornado Diagram

This model has a strategy table plus a deterministic model created from a spreadsheet. There are six variables on which to perform sensitivity analysis (all the value nodes except Profit).

▸ Select Analysis Expected Value Tornado Diagram.

DPL displays the Select Value for Sensitivity Analysis dialog box.

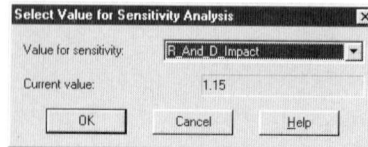

You have determined ranges of variation for each variable in the model. You can enter these ranges and evaluate the model's sensitivity to each.

▶ Select Annual Market Growth Rate from the drop-down dialog box.

DPL displays the value currently assigned to Annual Market Growth Rate.

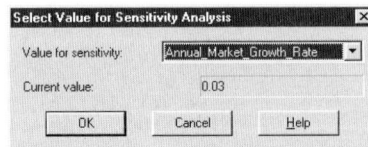

▶ Click OK.

DPL displays the Run Sensitivity Comparison dialog box.

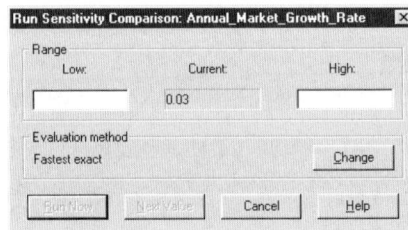

DPL asks you to enter values for the low and high ends of the uncertainty range for Annual Market Growth Rate.

▶ Enter –0.09 for the Low value and 0.15 for the High value.

The Evaluation method section of the dialog box allows you to change the evaluation method in the same manner as you would during a Decision Analysis run. Because Expected Value Tornado Diagrams are usually run on deterministic models, the run times are generally short. However, if your model is not deterministic or is linked to an extraordinarily large spreadsheet model, you may wish to run the tornado using Simulation. Because this model is deterministic and the spreadsheet model is small, accept the default of Fastest Exact.

▶ Click Run Now.

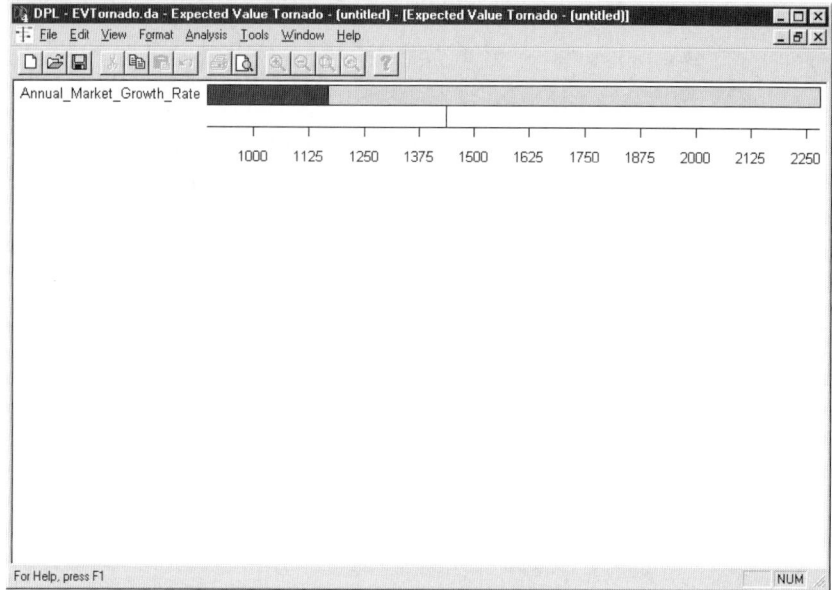

DPL has run the model three times, once each with Annual Market Growth Rate set to the low value and high value, and once at the current value. The Profit values for the high and low runs are represented by the ends of the bar. The vertical line between the x-axis and the bar represents the Profit value of the model when Annual Market Growth Rate is set to its current value. (This is the Profit value of the original model as defined in the Influence Diagram and is therefore the same for all variables. When you run subsequent variables, DPL will only run twice per variable.)

Viewing Values

You can have DPL display the values associated with the diagram.

▶ Select View Values.

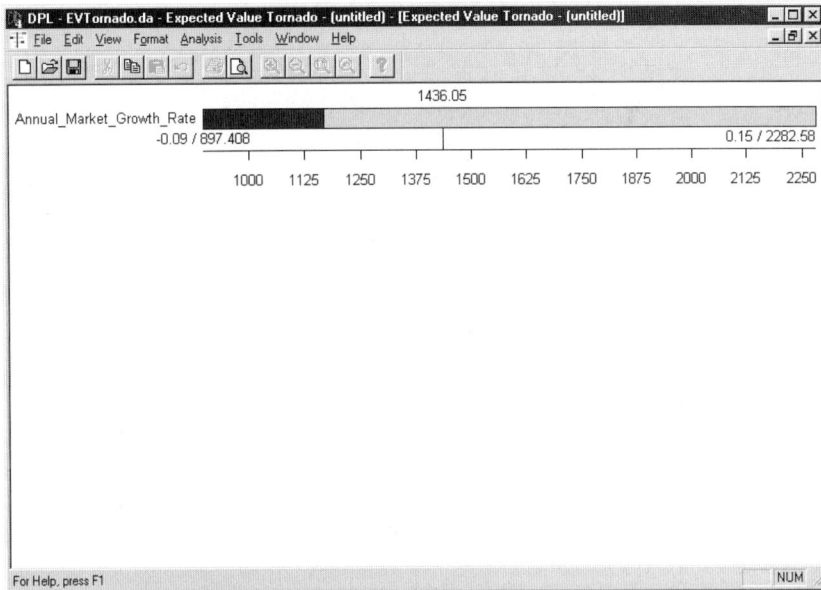

DPL displays the Profit value of the original model at the top of the tornado. In addition, the values used in the two runs are displayed at either end of the bar with the associated Profit value.

The variance in Annual Market Growth Rate appears to have a large effect on Profit, as it ranges from below $1 billion to over $2.2 billion.

Reading Color Changes

The color change in the bar indicates that the optimal strategy changes as the value of Annual Market Growth Rate changes. Annual Market Growth Rate is decision-sensitive.

The color change occurs at the midpoint between the line indicating the Profit value of the original model and the end of the bar; it does not indicate the precise point where the policy change occurs. Green indicates the original optimal policy; blue indicates that the low value causes a policy change; and magenta indicates that the high value causes a policy change. Based on this, you can say that the low value of Annual Market Growth Rate caused the policy change.

Adding Bars to an Expected Value Tornado Diagram

The true benefit of a Tornado Diagram comes from comparing the bars generated for several variables. Next you will add bars for the other variables.

▸ Select Analysis Add Bar.

As before, DPL displays the Select Value for Sensitivity Analysis dialog box.

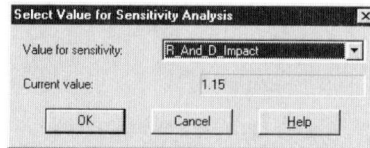

▸ Select Discounting Impact from the drop-down menu.
▸ Click OK.

DPL displays the Run Sensitivity Comparison dialog box.

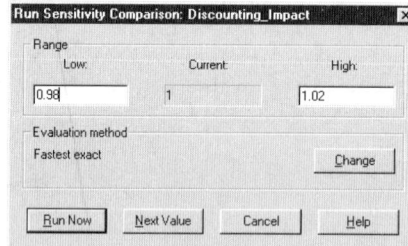

▸ Enter 0.98 for the Low value and 1.02 for the High value.

Adding Multiple Bars at Once

You could continue to run one bar at a time until you have generated bars for all the variables. Or, you can enter the data for the remaining variables and run them all at once.

▸ Click Next Value.
▸ Repeat the process of selecting a variable and defining ranges for Marketing Impact, Pricing Impact, and R And D Impact. Use the following ranges:
   Marketing Impact: 0.95 to 1.2
   Pricing Impact: 0.98 to 1.03
   R And D Impact: 0.97 to 1.15
▸ After you have entered the final range, click Run Now.

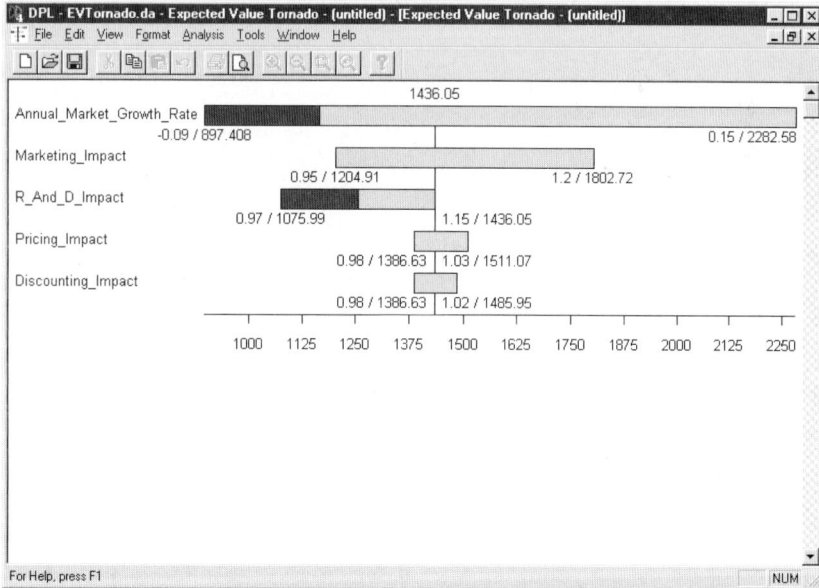

The tornado shows that Annual Market Growth Rate, Marking Impact, and R And D Impact have the greatest affect on Profit. In addition, Annual Market Growth Rate and R And D Impact are decision-sensitive. You might choose to model these three variables as uncertainties.

Hiding Bars

You might decide to simplify your chart by not displaying some of the bars, perhaps those to which the model is not very sensitive. (This may be particularly useful when showing the results to others, particularly if there are large number of bars.) You could hide the two variables you have decided not to model as uncertainties, Pricing Impact and Discounting Impact.

▸ Select Format Bars.

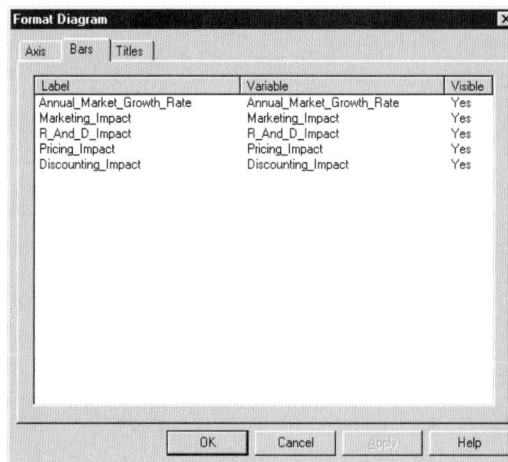

Format Diagram

Axis | Bars | Titles

Label	Variable	Visible
Annual_Market_Growth_Rate	Annual_Market_Growth_Rate	Yes
Marketing_Impact	Marketing_Impact	Yes
R_And_D_Impact	R_And_D_Impact	Yes
Pricing_Impact	Pricing_Impact	Yes
Discounting_Impact	Discounting_Impact	Yes

OK | Cancel | Apply | Help

▸ Right-click on Pricing Impact.
▸ Click on Hide/Unhide.
▸ Repeat the previous steps to hide Discounting Impact.
▸ Click OK.

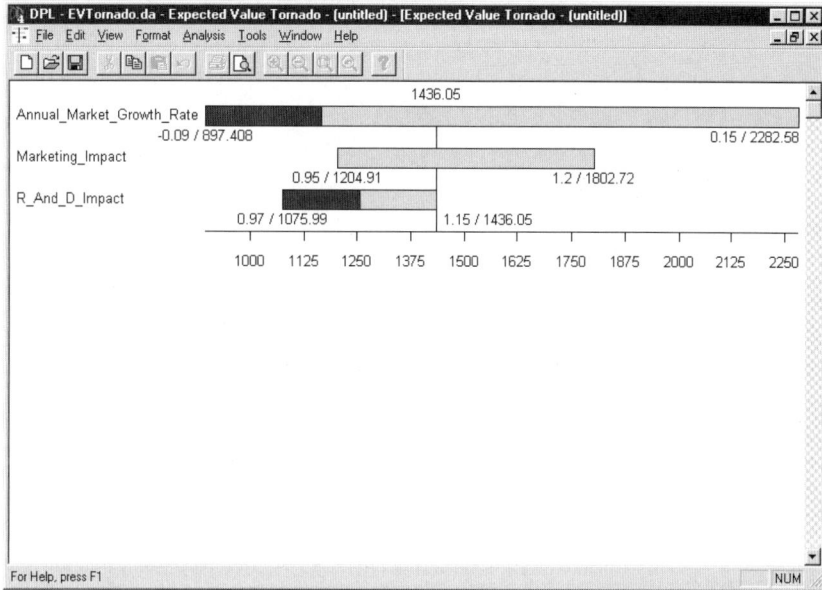

DPL has removed the bars for Pricing Impact and Discounting Impact from the diagram. You can replace them by repeating the procedure for removing them.

Index

SYMBOLS

A

B

E

F

T

Glossary

A

Always Gamble

Requires the use of the distribution of an event to determine its outcome, even in cases where the node would otherwise be given a "Dont Gamble" specification, such as during the generation of an event sensitivity comparison.

Array

A two-dimensional group of numbers. Arrays can be used to store related numbers, much as a look-up table is used in a spreadsheet. The syntax is arrayname[row #][column #]. The first element of a two-dimensional array is arrayname[0][0].

Array Formulas

Operations in DPL ordinarily take scalars (single values) as operands and return scalars as results. When an operand that is ordinarily required to be a scalar is replaced with an array, the containing formula is said to be an array formula. An array formula must be enclosed in the special symbols { = and } (the notation used by Microsoft Excel).

Array Subscripts

Numbers or formulas used to reference the elements of an array. For example, arrayname[1][3] has subscripts of 1 and 3. Array subscripts can also be formulas.

Arrowheads

In the influence diagram, the color of an arc's arrowhead indicates the level of conditioning the arc implies. Black arrowheads indicate timing only; blue arrowheads indicate values only; green arrowheads indicate probabilities only; burgundy arrowheads indicate both values and probabilities.

Asymmetric Node

A decision tree node with branches that lead to different nodes.

Attribute	A measure of value (Profit, Health Effects, Environmental Effects) tracked for each path of a decision tree during an evaluation. Usually combined at each endpoint by an objective function, generating a single measure of value that is used during tree rollback. Probability distributions for each attribute and the objective function may be graphed separately.

B

Base Case Tornado Diagram	Used with probabilistic models; a display generation by evaluating the outcome of a model when chance events are individually set to high and low states and compares the results to the outcome of the model when all events are set to their nominal state.
Bayes' Rule	A mathematical method for reversing the order of conditioning in chance events (e.g., used to assign probability when you have the probability of A given B, but need the probability of B given A). The formula for Bayes' Rule is:

$$P(A_i|B) = \frac{P(A_i)\,P(B|A_i)}{P(A_1)\,P(B|A_1) + P(A_2)\,P(B|A_2) + \ldots + P(A_n)\,P(B|A_n)}$$

Beta Distribution	A named distribution supported by DPL. For a detailed description see On-Line Help.
Binomial Distribution	A named distribution supported by DPL. For a detailed description see On-Line Help.
Branch	An element in a decision tree representing one state of an event. Decision node branches represent alternative choices and chance node branches represent possible outcomes.

Branch Control	A modeling option that forces the outcome (or branch) of an event to a particular state in a decision tree, overriding probabilities or optimal policy selection.

C

Certain Equivalent (CE)	The certain amount equivalent to the expected utility of the model, taking risk attitude into account. Specifically, the minimum guaranteed amount a decision-maker would accept in place of an uncertain lottery. When a utility function is specified, the certain equivalent is provided rather than the expected value.
Chance Nodes	Green ovals in the influence diagram or green circles in the decision tree representing events whose outcomes are uncertain and modeled with probability distributions.
Chi Distribution	A named distribution supported by DPL. For a detailed description see On-Line Help.
Chi-Square Distribution	A named distribution supported by DPL. For a detailed description see On-Line Help.
Code	A model, or part of a model, written in the DPL programming language. Can be viewed in a Program Window.
Command Window	A testing and debugging environment in which the analyst can display or temporarily change any value or calculation in a model.
Compilation	Before running a decision analysis, sensitivity analysis, or reduction on a DPL model, the model must be translated into a DPL program which is then automatically compiled. During

compilation, DPL checks for syntax errors in the program and prepares it for evaluation by the DPL engine.

Conditional Decision Policy

A policy containing decisions which will be made after current uncertainties are resolved. The optimal alternative may depend on the state of the uncertainties, and may therefore be different in different parts of the decision tree.

Constraint Function

An "if-then-else" expression that tests a condition and specifies different objective functions for true or false outcomes or prunes the tree, eliminating any further branches.

Control, Branch

See Branch Control.

Controlled Nodes

Gray rectangles in the influence diagram representing events whose state is set as an action on a branch in a decision tree. Controlled nodes do not appear in the decision tree.

Conversion

DPL can convert spreadsheet models to blocks of DPL code. The resulting code can be included in DPL models or edited and run as a stand-alone DPL program.

Cumulative Distribution

Displayed as a Risk Profile Chart, a graph generated by DPL whose X-coordinates represent outcome values and whose Y-coordinates represent the sum of the probabilities of all possible outcomes lower than or equal to the associated outcome value. In the Decision Analysis Options dialog box, the user can choose to have DPL generate and display cumulative distributions for the optimal policy or all initial decision alternatives. Also, if the analyst has defined more than one attribute for the model, a separate Risk Profile Chart for each of the attributes can be generated.

Cycle

A set of nodes and arrows creating a loop in which a node depends directly or indirectly on itself. An influence diagram may not include cycles.

D

DDE

See Dynamic Data Exchange.

Decision Nodes

Yellow rectangles in the influence diagram or yellow squares in the decision tree representing events with alternative choices. During evaluation, one alternative is determined to be optimal.

Decision Tree

A graphical representation of a decision model that displays the sequence of events and the range of possible scenarios. DPL has two views of decision trees:

1. Decision trees constructed in the Model Window which specify the structure of the model are called decision trees; they frequently are abbreviated and do not show every path explicitly.

2. The decision trees used to present the results of a decision analysis are called policy trees and are displayed in a Policy Tree Window; policy trees show every path explicitly, along with all probabilities and rollback values.

Definition Section

A section of a DPL program which specifies and initializes the elements of a decision model—decisions, uncertainties, variables, values, series, arrays, etc. The influence diagram is an example of a definition section.

Deterministic Model

A model that only has deterministic relationships—there are no probabilities. The model contains only value and decision variables, and yields a single output for a single set of inputs or setting of each variable. A spreadsheet is an example of a deterministic model.

Deterministic Sensitivity Analysis	A form of analysis in which a variable that is treated as a constant in a deterministic model is varied over a range to examine its impact on the decision policy. A Deterministic Sensitivity Analysis is used after the creation of a deterministic model to identify the critical variables that should be modeled probabilistically. The results of the analysis are displayed in an Expected Value Tornado Diagram.
Direct Conditioning	Occurs when the conditioned event or value has separate numbers or formulas for each state of the conditioning event.
Display Function	A function that enables the analyst to write text or formatted numbers to the Session Log during an analysis, allowing the writing of custom reports.
Distributed Sampling	A method of evaluating models based on Monte Carlo analysis. In Distributed Sampling, a variation of the Modified Monte Carlo method is used until the number of samples remaining at a node is small. At this point the Monte Carlo method is used for the remainder of the analysis.
Distribution	There are two types of Distributions in DPL: 1. Chance node data specified as set of discrete states and associated probabilities. 2. Chance node data specified as a named distribution. See Named Distributions.
DLL	See Dynamic Link Library.
Dont Gamble	A property assigned to a chance node that effectively reduces it to one branch. Subsequent variables depending on the state of this event will be assigned values calculated by taking the expected value over the event, where feasible.

Downstream Decision	A decision which will be made after current uncertainties are resolved.
DPL Program	A description of a decision model in DPL's own language (i.e. DPL Code). It presents all the structure and data necessary to analyze a model, and can be used in place of an influence diagram and decision tree, or as documentation for a graphic model.
Dummy Nodes	Blue rounded rectangles in the influence diagram included for clarity or communication only; they are ignored in the analysis. Dummy nodes usually represent intermediate calculations in a linked spreadsheet.
Dynamic Data Exchange	A Microsoft Windows communications protocol that allows Windows applications to communicate with each other. DPL uses DDE to communicate with Excel.
Dynamic Link Library	A file containing functions, instructions or program for use by DPL during analysis.

E

Enumerate Full Tree	An evaluation method which evaluates each path of the tree completely. This is in contrast to Fastest Exact.
Endpoint	The terminus of each path through a decision tree. The number of endpoints (or leaf nodes) of a decision tree equals the number of paths through the tree. Endpoints can be represented as blue triangles: 1. Each blue triangle in a policy tree is an endpoint. 2. Because decision trees are schematic, blue triangles can represent more than one endpoint. In the decision tree an

	endpoint triangle serves as the point of connection when you add a node to the tree.
Erlang Distribution	A named distribution supported by DPL. For a detailed description see On-Line Help.
Event	A variable or node with a set of possible outcomes or alternatives. In DPL, decision nodes and chance nodes are events.
Event Sensitivity Comparison	See Event Tornado Diagram.
Event Tornado Diagram	Used with probabilistic models, evaluates each chance event in the model. There are two types of Event Tornado Diagrams: 1. Deterministic evaluates the outcome distribution of the model when the variable being analyzed is evaluated over the range of its probability distribution, with all other variables set to their expected values. The results are compared to the expected value of the probabilistic model. 2. Probabilistic evaluates outcome distribution of model when the variable being analyzed is reduced to its expected value and all other variables are evaluated probabilistically. The results are compared to the expected value of the probabilistic model.
Expected Value (EV)	The probability-weighted average of all possible outcomes.
Expected Value (EV) Tornado Diagram	Used with deterministic models, evaluates the outcome of the model when multiple variables are individually set to high and low values and compares the results to the deterministic outcome of the model. The diagram is used to help determine which variables have the greatest effect on the measure of value and/or are decision sensitive, if applicable. When used with probabilistic models, EV Tornado Diagrams can only be

run on individual states of events or conditioned values. For probabilistic models, EV Tornado Diagrams are typically not as useful as Event Tornado Diagrams or Base Case Tornado Diagrams.

Exponential Distribution A named distribution supported by DPL. For a detailed description see On-Line Help.

Export Variable A node linked to a specific cell in a spreadsheet to provide input for the spreadsheet model. Values exported by DPL will overwrite any information contained in the cells.

F

Fastest Exact The default method of calculation DPL uses for an analysis. This is the fastest method of evaluation. See Optimization.

Frequency Histogram A viewing option in the Risk Profile Chart. Plots probability of occurrence for each possible outcome value as a discrete vertical line.

Frequency X-Y A viewing option in the Risk Profile Chart. Plots probability of occurrence for each possible outcome value and draws a line through the resulting set of points.

G

Gamma Distribution A named distribution supported by DPL. For a detailed description see On-Line Help.

Gaussian Distribution (Also Normal Distribution.) A named distribution supported by DPL. For a detailed description see On-Line Help.

Geometric Distribution   A named distribution supported by DPL. For a detailed description see On-Line Help.

Get/Pay Expressions   Assigned to branches of the decision tree to associate values with branches and endpoints. "Get" expressions are added to the value function and "Pay" expressions are subtracted.

H

Hide Intermediates   An option which prevents Dummy Nodes from being imported during spreadsheet importation.

Hyperexponential Distribution   A named distribution supported by DPL. For a detailed description see On-Line Help.

I

Import Variable   A node linked to a specific cell in a spreadsheet to return the output of the spreadsheet model calculations to the DPL analysis. A DPL-Import Node does not contain any data in DPL.

Indirect Conditioning   Node B is indirectly conditioned by a node A if one of the value expressions for B includes A, but B does not have a separate value expression for each state of A. See Direct Conditioning.

Influence Arc   Indicates timing or conditional dependence of nodes in the influence diagram. An arc from node A to node B means A influences B.

Influence Diagram	Represents all the components of a decision problem—decisions, uncertainties, and values—and the relationships among them. Comprised of nodes and influence arcs.
Interval	A subset within a Series in which (one or more) elements are defined by the same expression. For example, a series defining the number of days in each year from 1992–1997 would have an Interval (from 1993–1995) in which each element would have a definition of 365. The element 1996 would be an Interval with a definition of 366.

J

Joint Probability	The probability associated with a particular set of outcomes for multiple independent events. The Joint Probability is calculated by taking the Marginal Probability for the selected outcome of each event and multiplying them together.

L

Laplace Distribution	A named distribution supported by DPL. For a detailed description see On-Line Help.
Links	Connections to external documents (such as spreadsheets) which allow DPL to perform calculations or obtain data using other programs. See Dynamic Data Exchange.
Local Variables	Nodes which are not linked to cells in a spreadsheet.
Logistic Distribution	A named distribution supported by DPL. For a detailed description see On-Line Help.

Lognormal Distribution	A named distribution supported by DPL. For a detailed description see On-Line Help.
Lottery	An event whose outcome depends on chance.

M

Marginal Probability	The probability that a given outcome of a chance event will occur. For example, if there is an even chance of sun or rain, the Marginal Probability of rain is 0.5.
Maximize	Instructs DPL to choose the decision policy which returns the maximum expected value for the Objective Function.
Maxwell Distribution	A named distribution supported by DPL. For a detailed description see On-Line Help.
Minimize	Instructs DPL to choose the decision policy which returns the minimum expected value for the Objective Function.
Model Window	Provides graphical interface for designing decision models and running analyses. Contains both the Influence Diagram and the Decision Tree.
Modified Monte Carlo	Simulation method of evaluating models based on the Monte Carlo method. In Modified Monte Carlo, samples are first allocated to branches of chance events using the branch probabilities rather than completely random sampling.
Monte Carlo	Simulation method of evaluating models utilizing random sampling with a set sample size.

Multiattribute Utility Analysis (MUA)	A method for making decisions where the value function depends on more than one factor, such as cost and schedule. DPL allows you to track multiple attributes using independent value expressions and combine them in a single utility function.

N

Name	A string of characters used to identify a node, string or constant. Nodes will be referenced by their names when used as variables.
Named Distributions	Pre-defined probability distributions which can be assigned to chance events in DPL based on input for one or more parameters. DPL supports 21 different Named Distributions, such as the Normal Distribution and the Beta Distribution.
Negative Binomial Distribution	A named distribution supported by DPL. For a detailed description see On-Line Help.
Node	A graphical representation of a decision, value, uncertainty, strategy table, or controlled event.
Nominal Sensitivity Comparison	See Base Case Tornado Diagram.
Normal Distribution	(Also Gaussian Distribution.) A named distribution supported by DPL. For a detailed description see On-Line Help.

O

Objective Function	An expression defining the quantity to be optimized during the analysis. Generally used to express the relationship between multiple attributes.

Optimal Policy	A set of decision choices which optimize the user-defined value function.
Optimization	Allows DPL to take advantage of special structural properties of the model to reduce computation time using the Fastest Exact calculation method. Certain model structures are not suitable for Optimization, and should instead be analyzed using Enumerate Full Tree.

P

Pascal Distribution	A named distribution supported by DPL. For a detailed description see On-Line Help.
Perform Subtree	Occur when a subtree is only drawn once and referred to elsewhere. This technique is called "performing a subtree."
Poisson Distribution	A named distribution supported by DPL. For a detailed description see On-Line Help.
Policy Summary	An output generated by DPL which displays each event and its associated distribution under the optimal policy resulting from the analysis. The Policy Summary also offers an opportunity to compare these distributions with distributions associated with subsets of the optimal policy (e.g., for negative values of Profit).
Policy Tree	A setting for every decision in a decision tree. Shows all possible paths through a decision tree and indicates the value of all expressions in the model, the probabilities associated with each chance event outcome, the rollback values (EV or CE) for each node, and the optimal policy choices for each decision.

Prob Function

A function that returns the probability of an event at any point in an analysis at which the state of the event is unknown. This function is useful in decision problems in which the probability of an event is itself a major consideration in making decisions.

Probabilistic Model

A model in which some inputs are described with probability distributions and which generates probability distributions as outputs.

Probabilistic Sensitivity Analysis See Event Tornado Diagram.

Program Window

Provides interface for creating, editing and compiling DPL program files. Converted spreadsheets can be viewed in this window.

Project Manager

Window in DPL interface which lists all open windows and stored data associated with the current project. The Project Manager can be used to rename or delete items, access saved data, or switch to another window.

Promoting Terms

Moving elements of get/pay expressions to nodes which are encountered earlier in the evaluation. The advantage to this approach is that terms can be calculated in the tree as soon as all conditioning events are in known states, rather than waiting until the end of the path to calculate all the terms at once.

R

Rainbow Diagram

An analysis tool which varies a single value over a user-specified range and displays the impact on the expected value or certain equivalent and optimal decision policy.

Rayleigh Distribution

A named distribution supported by DPL. For a detailed description see On-Line Help.

Reduction	A DPL tool used to reduce an output probability distribution (risk profile) to a single chance event. The reduced event will be described by a discrete probability distribution with a user-specified number of states. The output is in the form of DPL code, and can be used as input to another model.
Risk Attitude	Decision-maker's willingness to gamble relative to the expected value. One who is not willing to gamble in the face of a positive expected value is "risk averse," while one who is willing to gamble in the face of a negative expected value is "risk seeking." One who makes decisions based on expected value is "risk neutral."
Risk Profile	The set of outcomes possible under a decision policy (optimal or other) and their associated probabilities. DPL can display graphically in a cumulative distribution or a frequency histogram. DPL can export a distribution in .CSV format for transfer to a spreadsheet or other application.
Risk Profile	A cumulative probability distribution for a specified policy. The Risk Profile of the optimal policy is a typical output of a DPL decision analysis.
Risk Tolerance Coefficient	Measure of the decision-maker's attitude towards risk. Common definition for risk tolerance r is the highest value for which you would accept a gamble in which you could win r or lose $r/2$ with equal probability. In DPL, the Risk Tolerance Coefficient describes the incorporation of risk when the built-in exponential utility function is used.
Roll Forward	The first phase of a DPL decision analysis run, in which DPL calculates an outcome value and joint probability for each path in the decision tree. As each path is traversed, get or pay

expressions are evaluated and combined to provided the path outcome value. If the model uses multiple attributes, each endpoint will have multiple outcome values.

Rollback
The third phase of a DPL decision analysis run, in which DPL uses the outcome values and joint probabilities of each endpoint to calculate expected values or expected utilities at each node in the decision tree. At each chance node in the Rollback procedure, DPL determines the expected utility/ value by calculating the probability-weighted average of outcome values. At each decision node in the Rollback, DPL determines the alternative providing the maximum or minimum value, as appropriate. The aggregate of these optimal decision alternatives comprise the optimal decision policy.

Root Node
The node on the decision tree that does not have any branches leading into it (the first node on the far left of the decision tree). Every tree contains a root node.

S

Sampling
Approximating a full range of outcomes by evaluating a subset number of paths.

Scenario
A path through the tree, from the root to an endpoint, that is a combination of specific decision alternatives and chance event outcomes. This path, a Scenario, represents a single possible state of the world.

Schematic Diagram
A type of decision tree diagram that does not show a full representation of the tree. Most decision trees appear as schematic diagrams in the Model Window.

Sensitivity Analysis   Allows the analyst to investigate the impact of a node or group of nodes on the Expected Value and/or Optimal Policy of a model. There are many types of Sensitivity Analyses. See Rainbow Diagram, Expected Value Tornado Diagram, Base Case Tornado Diagram, and Event Tornado Diagram.

Separable Value Model   A model whose value function contains terms which are combined by addition and subtraction. It is possible to promote terms in these models, which may allow for faster calculations.

Sequence Section   Systematic English-language descriptions of decision tree structures written in DPL code. The Sequence Section tells DPL in what order to evaluate nodes and get/pay expressions.

Series   A type of variable that allows you to define a one-dimensional group of values, all of which may depend, indirectly, on the states of events.

Session Log   A window which maintains a record of the DPL session. Error messages and DPL commands performed in the Model Window will be automatically written to the Session Log. Information from an analysis can also be written to the Session Log (see Display Function).

Simulation   Reduces the run-time of very large decision trees by evaluating a subset of the paths in the tree. There are two techniques used in DPL simulations: Monte Carlo and Distributed Sampling.

Spreadsheet Model   A spreadsheet which contains a set of calculations which evaluates one scenario of a decision analysis. DPL links to the spreadsheet model in order to rapidly evaluate a wide range of scenarios based on specified levels of uncertainty.

State	An alternative of a decision or an outcome of an uncertainty. Each branch of an event in the decision tree represents a state.
State Function	A function that returns a numerical value representing the state of an event. It can be used to process and test any event's state.
Statename Function	Returns the name of a state of an event as a string ("High", "Low", etc.).
Strategy	A single set of alternatives for all of the decisions contained in a Strategy Node.
Strategy Node	Yellow rectangles in the influence diagram or yellow squares in the decision tree representing a Strategy Table.
Strategy Table	A collection of decision nodes and a set of defined strategies. In the influence diagram, a Strategy Node contains the name of the node and the names of the included decisions. During evaluation, one strategy is selected as optimal.
Subscript	Used to reference particular elements of a series which can range between the lower and upper boundaries of the series. For using subscripts with arrays see Array Subscripts.
Symmetric Node	A node whose branches all lead to the same next event.

T

Tolerance	See Risk Tolerance Coefficient.
Tornado Diagram	A sensitivity analysis that displays the value and policy impacts of varying input values. See Expected Value Tornado Diagram Base Case Tornado Diagram, and Event Tornado Diagram.

Triangular Distribution	A named distribution supported by DPL. For a detailed description see On-Line Help.

U

Uncertainty	See Chance Nodes.
Uniform Distribution	A named distribution supported by DPL. For a detailed description see On-Line Help.
User Library	A Dynamic Link Library (DLL) file which contains a set of functions or a program (to be run during a DPL analysis) and language that allows communication with DPL. User Libraries are called from commands in the decision tree or program file.
User Library Function	An external set of commands which perform a routine or calculation during a DPL analysis. User Library Functions are stored in a User Library, which has a file extension of .DLL.
Utiles	The unit of measurement for the transformation performed by the Utility Function. In a model incorporating risk tolerance, DPL selects the policy which maximizes the expected utility (i.e. greatest utile value).
Utility Function	A mathematical expression which captures the decision-maker's Risk Attitude. The Utility Function translates the outcomes of a value model into utiles. In DPL, the built-in exponential utility function is defined as: $u(x)=-e^{(-x/r)}$, where x is the value to be converted to utiles and r is the risk tolerance coefficient. Alternatively, the user may specify a user-defined utility function.

V

Value Function Expression defining quantity to be optimized during analysis.

Value Model A model that defines a particular value of interest (e.g., costs, profits, social welfare) from other data.

Value Nodes Blue rounded rectangles in the influence diagram representing a number or an expression.

Value of Control The difference between the expected value of the model and the expected value of the model when a particular uncertainty is changed into a decision.

Value of Information The difference between the expected value of the model and the expected value of the model when the outcome of a particular uncertainty is known before a decision is made.

Value Sensitivity Analysis See Rainbow Diagram.

Value Sensitivity Comparison See Expected Value Tornado Diagram.

W

Weibull Distribution A named distribution supported by DPL. For a detailed description see On-Line Help.

Bibliography

Books

Baird, B.F. *Managerial Decisions Under Uncertainty: An Introduction to the Analysis of Decision Making*, John Wiley and Sons, 1989.

Behn, R.D. and J.W. Vaupel. *Quick Analysis for Busy Decision Makers*, Basic Books, Inc., 1982.

Center for Chemical Process Safety. *Tools for Making Acute Risk Decisions with Chemical Process Safety Applications*, American Institute of Chemical Engineers, 1995.

Clemen, R.T. *Making Hard Decisions: An Introduction to Decision Analysis*, Duxbury Press, 1996.

Edwards, W. *Utility Theories: Measurement and Applications*. Kluwer Academic Publishers, 1992.

Holloway, C.A. *Decision Making Under Uncertainty: Models and Choices*, Prentice Hall, Inc., 1979.

Howard, R.A. and J. E. Matheson. *The Principles and Applications of Decision Analysis*, Strategic Decisions Group, 1989.

Kahneman, D., P. Slovic, and A. Tversky. *Judgment Under Uncertainty: Heuristics and Biases,* Cambridge University Press, 1990.

Keeney, R.L. *Value-Focused Thinking: A Path to Creative Decisionmaking,* Harvard University Press, 1992.

Keeney, R.L. and H. Raiffa. *Decisions with Multiple Objectives,* Cambridge University Press, 1993.

Kirkwood, C.W. *Strategic Decision Making: Multiobjective Decision Analysis With Spreadsheets,* Duxbury Press, 1997.

Marshall, K.T. and R.M. Oliver. *Decision Making and Forecasting: With Emphasis on Model Building and Policy Analysis,* McGraw-Hill, 1995.

Merkhofer, M.W. *Decision Science and Social Risk Management,* Reidel, 1987.

Oliver, R.M. and J.Q. Smith. *Influence Diagrams, Belief Nets, and Decision Analysis,* John Wiley and Sons, 1990.

Raiffa, H. *Decision Analysis: Introductory Lectures on Choices under Uncertainty,* Random House, 1968.

Samson, D. *Managerial Decision Analysis,* Irwin, 1988.

Schuyler, John R. *Decision Analysis in Projects.* Project Management Institute, 1996.

Von Winterfeldt, D. and W. Edwards. *Decision Analysis and Behavioral Research,* Cambridge University Press, 1986.

Watson, S. R. and D. M. Buede. *Decision Synthesis: The Principles and Practice of Decision Analysis,* Cambridge University Press, 1987.

Wright, G. and P. Ayton. *Subjective Probability.* John Wiley and Sons, 1994.

Journals

Decision Sciences, Decision Sciences Institute.

Interfaces, Institute for Operations Research and the Management Sciences.

Journal of Behavioral Decision Making, John Wiley and Sons.

Journal of Risk and Uncertainty, Kluwer Academic Publishers.

Medical Decision Making, Society for Medical Decision Making, Hanley & Belfus, Inc. Publishers.

Multi-Criteria Decision Analysis, John Wiley and Sons.

OR/MS Today, Institute for Operations Research and the Management Sciences.

6. TERMINATION. This license is terminated automatically if Customer fails to comply with any part of this Agreement. Upon termination, Customer must immediately destroy all copies of the Licensed Software and Documentation or return them. All provisions of this Agreement relating to proprietary rights, non-disclosure and limitation of liability shall survive the termination of this Agreement. In the event this Agreement is terminated pursuant to this paragraph, Customer will not be entitled to a refund of any fees paid to PwCD under this Agreement.

7. NOTICES. Any notice provided pursuant to this Agreement, if specified to be in writing, shall be in writing and shall be deemed given (i) if by hand delivery, upon receipt thereof, (ii) if mailed, three (3) days after deposit in the United States mails, postage prepaid, certified mail return receipt requested, or (iii) if by next day delivery service, upon such delivery. All notices to PwCD shall be addressed as follows (or such other address as either party may in the future specify in writing to the other): PricewaterhouseCoopers Distributions LLC, c/o Applied Decision Analysis, 2710 Sand Hill Road, Menlo Park, CA 94025.

8. GOVERNING LAW. This Agreement shall be governed by and construed in accordance with the laws of the State of California without regard to conflict of law provisions.

9. EFFECT OF INVALIDITY. The invalidity in whole or in part of any provision hereof shall not affect the validity of any other provision.

10. COMPLETE AGREEMENT. This Agreement supersedes all prior agreements and all contemporaneous agreements not required or contemplated hereby and all representations, warranties, undertakings and understandings made with respect to the same subject and is the entire agreement of the parties as to its subject matter. This Agreement may not be changed or modified in any manner, orally or otherwise, except in writing by PwCD.

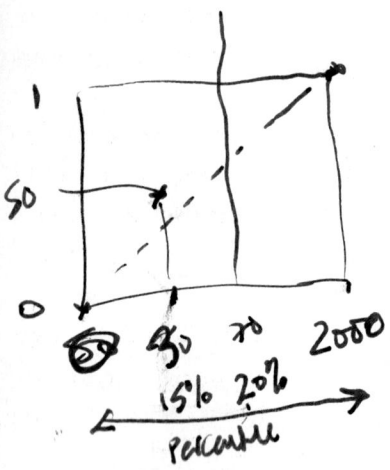